Lecture Notes Editorial Policies

Lecture Notes in Statistics provides a format for the informal and quick publication of monographs, case studies, workshops, and conference proceedings of theoretical or applied importance. Thus, in some instances, proofs may be merely outlined and results presented which will later be published in a different form. A subject index is expected for authored but not edited volumes. Conference proceedings must be focused and the contents refereed.

Publication of the Lecture Notes is intended as a service to the international statistical community, in that a commercial publisher, Springer-Verlag, can provide efficient distribution of documents that would otherwise have a restricted readership. Once published and copyrighted, they can be documented and discussed in the scientific literature.

Lecture Notes are reprinted photographically from the copy delivered in camera-ready form by the author or editor. Springer-Verlag provides technical instructions for the preparation of manuscripts. Volumes should be no less than 100 pages and preferably no more than 400 pages. Proposals for volumes should be sent to one of the series editors or to Springer-Verlag in Santa Clara.

Authors of monographs receive 50 free copies of their book. Editors receive 50 free copies and are responsible for distributing them to contributors. Authors, editors, and contributors may purchase additional copies at the publisher's discount. No reprints of individual contributions will be supplied and no royalties are paid on Lecture Notes volumes. Springer-Verlag secures the copyright for each volume.

Lecture Notes in Statistics **119**

Springer
New York
Berlin
Heidelberg
Barcelona
Budapest
Hong Kong
London
Milan
Paris
Santa Clara
Singapore
Tokyo

Masanao Aoki
Arthur M. Havenner
Editors

Applications of Computer Aided Time Series Modeling

Springer

Masanao Aoki
Department of Economics
University of California, Los Angeles
Los Angeles, CA 90024

Arthur M. Havenner
Department of Agricultural Economics
University of California, Davis
Davis, CA 95616

CIP data available.
Printed on acid-free paper.

Softcover reprint of the hardcover 1st edition 1997

Camera ready copy provided by the author.

9 8 7 6 5 4 3 2 1

ISBN-13: 978-0-387-94751-8 e-ISBN-13: 978-1-4612-2252-1
DOI: 10.1007/ 978-1-4612-2252-1

Preface

This book consists of three parts: Part One is composed of two introductory chapters. The first chapter provides an instrumental varible interpretation of the state space time series algorithm originally proposed by Aoki (1983), and gives an introductory account for incorporating exogenous signals in state space models. The second chapter, by Havenner, gives practical guidance in applying this algorithm by one of the most experienced practitioners of the method. Havenner begins by summarizing six reasons state space methods are advantageous, and then walks the reader through construction and evaluation of a state space model for four monthly macroeconomic series: industrial production index, consumer price index, six month commercial paper rate, and money stock (M1). To single out one of the several important insights in modeling that he shares with the reader, he discusses in Section 2ii the effects of sampling errors and model misspecification on successful modeling efforts. He argues that model misspecification is an important amplifier of the effects of sampling error that may cause symplectic matrices to have complex unit roots, a theoretical impossibility. Correct model specifications increase efficiency of estimators and often eliminate this finite sample problem. This is an important insight into the positive realness of covariance matrices; positivity has been emphasized by system engineers to the exclusion of other methods of reducing sampling error and alleviating what is simply a finite sample problem.

The second and third parts collect papers that describe specific applications. Part Two collects papers on linear models, and Part Three presents recurrent neural networks as useful nonlinear models. Applications in agriculture, macroeconomics, financial economics, and engineering are featured. More specifically, the chapter by Berwald and Havenner provides a good illustration of the potential of state space modeling in agricultural applications. They model soybean complex spot and futures prices (six-dimensional observation vectors). These series are volatile and difficult to model. They use both in-sample and out-of-sample forecasts to evaluate the model they estimate. Egan and Havenner compare state space, VAR, and futures markets of sequentially produced types of cattle. Arndt and Foster model monthly U. S. spot wheat prices in nine geographically separate markets. Their results tend to support the hypothesis that spot wheat markets are inefficient. Fiorito calculates impulse response functions from the estimated model for the U. S. labor markets. This chapter is a good illustration of how one might use the estimated models for policy analysis by calculating impulse responses of the economy to various exogenous shocks.

State space modeling of time series with exogenous signals has been sketched out in Chapter One. The applications reported in the next three papers all incorporate exogenous signals as parts of their models. Dorfman compares several exchange rate models reported in the literature. His study points to the usefulness of some macroeconomic variables as exogenous signals in forecasting exchange rates. Östemark examines the effects of Japanese stock markets on Finnish derivatives by building state space models of the series. DiRusio has an example of an exchange rate as an exogenous signal in his model.

Mendelssohn and Schwing tackle the very difficult problem of building time series models of climactic changes with both trend and seasonal components. Because of these components, they use an alternative method for building time series models due to Akaike, Kitagawa, and Gersch.

Part Three consists of two contributions on neural networks. State space time series models are linear models and may not work well with time series in which nonlinear effects are significant. For example, financial time series often contain significant nonlinear components. The chapter by Rhee reports on forecasting stock indices with recurrent neural networks, and the chapter by Saxen is a theoretical background piece that connects simple recurrent neural networks with familiar ARMA models.

Masanao Aoki
August 1996

Los Angeles, California

Table of Contents

Part I

Introduction to State Space Modeling

The SSATS[1] Algorithm and Subspace Methods

Masanao Aoki
Department of Economics
University of California
Los Angeles, California 90095-1477

Keywords: Instrumental Variables, State Space Modeling, Subspace Method.

Abstract

This chapter summarizes the development of the State Space Aoki Time Series algorithm since its inception in 1983 as an instrumental variables method, and relates it to subspace methods popular in the engineering literature from the viewpoint of orthogonal projections. Except for the numerical method used to calculate orthogonal projections these two classes of methods are conceptually equivalent.[2]

1 Background

As evidenced in the literatures on statistics, macroeconomic time series, econometrics, and control and systems, the "art" of time series modeling is practiced in many disciplines with slightly different emphases which are dictated by intended applications. For vector-valued time series, state space modeling algorithms seem superior to the ARMA modeling approaches. In the former category of modeling procedures we have the Aoki algorithm, originally announced in 1983 and modified subsequently in Aoki (1987, 1990) and in Havenner and Aoki (1988), and subspace methods which are popular in the control and systems literature.[3]

We observe some degree of duplication of effort in some parts of the engineering literature of what has been available in the instrumental variables version of the Aoki algorithm. More specifically, it appears that improvements and modifications of the original Aoki algorithm which were carried out by Aoki and others since 1987 appear

[1] State Space Aoki Time Series

[2] March 1996. The author benefited from seeing an unpublished manuscript by G. Picci and T. Katayama (1996a, b). The author also thanks A. Havenner for helpful discussion.

[3] See Viberg (1994) or van der Veen et al. (1993) for surveys of the subspace methods.

unknown to the control audience if one judges from the references in recent papers on subspace methods.[4]

The main purpose of this chapter is to provide a short and coherent summary of the development of the state space method for time series modeling since the 1987 publication by Aoki. This chapter discusses similarities and differences of these two classes of algorithms from the point of view of instrumental variables estimation, and may be of interest to the general reader of time series literature, and especially those who are interested in the class of algorithms called subspace methods. We show in this chapter that although these two classes of methods use different numerical methods to calculate orthogonal projections, they share the same conceptual framework, in particular, for series without exogenous signals.[5]

Aoki's state space modeling algorithm has been motivated by a desire to construct models for economic time series data. Economic time series are usually characterized by the absence of what the control literature calls "input" signals, although in some applications such as financial time series some macroeconomic variables may be used with advantage as exogenous signals. This part of Aoki's algorithm therefore corresponds to models of the stochastic part in the subspace literature.

Havenner and Aoki (1988, Sec.2.2) pointed out that at time t, the data vector at time $t-1$, y_{t-1}, can be interpreted as an instrument in Aoki's algorithm. Later in Aoki (1989) and Aoki (1990, Sec. 9.2) it has been pointed out that the stacked past data vector is an instrument, although inefficient, and that a more efficient instrument is the state vector itself, constructed as $z_t = \Omega R_-^{-1} y_{t-1}^-$, where Ω is the second half of the decomposition of the covariance between past and future stacks $E(y_t^+ y_{t-1}^{-\prime}) = \mathcal{O}\Omega$, and where $\mathcal{O}$ is the extended observability matrix in the notation of Aoki (1987, 1990).

As summarized below, this new vector of instruments is the projection on the subspace spanned by past data of an unobserved state vector of a finite dimensional linear data-generating processes which is modeled as

$$x_{t+1} = Ax_t + \xi_t; \; y_t = Cx_t + \eta_t, \tag{1}$$

where x_t is unobserved dynamic factors, and ξ's and η's are serially uncorrelated disturbances. Then, the orthogonal projection $z_t = E(x_t | y_s; s \le t-1)$ evolves with

[4] For example, van Overschee, de Moor, and Suykens (1992), van der Veen et al. (1993), and Verhaegen (1994) do not reference any work by Aoki, or Havenner. van Overschee and de Moor (1993), and van Overschee (1995) have inaccurate or misleading representations of Aoki (1987), and are apparently unaware of more recent developments by Aoki (1990, 1994) and others such as Havenner and Aoki (1988), and Havenner and Leng (1994), which have some overlap or close relations with their own work.

[5] This apparently independent development of algorithms in separate literatures may be because the control literature did not realize that the original algorithm in Aoki (1987) had been substantially modified and reinterpreted since introduction, and that the new algorithm has been interpreted and modified using the instrumental variables interpretations.

time and is related to the data series by

$$z_{t+1} = Az_t + Be_t; \; y_t = Cz_t + e_t,$$

where e_t is the innovation vector of y_t with respect to its own past. Let us call this an innovation model.

Aoki's algorithm has been extended later to model economic time series with exogenous signals which are observed on the assumption that the data vector y does not Granger-cause the exogenous signals, i.e., that exogenous signals have no feedback from observed data, Aoki (1990, Sec.9.7). See Hosoya (1977) or Aoki (1989, p.169-170) for the notion of Granger causality which lies behind this extension. Modeling of time series with exogenous signals are further discussed in Aoki (1994) and in Östermark and Aoki (1992) as well. See Picci and Katayama (1996a, b) for more complete treatment of feedback-free exogenous signals.

The chapter closes by mentioning the strict nestedness aspects of the Aoki algorithm which is not emphasized or mentioned in the control literature. This is a useful property of any modeling algorithm worthy of considertion in developing new algorithms. See Aoki (1987, p.76, 1990, p.112). This chapter also points out to the control audience the important distinction between model specification and model parameter estimation, a distinction which is not perhaps important in the traditional control literature because the forms of models to be estimated are prescribed by physical constraints or requirements of applications since the form of dynamics are a priori known. This is not the case with economic time series in which the form of dynamics is not known.

2 The Approach

It is fairly clear how we should proceed with modeling of time series with the help of exogenous signals when the exogenous signal is free of feedback from data series. We decompose the data series into two parts; the part which is influenced by exogenous signals and the part free of the effects of the exogenous signals which we call the (purely) stochastic part. We extract the effects of exogenous signals by the orthogonal projection onto appropriate subspaces of the data vectors, and perform modeling tasks separately for the two parts of the data series; one for the part of data influenced by exgoenous signals and the other for the stochastic part.[6] As we soon discuss, the innovation modeling algorithm of Aoki (1987, 1990) can be applied without any modification to the stochastic part of the data.

To conform with the convention in the control literature we use u_t to denote the

[6] This clean separation is contaminated by estimation error due to a finite data set as we mention later.

m-dimensional exogenous signal and denote

$$U_t^- = \bar{span}\{u_s; s \le t-1\},$$
$$U_t^+ = \bar{span}\{u_s; s \ge t\},$$

where by $\bar{span}$ we mean the closure of the linear span of the signals with respect to the inner product norm $\langle x, y\rangle = E(xy')$, where x and y are mean zero vectors with finite variances. Define Y_t^+ and Y_t^- similarly.

The assumption that there is no feedback from y to u is that conditional on U_t^-, U_t^+ and Y_t^- are orthogonal (uncorrelated). Lindquist and Picci (1985, Prop.2.1) have shown that this is equivalent to the statement that

$$E(y_t|U) = E(y_t|U_{t-1}^-),$$

for all t, where $U = U_t^+ + U_t^-$.

We decompose y_t into two components:

$$y_t = y_t^s + y_t^u,$$

where

$$y_t^u = E(y_t|U),$$
$$y_t^s = E(y_t|U^\perp) = y_t - y_t^u.$$

Here, $U^\perp$ is defined to be the orthogonal complement of U in the space $U \vee Y$, i.e., $U \oplus U^\perp = U \vee Y$.

The stochastic part of y_t is denoted by y_t^s and the second non-stochastic part by y_t^u. This second part is also called the deterministic part of y_t in the systems literature, but this name is misleading since exogenous signals need not be deterministic. It merely means that part of y_t which is influenced by the exogenous signals.

We denote by $\tilde{Y}$ the Hilbert subspace of $U \vee Y$ that is linearly generated by $\{y_t^s\}$ for all t.

We have an important fact that the innovation vector of y_t is the same as the innovation vector of y_t^s because

$$\begin{aligned}
e_t &= y_t - E(y_t|u_s, y_{s-1}, s \le t) \\
&= y_t - E(y_t|U_{t+1}^- \vee Y_t^-) \\
&= y_t - E(y_t|U_t^- \oplus \tilde{Y}_t^-) \\
&= y_t - E(y_t|U_{t+1}^-) - E(y_t|\tilde{Y}_t^-) \\
&= y_t^s - E(y_t^s + y_t^u|\tilde{Y}_t^-) \\
&= y_t^s - E(y_t^s|\tilde{Y}_t^-) \\
&= e_t^s,
\end{aligned}$$

since $y_t^u \perp \tilde{Y}_t^-$, where $\tilde{Y}_t^- = E(Y_t^-|U^\perp)$.

3 Model Specification

3.1 Innovation Model

We are left with the stochastic part of the time series after removing the influences of exogenous signals. Assume that the series of this stochastic part have rational spectral density matrix and satisfy positive realness. Then, modeling of this series is the same as in Aoki (1987, 1990). The model is taken to be

$$z_{t+1} = Az_t + Be_t;\ y_t^s = Cz_t + e_t,$$

for some finite dimensional vector z_t. It has been shown in Aoki (1990) that the model in the form (1) where x_t is a finite dimensional unobserved vector, and where ξ_t and η_t are serially uncorrelated in the sense that

$$E(\xi_t|\tilde{Y}_t^-) = 0,$$

and likewise for η_t, can be reduced to the above form by taking

$$z_t = E(y_t^s|y_k^s, k \leq t-1).$$

3.2 Model Driven by Exogenous Signal

We assume that the spectral density matrix of the exogenous signal process is strictly positive, which implies that the exogenous signals vary sufficiently often to reveal their stochastic structure. This is a natural assumption for most economic time series.

The content of this assumption is that the model is specified to be

$$s_{t+1} = Fs_t + Gu_t,\ y_t^u = Hs_t + Ku_t \tag{2}$$

where s_t is finite dimensional.

4 Orthogonal Projection Calculations

4.1 Effects of Finite Data Length

In the above discussions, we tacitly assumed that we have observations from the infinite past. Actually we only have a finite time span over which data are observed. We denote them by

$$U([0,T]) = \bar{span}\{u_s; 0 \leq s \leq T\},$$
$$Y([0,T]) = \bar{span}\{y_k; 0 \leq k \leq T\}.$$

We define $U^{\perp}([0,T])$ as the orthogonal complement of $U([0,T]) \vee Y([0,T])$. Since $U([0,T]) \subset U$, we have

$$E(y_t|U([0,T])) = E(y_t^u|U([0,T])) = \hat{y_t}^u,$$

and consequently

$$\begin{aligned} \hat{y_t}^s &= y_t - E(y_t|U([0,T])) \\ &= y_t^s + y_t^u - E(y_t^u|U([0,T])) \\ &= y_t^s + \tilde{y_t}^u \end{aligned}$$

where we denote the discrepancy

$$\tilde{y_t}^u = y_t^u - \hat{y_t}^u.$$

Although $\{s_t\}$ is governed by (2) with the initial condition s_0, the series $\{\hat{s}_t\}$ is generated by

$$\hat{s_{t+1}} = F\hat{s}_t + Gu_t,$$
$$\hat{y_t}^u = H\hat{s}_t + Ku_t,$$

with the initial condition $\hat{s}_0 = E(s_0|U([0,T]))$. We see that

$$\tilde{y_t}^u = HF^t\tilde{s}_0.$$

This error of the estimate of the initial condition state of the non-stochastic component causes $\hat{y_t}^s$ to differ from y_t^s. Picci and Katayama (1996a, b) discuss this source of error and propose a scheme for filtering this component out.

4.2 LQ Decomposition

Orthogonal projections which are basic to the algorithm can be implemented in several ways. For any two mean-zero random vectors with finite covariances we define the orthogonal projection by

$$E(a|b) = E(ab')E(bb')^{-1}b.$$

A convenient way of calculating orthogonal projections is to use the transpose of the QR factorization, known as the LQ decomposition which is a key step in many subspace algorithms. For example, let

$$\left(U \right) = \begin{pmatrix} L_u & 0 \\ L_{yu} & L_y \end{pmatrix} \begin{pmatrix} Q'_u \\ Q'_y \end{pmatrix},$$

where $Q = [Q_u \; Q_y]$ is an orthogonal matrix, which is premultiplied by a lower triangular matrix with elements L_u etc. See Householder (1958) or Paige (1985). In terms of the LQ factorization

$$\hat{Y} = E(Y|U) = L_{yu}Q'_u,$$

and

$$\tilde{Y} = E(Y|U^{\perp}) = L_yQ'_y.$$

4.3 Generalized Singular Value Decomposition

To calculate orthogonal projections numerically, we can use the LQ decomposition of Householder (1958) (transpose of QR factorization), where L refers to lower triangular matrices. For example, consider

$$\begin{pmatrix} Y_- \end{pmatrix} = \begin{pmatrix} L_- & 0 \\ L_{+-} & L_+ \end{pmatrix} \begin{pmatrix} Q'_- \\ Q'_+ \end{pmatrix}.$$

In this form,

$$E(Y_+ Y'_-) = L_{+-} L'_-,$$

where the matrices L_{+-} and L_- are decomposed by the generalized singular value decomposition of Paige into

$$L_{+-} = XCV',$$
$$L_- = XSW',$$

with V and W orthogonal, C and S diagonal, X non-singular, and $C^2 + S^2 = 1$. See Paige (1985) for proofs, and Picci and Katayama (1996a, b) on how to use this procedure to find a basis for the state space.

5 Iterative Improvement of Estimates

5.1 Model with Exogenous Signal

The matrices in the model with exogenous signal (2) are solved together by least squares, i.e., using the stacked vector $(s'_t\, u'_t)'$ as instruments. Here the state vector s_t is generated by the orthogonal projection of $\hat{y}_t$ on the subspace U_t^-, i.e.,

$$s_t = \Omega_u cov(u_t^-)^{-1} u_t^-,$$

where Ω_u is the second factor of the factorization of $H_{yu} = E(\hat{y}_t^+ u_t^{-\prime}) = O_u \Omega_u$, and O_u is the extended observability matrix for the model of the exogenous part. In other words, we multiply the model (2) from the right by the instrument vectors and form

$$\langle s_+, s \rangle = F \langle s, s \rangle + G \langle u, s \rangle$$
$$\langle y, s \rangle = H \langle s, s \rangle + K \langle u, s \rangle,$$

where we use the notation

$$\langle s_+, s \rangle = \frac{1}{T} \sum_t s_{t+1} s'_t,$$

$$\langle s, s \rangle = \frac{1}{T} \sum_t s_t s'_t,$$

$$\langle y, s\rangle = \frac{1}{T}\sum_t \hat{y}_y s_t',$$

$$\langle u, s\rangle = \frac{1}{T}\sum_t u_t s_t'.$$

Since $y_{t+1}^+ = O_u s_{t+1}$ we may use the orthogonal projection of y_{t+1}^- onto U_t^- to obtain s_{t+1} directly to be used in the above least squares estimation.

5.2 Stochastic Model

The initial estimates of matrices A, B, C and covariance matrix of the innovation vector may be obtained by the algorithm in Aoki (1987). The initial condition of the dynamics z_0 may be set to zero, or alternatively a method similar to backcasting may be used, or we can arrange the data to start model construction from date K to T by using the first K data vectors to construct the initial estimate z_K^0 from $y_0, \ldots y_{K-1}$ for some positive integer K. It is convenient to introduce a short-hand notation

$$\langle z, z\rangle = \frac{1}{T}\sum_{s=K}^{T+K} z_s z_s'.$$

Define $\langle y, y\rangle$ and $\langle y, z\rangle$ etc., analogously. When one of the factors has index $s+1$ rather than s, we write $\langle z_+, z\rangle$, and we write $\langle z, z_-\rangle$ to indicate that one of the $z's$ has time index $s-1$ rather than s. Denote the i-th iterates of matrix estimates by the superscript i, such as $A^{(i)}$. Then the estimates are updated by

$$A^{(i+1)} = \langle z^{(i)}, z_-^{(i)}\rangle\langle z^{(i)}, z^{(i)}\rangle^{-1},$$

$$C^{(i+1)} = \langle y, z^{(i)}\rangle\langle z^{(i)}, z^{(i)}\rangle^{-1},$$

$$e_t^{(i)} = y_t^s - C^{(i)} z_t^{(i)},$$

and so on. See Aoki (1994) for detail, and more recent related work by Havenner and Leng (1994) which has proofs of some convergence properties of this and related procedures and assesses the gain in efficiency of the proposed instrumental variable approach by running simulations for the estimators' effects on in- and out-of-sample predictions.

Alternatively we may generate an estimate of z_{t+1} separately from that of z_t by recalculating the orthogonal projection matrix separately from y_t^- rather than y_{t-1}^- and use it to calculate $\langle z_+, z\rangle$. There are several other possibilities in implementing the basic idea.

6 Concluding Remarks

Whenever models are constructed by retaining a subset of singular values the estimated matrices can possess the strict nestedness properties. In Akaike's canonical correlation method, orthogonality is imposed to produce state vectors with orthogonal components, i.e., when the dimension of the state vector is increased by one, the newly introduced state vector component is orthogonal to the existing state vector components by construction. In Aoki's algorithm the orthogonality property is used to ensure that the upper-left corners of estimated matrices are the correct estimates when the dimension of the state vector is reduced by one. This strict nestedness property is retained even in the case in which the estimated state vector is used as an instrument, even though one extra step is needed in the algorithm, see Aoki and Havenner (1991).

There are many ways of conducting constrained least squares estimations. The generalized singular value decomposition may be used to carry out these as discussed by van Loan (1976) who list four such problems. Thus, the key to modeling is model specification since there are many alternative ways of carrying out the estimates of the parameters, once the form of the model is specified. Modeling experience has established the superiority of state space specification over that of autoregressive moving average specification for vector valued data series, with and without exogenous input series. See the examples in this volume. Subspace methods are but one of many estimation/computation procedures for state space models in the same model specification framework as Aoki's method.

References

Aoki, M. (1983) *Notes on Time Series Modeling: System Theoretic Approach* Lecture Notes in Systems and Economic Dynamics, Springer, Heidelberg

——. (1987) *State Space Modeling of Time Series* (Springer-Verlag, New York).

——.(1989) *Optimization of Stochastic systems: Topics in Discrete Time Dynamics, Second Edition* (Academic Press, New York).

———. (1989b) Instrumental Variable Estimators for State Space Models, Discussion paper No.19, Institute for Empirical Macroeconomics, Federal Reserve Bank of Minneapolis, and University of Minnesota

————. (1990) *State Space Modeling of Time Series, Second, Revised and Enlarged Edition* (Springer-Verlag, New York).

————. (1991) Instrumental Variables Estimators for State Space Models of Time Series, pp.73-80, G.Conte et al. (ed) *New Trends in System Theory* Proceedings Univ. Genova-the Ohio State University Joint Conf. (Birkhauser, Boston).

——. (1994) Two Complementary Representations of Multiple Time Series in State-Space Innovation Forms *Journal of Forecasting* **13**, 69-90.

——, and J.H.Dorfman (1992) Statistical Analysis of Some Identification Procedures for State Space Innovation Models, pp.619-624 *Recent Advances in Mathematical Theory of Systems, Control, Networks and Signal Processing I*, Proc. International Symposium MTNS -91, H. Kimura and S. Kodama (eds), Mita Press, Osaka

——, and A. Havenner (1991) *Econometric Reviews* **10**, 1-59.

——,and ——. (1989). A Method for Approximate Representation of Vector-Valued Time Series and Its Relation to Two Alternatives *Journal of Econometrics* **42**, 181-199.

Havenner, A., and M. Aoki (1988) An Instrumental Variable Interpretation of Linear Systems Theory Estimation *Journal of Economic Dynamics and Control* **12** 49-54.

——, and Z.Leng (1994) Improved Estimates of the Parameters of State Space Time Series Models Working Paper, Department of Agricultural Economics, U. C. Davis. Published in *J. Economic Dynamics Control* **20** 767-789 (1996).

Hosoya, Y. (1977) On the Granger Condition for Non-Causality *Econometrica* **45** 1735-1736.

Householder, A. S. (1958) Unitary Triangularization of a Nonsymmetric Matrix *Journal ACM* **5**, 339-342.

Katayama, T., and G. Picci (1995) Stochastic Realization with Exogenous Inputs and "Subspace Methods" Identification Mimeo, Department of Applied Mathematics and Physics, Kyoto University

Lindquist, A., and G. Picci (1985) Realization Theory for Multivariate Stationary Gaussian Processes *SIAM J. Control and Optimization* **23** 809-857

Östermark, R., and M. Aoki (1992) Time Series Evidence of Impacts of the US Economy on the Scandinavian Economy *IFAC Workshop on Economic Time Series Analysis and System Identification*, Vienna

Paige, C. C. (1985) The Generalized Linear Model and the Generalized Singular Value Decomposition *Linear Alg. Appl.* **70** ,269-284.

——, and M. A. Saunders (1976) Towards a Generalized Singular Value Decomposition *SIAM J. Numer. Anal.* **18**, 398-405

G. Picci and T. Katayama: A Simple "Subspace"Identification method with Exogenous Inputs, *presented at 1996 IFAC (international federation of automatic control) Meeting*, San Francisco 1996.

G. Picci and T. Katayama, Stochastic realization with exogenous inputs and "subspace-methods" identification, *1996 mimeo (in review at Signal Processing).*

van der Veen, A.-J., E. F. Deprettere, and A. L. Swindlehurst (1993) Subspace-Based Signal Analysis Using Singular Value Decomposition *Proc. IEEE* **81**, 1277-1308.

van Loan, C. F. (1976) Generalizing the Singular Value Decomposition *SIAM J. Numer. Anal.* **13**, 76-83.

van Overschee, P. (1995) Subspace Identification: Theory-Implementation-Application Ph.D. thesis, Catholic University Leuvan, Department of Electrotechnique

van Overschee, P., and B. de Moor (1993) Subspace Algorithms for the Stochastic Identification problem *Automatica* **29**, 649-660.

—, B. de Moor, and J. Suykens (1992) Subspace Algorithms for System Idendification and Stochastic Realization *Proc. Mathematical Theory of Systems, Control, Networks and Signal Processing*, H. Kimura and S. Kodama (eds), 589-594, Mita Press, Osaka

Verhaegen, M. (1991) A Novel Non-iterative MIMO State Space Model Identification Technique *Proc. 9the IFAC/IFORS Symposium on Identification and System Parameter Estimation*, 1453-1458.

Viberg, M. (1994) Subspace Methods in System Identification *Proceeding SYSID* , 1-12.

A Guide to State Space Modeling of Multiple Time Series

Arthur Havenner
University of California
Davis, California 95616

I Introduction

In teaching time series analysis over the past several years I have become aware of the gap between the often elegant theory of the methods developed and the sometimes crude issues that arise in their application to real problems. A substantial amount of important practical information is relegated to a minor position and is either haphazardly conveyed in long sessions in the computer laboratory, or not at all. This guide is an attempt to organize and formalize that information as it applies to the use of Aoki's state space time series procedure, employing monthly data on four macroeconomic series as an illustration. Here I am taking the role of the expert witness, and offering opinions where relevant; the judgmental decisions embodied in the discussion reflect my experience, both from my own personal projects and from those of my students.

State space methods for multiple time series analysis emerged in the middle 1970's, based on multivariable systems/matrix transfer function extensions of scalar systems theory [Akaike (1974, 1975, 1976), Popov (1969), Rosenbrock (1970)]. Work accelerated in the last decade [Aoki (1983, 1987, 1990), and Hannan and Deistler (1988), *inter alia*], and crossed over from linear systems theorists to statisticians and econometricians: see, for example, Harvey's *Forecasting, Structural Time Series Models and the Kalman Filter* (1989), Shumway's *Applied Statistical Time Series Analysis* (1988), Wei's *Time Series Analysis* (1990), and Hamilton's *Time Series Analysis* (1994).

The state and observation equation pair underlies all state space methods. Arrange the series to be modeled in the $(m \times 1)$ vector y_t. The $(n \times 1)$ vector z_t denotes the unobservable states, while the m-element vector e_t indicates serially uncorrelated but perhaps contemporaneously correlated error terms. The notation $t \mid t-1$ indicates the orthogonal projection on the information set available at time *t-1*. Then a state space time series model can be written in innovation form as:[1]

$$\underset{(n\times 1)}{z_{t+1|t}} = \underset{(n\times n)(n\times 1)}{A z_{t|t-1}} + \underset{(n\times m)(m\times 1)}{B e_t} \tag{1}$$

$$\underset{(m\times 1)}{y_t} = \underset{(m\times 1)}{\mu} + \underset{(m\times n)(n\times 1)}{C z_{t|t-1}} + \underset{(m\times 1)}{e_t} \tag{2}$$

where the matrices A, B, and C are coefficients to be estimated.

While they all employ a matrix structure based on an augmentation to first order popularized in the Kalman filter, the state space time series analyses cited above differ dramatically in their reliance on the concepts of linear systems theory. A critical distinction centers on the process of model identification: if a procedure first identifies the model (typically an ARMA form) and then augments the model to first order, it implicitly discards the identification insights of systems theory. Although dynamic analysis is facilitated by the augmentation of the previously identified model, a number of other advantages of modern systems theory are lost.

The state space time series procedure proposed by Aoki[2] uses the concepts of systems theory at all stages of model specification and estimation, and, as a result, has a number of advantages over other apparently similar procedures. Before investigating how to effectively use the method, it is worthwhile to briefly consider six of these advantages:

1 Separate ARMA orders unnecessary

The n states in equation 1 are capable of representing any ARMA model up to and including an ARMA(n,n) by suitable augmentation to first order; see Aoki for details. Given the series to be modeled y_t, state space model specification consists of determining the number of states, n, rather than the two separate parameters p and q of an ARMA(p,q) process. In practice, accurate empirical determination of both p and q can be difficult, often requiring subjective evaluation of models generated by a two-dimensional scan over values of p and q. Errors in determining either p or q or both can seriously affect model performance. This problem is avoided by procedures that directly determine the number of states from the data (see below) rather than augmenting an ARMA model into state space form.

1 Sometimes the series to be modeled are assumed to be centered with compatible units of measurement in equations 1 and 2 above. This is accomplished by subtracting the mean vector μ and transforming by a matrix D. In terms of the original data, the observation equation becomes $D(y_t - \mu) = DCz_t + De_t$ with the moving average matrix B in the state equation redefined to include D. One adjustment of units of measurement sets the matrix D to a diagonal matrix of standard errors so that the Hankel matrix developed below is in terms of autocorrelations. (The program described in Appendix 2 defaults to this form automatically.) See Havenner and Criddle for a discussion of the effects of this measurement choice.

2 Henceforth references to Aoki will refer to the 1990 book unless otherwise specified.

2 The Kronecker Theorem

The Kronecker Theorem is the key model identification result from linear systems theory.[3] It says that the number of states required to characterize the dynamics of a stationary multivariate system is the rank of the Hankel matrix[4]

$$H = E(y_t^+ - Ey_t^+)(y_{t-1}^- - Ey_{t-1}^-)'$$

$$= \begin{pmatrix} \Gamma_1 & \Gamma_2 & \Gamma_3 & \Gamma_4 & \cdots & \Gamma_{N_p} \\ \Gamma_2 & \Gamma_3 & \Gamma_4 & \Gamma_5 & \cdots & \Gamma_{N_p+1} \\ \Gamma_3 & \Gamma_4 & \Gamma_5 & \Gamma_6 & \cdots & \Gamma_{N_p+2} \\ \Gamma_4 & \Gamma_5 & \Gamma_6 & \Gamma_7 & \cdots & \Gamma_{N_p+3} \\ \cdot & \cdot & \cdot & \cdot & \cdots & \cdot \\ \cdot & \cdot & \cdot & \cdot & \cdots & \cdot \\ \cdot & \cdot & \cdot & \cdot & \cdots & \cdot \\ \Gamma_{N_f} & \Gamma_{N_f+1} & \Gamma_{N_f+2} & \Gamma_{N_f+3} & \cdots & \Gamma_{N_f+N_p-1} \end{pmatrix} \tag{3}$$

where

$$y_t^{+\prime} = (y_t', y_{t+1}', y_{t+2}', \ldots, y_{t+N_f-1}'); \tag{4}$$

$$y_{t-1}^{-\prime} = (y_{t-1}', y_{t-2}', y_{t-3}', \ldots, y_{t-N_p}'); \tag{5}$$

$$\Gamma_l = E(y_{t+l} - Ey_{t+l})(y_t - Ey_t)', \quad l = 0, 1, 2, 3, \ldots; \tag{6}$$

and N_f and N_p are sufficiently large that they do not constrain the choice of n.

Choosing the number of states based on the rank of the Hankel matrix brings a direct formal mathematical result to bear on the main model specification issue in time series modeling. Procedures that simply augment to state space form do not use the important information available in this arrangement of the sample information.

3 Data based restrictions

Any time series model identification method consists of finding or arbitrarily imposing restrictions to eliminate excess sampling error in the parameter estimates and forecasts. For example, the early Box and Jenkins univariate time series identification procedure can be interpreted as a prescription for what to calculate from the data -- the autocorrelations and partial autocorrelations -- in order to determine the orders p and q of the ARMA process, thus zeroing coefficients beyond these lags. Similarly, VAR restrictions exclude moving average coefficients, and Litterman's procedure imposes stochastic restrictions or Bayesian priors on VAR coefficients to reduce sampling error.

---

3 See, *e.g.*, Aoki and Havenner (1991).

4 Individual block components of the matrix H are autocovariances, or if suitably scaled, autocorrelations.

The arbitrary near random walk restrictions proposed by Litterman are often appropriate in areas such as finance and thus can be highly successful; in other areas, such as natural resources, they are less appropriate and less successful. (No one would seriously propose this method to model river flooding like that studied in Havenner and Tracy, for example.) The data-based restriction methodology of Box and Jenkins, while very successful in the univariate case, has not been adequate for the complex interactions of multivariate series.

The states constructed in a true systems theory method such as Aoki's can be interpreted as the n most important dynamic factors in approximating the autocovariance sequence of the vector y_t. In this case, the restrictions that reduce sampling error are imposed on the singular values of the Hankel matrix: those below a bound are assumed to be zero apart from sampling error and thus are set to zero, embodying the model specification step. These restrictions are based on the data, and are capable of characterizing any type of stochastic process. They are shared over the multiple series in the model. In addition, the restrictions are imposed directly on the determinants of the dynamic approximation, rather than on unstable coefficients of highly collinear "regressors" in a changing model specification.

4 Common cycles

The m series in y_t may require more or less than m states to characterize their dynamics. When $n < m$ states are required the series share common cycles, that is the C matrix in equation 2 has more rows than columns and the fundamental dynamics of just n states are disaggregated into all m multiple series. The ability of true systems theory procedures to find these common cycles (based on points 1, 2, and 3 above) allows a smaller specification for n and therefore less sampling error in the estimated parameter matrices A, B, C, and the associated covariance matrices. The possibility of common cycles can be extremely important in modeling large closely related multivariate systems (see, *e.g.*, the seven weekly live cattle series in Foster, Havenner, and Walburger, or the six monthly soybean complex series in Berwald and Havenner; recent unpublished work has involved even larger numbers of series).

5 Sequence-independent model choice

The true systems theory approach to time series modeling treats model specification as a problem of formally approximating the autocovariance (or autocorrelation) sequence of the series to be modeled. This is in contrast to other methods which attempt to discover the characteristics of the model that must have generated the autocorrelations, or impose arbitrary restrictions. In Aoki's procedure, states enter the model in order based on their contribution to the approximated autocovariance sequence. This natural order reduces the sequence dependence of many identification schemes, in which the ultimate model chosen depends on the order of the preliminary models examined.[5]

5 It also makes computationally feasible Bayesian analysis possible, see Dorfman and Havenner (1991).

6 Computational stability and model robustness

Since the states are unobservable latent variables, their measurement is arbitrary up to a nonsingular matrix transformation, *i.e.*, the state coordinate system is open. Modern systems theory has established the "balanced representation" as the coordinate system of choice for computational reasons (see, *inter alia*, Kailath or Aoki). Aoki's parameter estimates embody this scaling of the states, but state space forms based simply on augmentation to first order do not. While statisticians and econometricians have not been much concerned with computational considerations, they are important nonetheless. As Kailath notes in his preface (page *xv*): "The significance of this stability is that the effect of numerical errors in computation, *e.g.*, round off errors, does not build up and destroy the calculation -- obviously a very important practical consideration."

An implication of this choice of coordinate system is that some of the parameter estimates are "strictly nested." This means that the leading principal submatrices of certain parameter estimates are unchanged when additional states are added to the model. For instance, the northwest corner of the dynamics matrix A in equation 1 is unchanged when additional states are added; see, *e.g.*, Aoki and Havenner (1991). This adds an additional element of stability to the process of model identification. (Compare to VAR model specification, for example, where multicollinearity often radically alters the preceding coefficient estimates when an additional lag is included in the model.)

Based on the six points elaborated above, the main relative advantage of systems theory based methods lies in their application to multivariate systems. With multivariate systems, the dynamic interactions between multiple series are subtle and difficult to accurately characterize without powerful methods. The identification, common dynamics, and computational considerations of systems methods often lead to better specified, more accurate models of larger multivariate systems.

In the sections below we will go through the process of fitting time series models using Aoki's state space method. The discussion will be framed in terms of a sample data set of 177 seasonally adjusted observations on the U.S. monthly industrial production index (IPI), consumer price index (CPI), four to six month commercial paper rate (RCP), and money stock (M1). Although the monthly frequency makes the one period forecasts more accurate than usual, this dataset is appropriate because it involves many of the issues that occur in practical applications of the method to other series. These include problems such as nonstationarity, calculation instability, and choice of series to be modeled, for example.

II Preliminary Considerations

Before fitting a state space model, we must determine which series will be modeled together, and whether or not they embody deterministic or stochastic nonstationarities. The single most important tool is a simple plot of each series, provided for the macroeconomic data example in Figures 1-4. The idea is to get an overall impression of the shape and volatility of the series.

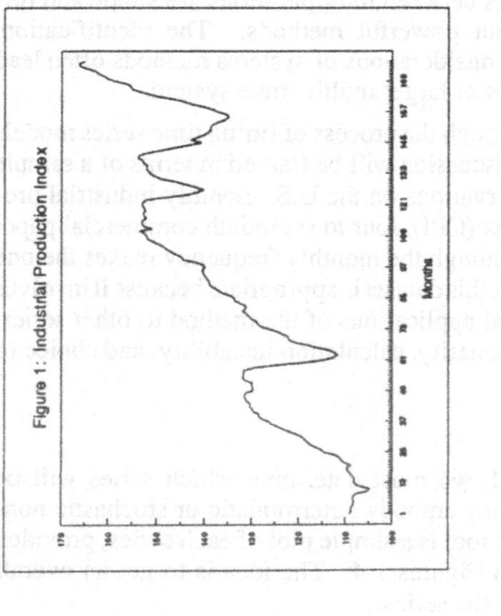

Figure 1: Industrial Production Index

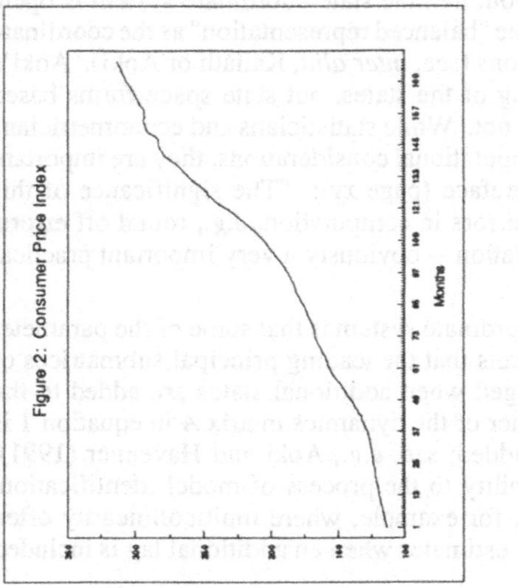

Figure 2: Consumer Price Index

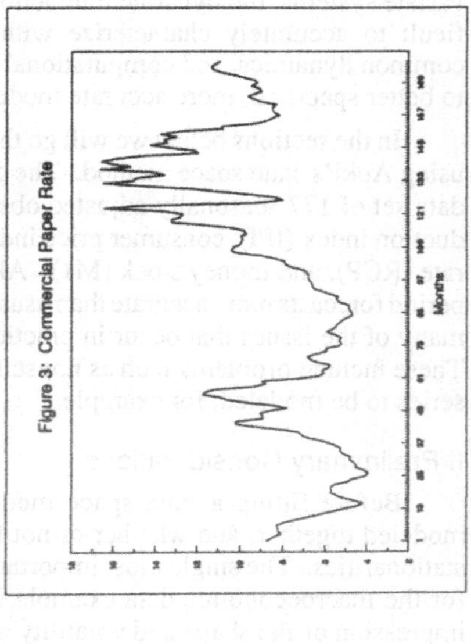

Figure 3: Commercial Paper Rate

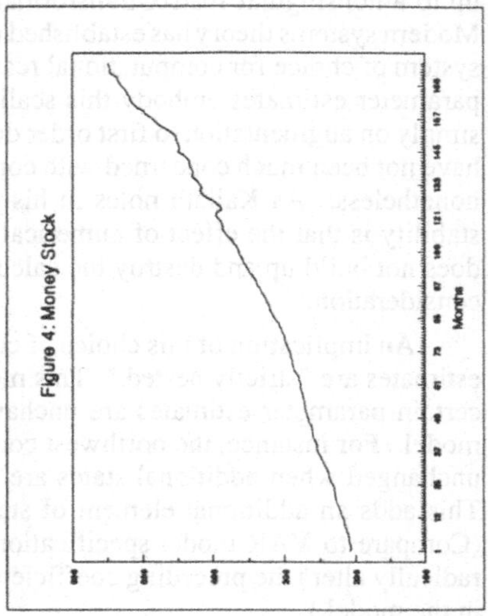

Figure 4: Money Stock

1 Series

Choosing which series to model together is a far less trivial undertaking than might be expected *a priori*. Obviously variables to be modeled by time series methods should have important autocorrelations; data that we think of as dynamic, such as salmon river runs, can be surprisingly nonautocorrelated, however. Also, the dynamic characteristics of series that theory groups together can differ dramatically.[6] The things we think we know are related sometimes are not, and the resulting multivariate models don't always predict better than univariate or smaller multivariate models. For example, expenditure line items such as for defense, agriculture, health and welfare, *etc.*, are grouped together by the Department of Commerce and may share a political base but, at least to some extent, the first depends on international events, the second on crop failures, the third on the business cycle, *etc.*, so that their common components are minimal compared to their differences. Even series that seem closely related can be detrimental to model accuracy -- see the brief discussion of the effect of hay prices on the cattle price model in the paper by Egan and Havenner in this volume.

Examining the plots of the macroeconomic data (Figures 1-4), we see marked differences in dynamic characteristics despite the fact that we certainly expect industrial production, consumer prices, interest rates, and the money stock to be related. For example, IPI has a lot of variation[7] between its maximum and minimum sample values relative to CPI and M1, which are comparatively smooth upward trends. The commercial paper rate RCP also has a lot of variation, with substantially less upward trend. Taking the logarithms of the series decreases their apparent differences (see Figure 5 in section II) and is consistent with proportionate changes affecting each other; the logarithms will be used in the subsequent analysis. Even in logarithms, RCP remains marked different in appearance. While these series don't have to look exactly alike -- the time series procedure can smooth them, for example -- the conspicuous differences in their graphs are a warning to examine models with different combinations of series.

2 Nonstationarity

Inexperienced analysts have the most difficulty with the question of nonstationarity in the series to be modeled. In the not so distant past, the accepted procedure was to difference every series until the graph against time looked the same upside down as right side up, *i.e.*, there was no noticeable trend remaining. In addition, the autocorrelations were calculated, and the series differenced until these tended toward zero at long lags. We now know that stationary series are not required for forecasting and hypothesis testing in many cases, and these differencing procedures can lose a lot of information.

6 Consider the series in a natural rate of unemployment model, or the commercial paper rate versus the other series in the example pursued below.

7 Plotting these with zero on the vertical axis makes them look a lot more alike, emphasizing the trend and reducing the relative variation. There is no natural zero for IPI or CPI, however.

Nelson and Plosser established the classic separation of nonstationarities into two categories, deterministic (*e.g.*, time trends[8]) and stochastic (unit root processes), and proposed a test to distinguish between them. In addition, they investigated the implications of misclassification and characterized the essential features of the two classes. Chapter 15 of Hamilton provides an excellent summary and integrates later work in the area.

Models with a mixture of stationarity assumptions present special problems. Combining a nonstationary series with a stationary series has certain implications -- the nonstationary series can depend on the stationary series, but not *vice versa*, although the stationary series can depend on the appropriate combination of cointegrated nonstationary series. Series that grow with time can depend on those that don't, but not *vice versa*, except in combinations that perfectly eliminate time effects. (Series that increase with time at different rates require the same relations.) These relationships should be consistent with theory if they are used as the basis of a model.

Although there are formal tests for the different types of nonstationarity, and an informal examination of the autocorrelations at long lags is useful, theoretical considerations (common sense?) combined with time plots of the series should be the first consideration. Prior information about the series is sometimes overlooked, and in any event there is no point in testing something we have to believe is true.[9]

Turning to the macroeconomic data example, industrial production and the money stock are obviously nonstationary -- they are not in per capita terms, and even if they were they would be expected to increase with time. Similarly, the consumer price index is expected to grow through time. These series either have deterministic trends or are random walks with drift. The commercial paper rate is expected *not* to increase with time; the only reasonable form of nonstationarity for this series is a random walk without drift, and stationarity remains a strong possibility.

Calculation[10] of the autocorrelations of all four series shows the slow (less than geometric) decay of nonstationary series. Although the RCP autocorrelations decline very slowly, they do tend toward zero by lag 30, and cycle toward zero at later lags. The other three series grow with time, and their autocorrelations show the classic slow decline to zero and then slow decrease (increasing negative, away from zero) at long lags that comes from calculation around a fixed mean when the series is trending.

Thus all evidence points to either a time trend or a random walk with drift for IPI, CPI, and M1, with RCP uncertain. The Nelson and Plosser test can be used to distinguish between trend and difference stationary series based on the regression

$$y_t = \mu + \rho y_{t-1} + \gamma t + \varepsilon_t \qquad (7)$$

---

8 Any known nonstochastic conditioning variable could be a deterministic trend (for example, irrigation in agricultural applications); for expositional convenience discussion centers only on time trends. The statistical properties -- such as superconsistency -- will depend on the dynamic attributes of the conditioning variable.

9 Testing for cointegration of regional agricultural market prices comes to mind. For example, we know regional live cattle prices cannot diverge infinitely due to the possibility of cross shipping and therefore must be cointegrated. Conclusions in cases such as these tell more about the power of the test than about the underlying economic phenomena.

10 All calculations are based on 153 in sample observations, with 24 reserved for out of sample model validation.

and the null hypothesis H_0: $\rho = 1$ and $\gamma = 0$; rejecting implies a trend stationary series, while not rejecting implies a difference stationary series. The test statistic is the usual calculated F, but its distribution is no longer the standard F distribution since $\rho = 1$ under the null hypothesis. The correct distribution has been tabled by Dickey and Fuller. At the five percent significance level, the critical value is 6.39 (by interpolation). The calculated value for IPI is 1.57, failing to reject the null hypothesis and implying difference stationarity. For CPI the calculated value is 52.48, implying trend stationarity, and for M1 the calculated value is 6.26, implying difference stationarity (although it is close). Neither of the models in the Nelson and Plosser test -- the random walk with drift, or a model with a time trend -- is plausible for RCP. To formally test for nonstationarity in this case, the random walk without drift was tested against a stationary process with an intercept (case two in Dickey-Fuller tests). The critical value of the t statistic analog is -2.88, and the calculated value is -1.59. Since the calculated value is not to the left of the critical value, the null hypothesis that $\rho = 1$ cannot be rejected in favor of $\rho < 1$.

All of this places us in a bit of a bind. By the test results, we have two random walks with drift, a series with a time trend, and a random walk without drift (that we think from theory might well be stationary). This is a somewhat odd combination of dynamics to collect in a model. The random walk with drift is not a bad model for IPI -- we can imagine population, capital, and technology shifting industrial production with its mean remaining at the shifted level rather than returning to the time trend as required by the trend stationary model. The same argument based on population and financial innovations fits M1. Finding CPI to be trend stationary is difficult to reconcile with its relation to M1. It seems reasonable to model CPI as a random walk with drift on economic grounds, although the strength of the Nelson Plosser test result is troubling. The net conclusion is that these series may not be as useful in predicting each other as we would expect *a priori*. The interest rate, RCP, is particularly suspect as a useful series in this multivariate model.

i Deterministic nonstationarity

If there are deterministic trends in the series, they can be removed by regressions and the errors used as input in the state space time series procedure.[11] Add the regression equation 8 to the original state equation 1 and a slight modification of the observation equation 2:

$$z_{t+1|t} = A z_{t|t-1} + B e_t \tag{1}$$

$$\varepsilon_t = C z_{t|t-1} + e_t \tag{2'}$$

$$y_t = \beta x_t + \varepsilon_t \tag{8}$$

11 This includes ordinary regression analysis with autocorrelated errors as a (possibly stationary) subcase, using the state space procedure as a multivariate error serial correlation model with strong identification results. The coefficient matrix β can be restricted if exogenous variables are absent from some regressions. See Dorfman and Havenner (1992) or Criddle and Havenner for applications.

where β is the $(m \times k)$ matrix of coefficients relating the k regressors in x_t to the m series in y_t to be modeled, and ε_t is the multivariate serially correlated error term that is input to the state space model of equations 1 and 2′ above. The regression and time series model estimates can be iteratively revised in the usual generalized least squares (GLS) manner. The error covariance matrix is easily developed [see Havenner and Aoki (1990)] even in the general multivariate state space model corresponding to a vector ARMA(*p*,*q*), but the $(mT \times mT)$ size is frequently unworkably large for the usual GLS calculations. An alternative to the customary GLS calculations is based on recognizing that since

$$\varepsilon_t - Cz_{t|t-1} = e_t \tag{9}$$

and e_t is serially uncorrelated, subtracting $Cz_{t|t-1}$ from each side of equation 8 eliminates the serial correlation and therefore the troublesome T dimension of the multivariate error covariance matrix:

$$y_t - CZ_{t|t-1} = \beta x_t + \varepsilon_t - C_{t|t-1} \tag{10}$$

or

$$y_t^* = \beta x_t + e_t. \tag{11}$$

The state space error e_t is in general contemporaneously correlated. Define its covariance matrix to be $\Psi = Ee_te_t'$, with factors $P'P \equiv \Psi^{-1}$. Then the error in

$$Py_t = P(x_t' \otimes I_m)vec(\beta) + Pe_t \tag{12}$$

is $iid\ (0, I_m)$. Substituting appropriate sample estimates and iterating produces asymptotically efficient parameter estimates.

If (as is often the case) the regressors in x_t are identical for all series, then the data can be reordered from the current T blocks of m into m blocks of T and the well known demonstration that the contemporaneous error covariance matrix drops out of the calculations is valid. The algebra is as usual, but it is worth noting that it only holds because of the manner in which the serial correlation is eliminated -- $x_t' \otimes I_m$ is unaffected by the subtraction of $Cz_{t|t-1}$, unlike, *e.g.*, the case with the Orcutt transformation. Thus for the same set of deterministic trends for all series, ordinary least squares on the serial correlation corrected data is asymptotically efficient, and P can be neglected.

If the exogenous variables are a constant and time trend, *i.e.*, $x_t' = (1, t)$, then the coefficients can be estimated by ordinary least squares and the coefficient on t will be superconsistent, converging at $T^{3/2}$, where T is the sample size. The intercept is $\sqrt{T}$ consistent, as usual. This provides some justification for ignoring possible efficiency gains in estimating the parameters by not iterating at all, and simply using the time series model error predictions to improve the forecasts.

ii Stochastic nonstationarity

When the series exhibit stochastic nonstationarity, a two step application of Aoki's procedure can be used to identify common long run features and to separate long and short run dynamics; see Aoki, or, especially, section 4 of Aoki and Havenner (1991). For applications, see among others Arndt and Foster in this volume, or Cerchi and Havenner, or Foster, Havenner, and Walburger.

The idea is to decompose the vector of observed series y_t into long run dynamics [corresponding to large or unit eigenvalues] summarized in a set of states $\tau_{t|t-1}$, and a stationary transform of y_t that retains the higher frequency dynamics, denoted y_t^*. This stationary transform of y_t takes the place of the error in the observation and state equation of the state space model. The observation equation (analogous to equation 2) that embodies this decomposition is

$$y_t = \mu_\tau + C_\tau \tau_{t|t-1} + y_t^*, \tag{13}$$

with the corresponding state equation

$$\tau_{t+1|t} = A_\tau \tau_{t|t-1} + B_\tau y_t^*. \tag{14}$$

The stationary residual from this model, y_t^*, is used as input into another state space model designed to capture the remaining (high frequency) cycles:

$$y_t^* = \mu_\eta + C_\eta \eta_{t|t-1} + e_t \tag{15}$$

with state equation[12]

$$\eta_{t+1|t} = A_\eta \eta_{t|t-1} + B_\eta e_t \tag{16}$$

where e_t is a serially uncorrelated error vector. Thus the two models, the trend model of equations 13 and 14 and the cycle model of equations 15 and 16, reduce the series y_t to serially uncorrelated errors e_t, although the output from the trend model alone y_t^* is serially correlated at the high frequencies.

The separate trend and cycle models can be combined into one stacked model for ease in analysis and forecasting. Substitute equation 15 for y_t^* in both 13 and 14, rearrange and stack with equations 15 and 16 to obtain

$$\begin{pmatrix} \tau_{t+1|t} \\ \eta_{t+1|t} \end{pmatrix} = \begin{pmatrix} B_\tau \\ 0 \end{pmatrix} \mu_\eta + \begin{pmatrix} A_\tau & B_\tau C_\eta \\ 0 & A_\eta \end{pmatrix} \begin{pmatrix} \tau_{t|t-1} \\ \eta_{t|t-1} \end{pmatrix} + \begin{pmatrix} B_\tau \\ B_\eta \end{pmatrix} e_t \tag{17}$$

$$y_t = (\mu_\tau + \mu_\eta) + (C_\tau \quad C_\eta) \begin{pmatrix} \tau_{t|t-1} \\ \eta_{t|t-1} \end{pmatrix} + e_t \tag{18}$$

the combined state and observation equations. These equations constitute a state space model in their own right,[13] and can easily be solved for forecasts of the original series regardless of the specific details of the particular trend and cycle models.[14]

---

12 Although the intercept μ_η in the cycle observation equation is expected to be small, there is no first order condition guaranteeing it will be zero in general. Experiments with the iterative estimator described in section IV below suggest that good estimates of these constants are important to accurate forecasting.

13 Indeed, the eigenvalues that characterize the dynamics of the series are the eigenvalues of the dynamics matrix from equation 18, see Aoki and Havenner (1991).

14 In the special case of differencing each series individually, these equations automatically sum back to the original observations.

In practice, the procedure is to fit a model using state space identification methods to the long term components of y_t, obtain the errors y_t^*, and then use state space methods again to fit a model to the high frequency components in the error. The lag parameter N_p is used to separate the dynamics. Heuristically, if we regress a series on just its single lagged value, we obtain a long run average effect, while if we regress a series on multiple lagged values we can get some cyclic detail. Thus in fitting the trend model we choose the lag parameter N_p to be very small, typically one for multivariate series (since the implied determinantal polynomial for each series has order up to mN_p and we want the equivalent of a regression on only immediate lags in order to capture the trends). For the cycle model the lag parameter is increased, with the particular value depending on the series being modeled. Since the input to the cycle model has had the trends removed, the autocorrelations are implicitly calculated around the trend values rather than around a fixed mean. This keeps the slow dynamics from obscuring the faster dynamics, and allows more accurate models of the higher frequency effects. A good job of trend removal captures the slow dynamics by fitting a model that smoothly approximates the low frequency dynamics, *i.e.*, by laying a smooth fitted line through the long easy curves of the graphed data.

In both the trend and cycle model formulation steps, the Hankel matrix of autocovariances or autocorrelations can be used to determine the required number of states. The specification advantages of the balanced representation carry over as well. Whether or not the magnitudes of the eigenvalues of A_τ are less than one, the long run dynamics are set by the data through the modeling process, permitting complex trends if appropriate. If there are more states in $\tau_{t|t-1}$ than series in y_t, a trend model in excess of first order was required, and has been implicitly augmented to a first order matrix state equation. If there are fewer trend states than series then there are "common trends," *i.e.*, the long run stochastic components are shared by the series being modeled.[15] These states may or may not embody nonstationarities, depending on the eigenvalues of A_τ; in any case, $y_t^* = y_t - C_\tau \tau_{t|t-1}$ has the long run dynamics removed.

Since three of the macroeconomic data series [IPI, CPI (forcing it difference stationary rather than trend stationary, to match the others), and M1] are suspected random walks with drift, and the RCP series is a suspected random walk without drift, the two step stochastic trend removal procedure just described is appropriate. Tables 1-6 summarize information from six state space models of the macroeconomic data: a multivariate model of all four series, a multivariate model without the mismatched commercial paper rate RCP, and separate univariate models for IPI, CPI, RCP, and M1. The statistics reported in the tables are the mean absolute percentage error (MAPE), the root mean squared error (RMSE), the correlation of each series' actual and forecast values (RSQ), the average error (AVGERR), the mean absolute deviation

15 When there are fewer states in the trend model than series modeled, any matrix in the null space of C_τ' annihilates the nonstationary components when applied to equation 13, and thus is a cointegrating matrix. Denote such a matrix α, where $C_\tau'\alpha = 0$. Then multiplying the observation equation 13 by α' gives

$$\alpha' y_t = \alpha' C_\tau \tau_{t|t-1} + \alpha' y_t^* = \alpha' y_t^*$$

with α indexing the transformations of y_t that are stationary.

(MAD), the confidence level for the Henriksson-Merton nonparametric test of direction[16] (HM), and the error autocorrelations (each series is a column). In addition to the unconditional series means (MEANS), the estimated intercept (CONSTANT) is also reported when the iterative estimator proposed by Havenner and Leng (discussed in section IV below) is used.

We will examine the multivariate model with all four series first. Setting the lag parameter N_p to one to focus on long run dynamics, the calculated singular values (Table 1) suggest that the rank of the (4×4) Hankel matrix is one, *i.e.*, there is one common stochastic trend in all of the series. Figures 6 and 7 provide, for IPI and RCP respectively, graphs of the actual and forecast values and the "errors" y_t^*, the stationary transform of y_t to be the input into the cycle model. The other two series, CPI and M1, are similar to the IPI plot. For using just one state to predict all four series, the model is surprisingly accurate on three of them (all but RCP) in the sense that the errors are small relative to the levels. Of course, the errors are highly serially correlated, and the cycle model would be expected improve the fit even more. As a trend run, the multivariate model falls short, however: the RCP trend does not smoothly track the series, and even the IPI errors (and the CPI and M1 errors not shown) exhibit slower dynamics than desired after detrending.

The statistical mismatch of RCP and the rest of the series is a problem. A univariate RCP trend run (with $N_p = 1$, see Table 5) produced Figure 8. Without the other series to fit, the trend forecasts are not as smooth, and the trend model is nearly a random walk. The separation into long run and short run dynamics is better, however. In the four series model (see Table 1) the second singular value is a little larger than typically seen in trend runs. Since the RCP dynamics are known to differ from the other series, it is reasonable to try a trend model with two states for these four series. (The additional state would not be expected to result in a second order trend, but rather to separately fit the RCP series.) Such a model does in fact fit the RCP series very much better. However, the resulting cycle models have symplectic matrix problems (see section III.3.ii below) at all reasonable values of the lag parameter ($N_p = 2, 3, 4$) and states ($n=1, 9$). Better models treat RCP separately. A trend model based on IPI, CPI, and M1 (without RCP) was estimated (Table 2); it was an improvement over the four series model. In addition to the separate univariate model for RCP, univariate IPI, CPI, and M1 trend models were also fitted (Tables 3, 4, 6), based on the initial observation that the IPI dynamics appeared different (in graphs) from those of the other series and CPI tested different from the others. These trend models will be kept in mind as various cycle models are developed.

16 See Henriksson and Merton for the statistical development. The test statistic itself is discussed in section III.4.ii. below.

Figure 5: Macroeconomic Data
In Logarithms

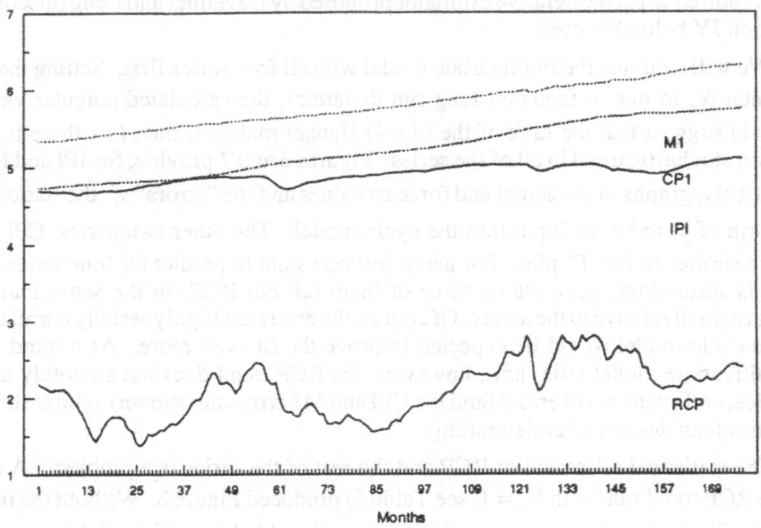

Figure 6: Multivariate Industrial Production Index Trend
In Logarithms In Sample

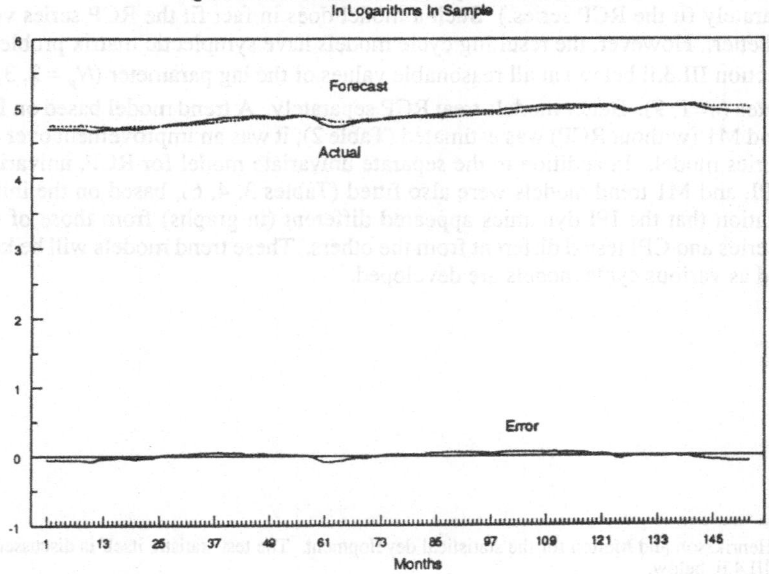

Table 1: Four Series Model

---------- Trend Model ----------
Np = Nf = 1

SINGULAR VALUES AND RATIO TO FIRST SINGULAR VALUE

SVS	NSVS	RATIO
3.3249	1	1
.40397	2	.1215
.16771	3	.050441
.0021954	4	6.6029E-4

C O E F F I C I E N T   E S T I M A T E S

A	B	C
.9836	-3.261	-.10463
	-1.1592	-.2813
	-.61311	-.31433
	-.93806	-.22776

TREND MODEL
NP = NF = 1, NSTAR = 1
IN SAMPLE RESULTS

SERIES	MEANS	MAPE	RMSE	RSQ	AVGERR	MAD	COVCHEKE	HM
IPI	4.88	.792	.0437	.858	.00158	.0386	1.32	1
CPI	5.15	1.25	.0776	.93	.00426	.0643	1.19	-
RCP	2.07	9.64	.23	.644	.00476	.19	.962	.986
M1	5.73	.819	.0549	.946	.00345	.0467	1.42	.091

LAG	ERROR AUTOCORRELATIONS			
1	.961	.948	.928	.92
2	.908	.868	.832	.798
3	.848	.778	.752	.677
4	.785	.684	.684	.558
5	.721	.597	.62	.454
6	.657	.515	.552	.357
7	.595	.446	.505	.283
8	.539	.385	.485	.228
9	.479	.327	.462	.172
10	.414	.266	.427	.116
11	.351	.213	.387	.0754
12	.294	.164	.339	.0436

Table 1 (continued): Four Series Model

---------- Cycle Model ----------
Nf = Np = 3

SVS	NSVS	RATIO
5.5162	1	1
3.8799	2	.70336
.13424	3	.024336
.080417	4	.014578
.030279	5	.0054891
.016135	6	.002925
.0064421	7	.0011678
.0058861	8	.001067
.0022615	9	4.0997E-4
.001283	10	2.3259E-4
3.7734E-4	11	6.8406E-5
2.1787E-4	12	3.9495E-5

SINGULAR VALUES AND RATIO TO FIRST SINGULAR VALUE

Symplectic Matrix Roots (Four States)

Eigenvalue	Absolute Value	
-.20812+.1488i	.25585	
-.20812-.1488i	.25585	
.39607	.39607	
.34006+.94041i	1	<-- Indication of
.34006-.94041i	1	<-- a problem
2.5248	2.5248	
-3.1795+2.2732i	3.9086	
-3.1795-2.2732i	3.9086	

(Unit roots in the symplectic matrix occurred with three to eight states.)

Aoki's C O E F F I C I E N T   E S T I M A T E S, Nf = Np = 3, NSTAR = 2

A		B				C	
.88283	-.07324	-15.736	6.3912	-3.8207	-4.5194	-.0017654	.042609
.093648	.95976	21.525	-.029583	-.79166	9.6048	.069456	-.029302
						-.16733	-.12647
						.05391	5.8614E-4

A: Eigenvalues	A: Absolute Value	A - BC: Eigenvalues	A - BC: Absolute Value
.92129+.073342i	.92421	-.024279+.39427i	.39502
.92129-.073342i	.92421	-.024279-.39427i	.39502

******** NEGATIVE EIGENVALUE IN COVARIANCE MATRIX ********

CYCLE MODEL
IN SAMPLE

SERIES	MEANS	MAPE	RMSE	RSQ	AVGERR	MAD	COVCHEKE	HM
IPI	.00158	29.1	.00962	.952	-3.16E-4	.00721	1.57	.974
CPI	.00426	131	.013	.973	.0015	.0099	-1.36	1
RCP	.00476	129	.0835	.87	-.00231	.0617	1.61	.901
M1	.00345	87.5	.0151	.926	.00101	.0125	-.247	1

Table 1 (continued): Four Series Model

---------- Cycle Model ----------
Np = Nf = 3

Iterated C O E F F I C I E N T		E S T I M A T E S, Nf = Np = 3, NSTAR = 2					
A		B				C	
.94099	-.10637	-14.468	6.3751	-3.6978	-3.722	-.0020846	.043125
.078341	.9784	21.535	.35927	-.85556	9.5073	.067753	-.029597
						-.18174	-.12177
						.051846	.0017545

A		A - BC	
Eigenvalues	Absolute Value	Eigenvalues	Absolute Value
.95969+.089349i	.96384	-.030338+.37862i	.37983
.95969-.089349i	.96384	-.030338-.37862i	.37983

CYCLE MODEL
IN SAMPLE

SERIES	MEANS	CONSTANT	MAPE	RMSE	RSQ	AVGERR	MAD	HM
IPI	.00158	9.67E-4	29.4	.00965	.951	-3.44E-4	.00724	.927
CPI	.00426	.0072	131	.013	.972	.00164	.00982	1
RCP	.00476	-5.66E-4	148	.0832	.869	-.00297	.061	.684
M1	.00345	.00539	77.3	.0147	.928	.00108	.0116	1

Aoki's Coefficient Estimates ERROR AUTOCORRELATIONS				LAG	Iterated Coefficient Estimates ERROR AUTOCORRELATIONS			
.402	.646	.451	.866	1	.382	.717	.423	.879
.262	.43	.0834	.708	2	.245	.509	.0318	.712
.286	.3	-.0206	.53	3	.283	.369	-.0558	.526
.208	.197	-.0544	.392	4	.219	.262	-.0762	.374
.211	.187	-.0365	.282	5	.225	.241	-.0555	.251
.223	.171	-.188	.219	6	.242	.233	-.221	.174
.104	.123	-.167	.16	7	.132	.201	-.202	.106
.082	.204	.0523	.125	8	.112	.261	.0287	.0599
.144	.331	.124	.106	9	.178	.364	.103	.0244
.043	.325	.11	.0784	10	.0902	.37	.0868	-.00448
.035	.307	.149	.0623	11	.0831	.344	.131	-.0173
-.015	.258	.0918	.0613	12	.0214	.3	.0776	-.00691

Table 1 (continued): Four Series Model

---------- Combined Trend and Cycle Model ----------

A			B				C		
.9836	-.00939	-.03313	-3.261	-1.159	-.613	-.93806	-.1046	-.00230	.04317
0	.94407	-.10729	-15.54	4.918	-3.57	-4.3922	-.2813	.06828	-.02976
0	.07633	.97805	22.38	1.31	-.935	9.9874	-.3143	-.18133	-.12178
							-.22776	.05217	.00161

COMBINED TREND AND CYCLE MODELS
IN SAMPLE RESULTS

SERIES	MEANS	CONSTANT	MAPE	RMSE	RSQ	AVGERR	MAD	HM
IPI	4.88	4.88	.148	.00955	.994	-3.87E-4	.0072	1
CPI	5.15	5.16	.187	.0129	.999	.00167	.00976	.757
RCP	2.07	2.06	2.97	.0832	.954	-.00286	.061	.994
M1	5.73	5.73	.203	.0147	.996	.00108	.0116	.998

LAG	ERROR AUTOCORRELATIONS			
1	.385	.728	.423	.881
2	.244	.515	.0317	.712
3	.283	.373	-.0556	.526
4	.219	.265	-.0754	.374
5	.225	.242	-.0546	.251
6	.238	.233	-.221	.173
7	.128	.204	-.202	.106
8	.114	.266	.0282	.0602
9	.189	.368	.103	.0243
10	.0857	.367	.0861	-.00518
11	.0681	.344	.13	-.0174
12	.0207	.31	.0755	-.00604

COMBINED TREND AND CYCLE MODELS
FULL SAMPLE (ALL 177 OBSERVATIONS)

SERIES	MEANS	CONSTANT	MAPE	RMSE	RSQ	AVGERR	MAD	HM
IPI	4.88	4.88	.221	.0156	.99	-.005	.0109	1
CPI	5.15	5.16	.204	.0139	.999	.00378	.0109	.945
RCP	2.07	2.06	2.91	.0812	.952	.00391	.0603	.995
M1	5.73	5.73	.347	.0323	.993	.0116	.0208	.997

Table 2: Three Series Model

---------- Trend Model ----------
Np = Nf = 1

SINGULAR VALUES AND RATIO TO FIRST SINGULAR VALUE

SVS	NSVS	RATIO
2.7598	1	1
.16803	2	.060887
.0027457	3	9.9489E-4

C O E F F I C I E N T   E S T I M A T E S

A	B	C
.98324	-3.6596	-.10726
	-1.5579	-.28416
	-.85198	-.23207

IPI, CPI, M1 TREND MODEL
IN SAMPLE RESULTS

SERIES	MEANS	MAPE	RMSE	RSQ	AVGERR	MAD	COVCHEKE	HM
IPI	4.88	.66	.0384	.89	.00191	.0322	1.36	1
CPI	5.15	1.11	.067	.948	.00507	.0571	1.3	-
M1	5.73	.443	.0351	.978	.00414	.0255	2	.477

LAG	ERROR AUTOCORRELATIONS		
1	.954	.967	.927
2	.892	.916	.819
3	.82	.853	.697
4	.739	.782	.566
5	.658	.71	.444
6	.58	.64	.336
7	.508	.572	.244
8	.445	.501	.159
9	.377	.434	.0817
10	.308	.369	.0241
11	.244	.307	-.0157
12	.185	.25	-.0383

Table 2 (continued): Three Series Model
---------- Cycle Model ----------
Np = Nf = 3

SINGULAR VALUES AND RATIO TO FIRST SINGULAR VALUE

SVS	NSVS	RATIO
6.782	1	1
.41088	2	.060584
.10689	3	.015761
.01976	4	.0029136
.004380	5	.0006459
.002664	6	3.9281E-4
.0017626	7	2.5989E-4
7.5781E-4	8	1.1174E-4
2.9365E-5	9	4.3298E-6

IPI, CPI, M1 CYCLE MODEL
Np = Nf = 3, Five States

Based on Aoki's Coefficient Estimates

A: Eigenvalues	A: Absolute Values	A - BC: Eigenvalues	A - BC: Absolute Values
.22941	.22941	-.36459	.36459
.75449+.22303i	.78677	-.13739	.13739
.75449-.22303i	.78677	.20619-.34749i	.40406
.94535+.027387i	.94575	.20619+.34749i	.40406
.94535-.027387i	.94575	.92576	.92576

IPI, CPI, M1 CYCLE MODEL
IN SAMPLE RESULTS

SERIES	MEANS	MAPE	RMSE	RSQ	AVGERR	MAD	COVCHEKE	HM
IPI	.00191	81.3	.008	.957	-1.83E-4	.00593	1.56	.5
CPI	.00507	18.3	.00647	.992	9.58E-5	.00517	3.69	1
M1	.00414	54.4	.006	.975	4.09E-4	.00438	2.99	1

Based on Iteration Two Coefficient Estimates

A: Eigenvalues	A: Absolute Values	A - BC: Eigenvalues	A - BC: Absolute Values
.18312	.18312	-.40125	.40125
.8068-.19904i	.83099	.19302+.32032i	.37398
.8068+.19904i	.83099	.19302-.32032i	.37398
.95434-.024717i	.95466	.51327	.51327
.95434+.024717i	.95466	.71551	.71551

SERIES	MEANS	CONSTANT	MAPE	RMSE	RSQ	AVGERR	MAD	HM
IPI	.00191	.00163	92.9	.00772	.959	-3.99E-6	.00565	.928
CPI	.00507	.00545	8.47	.00424	.996	1.13E-5	.00304	1
M1	.00414	.00497	60.2	.00533	.977	2.87E-4	.00392	1

Table 2 (continued): Three Series Model

LAG	ERROR AUTOCORRELATIONS			LAG	ERROR AUTOCORRELATIONS		
1	.123	.597	.148	13	-.00541	.119	-.0969
2	.0994	.197	-.0241	14	-.166	.117	.113
3	.0746	-.0639	-.1	15	-.0468	.123	.0844
4	.00599	-.15	-.294	16	-.0724	.0961	.0588
5	-.0571	-.0668	-.0125	17	-.0484	-.0188	.0918
6	-.0269	.0114	.0228	18	.0569	-.111	-.00327
7	-.0357	.0315	.0269	19	-.086	-.161	-.116
8	-.0169	.0181	.275	20	.0905	-.124	-.0952
9	.188	.0683	.166	21	.0184	-.0932	-.186
10	.098	.0627	.00745	22	-.0625	-.02	-.0132
11	.119	.05	-.0413	23	.00751	.0441	.145
12	-.0075	.0855	-.23	24	-.0754	-.00829	.028

---------- Combined Trend and Cycle Model ----------

IPI, CPI, M1 COMBINED TREND AND CYCLE MODEL
IN SAMPLE RESULTS

SERIES	MEANS	CONSTANT	MAPE	RMSE	RSQ	AVGERR	MAD	HM
IPI	4.88	4.88	.116	.00769	.996	-4.81E-5	.00563	1
CPI	5.15	5.16	.0602	.00423	1	1.27E-4	.00311	-
M1	5.73	5.73	.0678	.00525	1	3.6E-4	.00392	.801

LAG	ERROR AUTOCORRELATIONS			LAG	ERROR AUTOCORRELATIONS		
1	.116	.607	.13	13	-.00583	.0956	-.0981
2	.0947	.201	-.0336	14	-.167	.122	.122
3	.0724	-.0659	-.11	15	-.0454	.137	.0937
4	.00457	-.157	-.306	16	-.0692	.0986	.0627
5	-.057	-.0753	-.0133	17	-.0449	-.0337	.0887
6	-.024	.00565	.0246	18	.0592	-.138	-.0148
7	-.0321	.0282	.0253	19	-.088	-.186	-.133
8	-.0169	.0131	.273	20	.0887	-.138	-.109
9	.184	.0535	.155	21	.0176	-.0926	-.196
10	.0891	.0334	-.011	22	-.0655	-.00816	-.00853
11	.114	.00951	-.0586	23	.00597	.0613	.158
12	-.00744	.0419	-.243	24	-.0751	-.00207	.0374

IPI, CPI, M1 COMBINED TREND AND CYCLE MODEL
FULL SAMPLE (ALL 177 OBSERVATIONS)

SERIES	MEANS	CONSTANT	MAPE	RMSE	RSQ	AVGERR	MAD	HM
IPI	4.88	4.88	.139	.0091	.995	-.00197	.00681	1
CPI	5.15	5.16	.0695	.00491	1	8.74E-4	.00367	-
M1	5.73	5.73	.0797	.00633	1	.00163	.0047	.782

Table 3: Univariate Industrial Production Index Model

---------- Trend Model ----------
Np = Nf = 1

SINGULAR VALUES AND RATIO TO FIRST SINGULAR VALUE

SVS	NSVS	RATIO
.98022	1	1

COEFFICIENT ESTIMATES

A	B	C
.97991	10.934	.11471

UNIVARIATE IPI TREND
IN SAMPLE RESULTS

SERIES	MEANS	MAPE	RMSE	RSQ	AVGERR	MAD	COVCHEKE	HM
IPI	4.88	.138	.00936	.994	.00117	.00674	3.75	1

LAG	ERROR AUTOCORRELATIONS	LAG	ERROR AUTOCORRELATIONS
1	.495	13	-.0393
2	.364	14	-.127
3	.195	15	-.0464
4	.068	16	-.0382
5	-.04	17	-.0256
6	-.0685	18	.0164
7	-.0775	19	-.0277
8	-.0952	20	.038
9	.042	21	-.0145
10	-.00456	22	-.0713
11	.0063	23	-.0501
12	-.0505	24	-.106

---------- Cycle Model ----------
Np = Nf = 6

SINGULAR VALUES AND RATIO TO FIRST SINGULAR VALUE

SVS	NSVS	RATIO
.81072	1	1
.25174	2	.31051
.10235	3	.12625
.088411	4	.10905
.054022	5	.066635
.0096734	6	.011932

Table 3 (continued): Univariate Industrial Production Index Model

UNIVARIATE IPI CYCLE
Np = Nf = 6, Four States

COEFFICIENT ESTIMATES

A				B	C
.5809	-.37937	-.016242	8.353E-4	69.049	.0066654
.37937	.65797	-.033702	-.021918	-40.195	.0014777
-.016242	.033702	-.87985	-.40627	5.9653	2.5693E-4
-8.353E-4	-.021918	.40627	-.18975	.39139	-3.7326E-5

A		A - BC	
Eigenvalues	Absolute Value	Eigenvalues	Absolute Value
-.53473+.21412i	.57601	-.53671+.21052i	.57652
-.53473-.21412i	.57601	-.53671-.21052i	.57652
.61937+.37793i	.72557	.42017+.47327i	.63287
.61937-.37793i	.72557	.42017-.47327i	.63287

UNIVARIATE IPI CYCLE
IN SAMPLE RESULTS

SERIES	MEANS	MAPE	RMSE	RSQ	AVGERR	MAD	COVCHEKE	HM
IPI	.00117	223	.0079	.277	-3.85E-6	.00577	1	1

LAG	ERROR AUTOCORRELATIONS	LAG	ERROR AUTOCORRELATIONS
1	.00856	13	.0185
2	-.00801	14	-.142
3	-.0179	15	.0221
4	.0302	16	.00562
5	.00165	17	-.0187
6	.0221	18	.0294
7	-.0135	19	-.0811
8	-.126	20	.08
9	.115	21	-.00504
10	.00152	22	-.0711
11	.04	23	.0302
12	-.0524	24	-.0502

Table 3 (continued): Univariate Industrial Production Index Model

---------- Combined Trend and Cycle Model ----------

COMBINED IPI TREND AND CYCLE MODEL
IN SAMPLE RESULTS

SERIES	MEANS	CONSTANT	MAPE	RMSE	RSQ	AVGERR	MAD	HM
IPI	4.88	4.88	.118	.0079	.995	-1.39E-5	.00576	1

LAG	ERROR AUTOCORRELATIONS	LAG	ERROR AUTOCORRELATIONS
1	.00935	13	.021
2	-.00803	14	-.142
3	-.0171	15	.0211
4	.0304	16	.00507
5	.001	17	-.0182
6	.0215	18	.0309
7	-.0115	19	-.0798
8	-.123	20	.0776
9	.118	21	-.00351
10	-.00178	22	-.0709
11	.0356	23	.0288
12	-.0498	24	-.052

COMBINED IPI TREND AND CYCLE MODELS
FULL SAMPLE (ALL 177 OBSERVTIONS)

SERIES	MEANS	CONSTANT	MAPE	RMSE	RSQ	AVGERR	MAD	HM
IPI	4.88	4.88	.117	.00777	.996	4.69E-4	.00573	1

Table 4: Univariate Consumer Price Index Model

---------- Trend Model ----------
Np = Nf = 1, A=.98166

UNIVARIATE CPI TREND MODEL
IN SAMPLE RESULTS

SERIES	MEANS	MAPE	RMSE	RSQ	AVGERR	MAD	COVCHEKE	HM
CPI	5.15	.145	.00931	1	.00594	.00767	34.5	-

---------- Cycle Model ----------
Np = Nf = 6, Five States

SINGULAR VALUES AND RATIO TO FIRST SINGULAR VALUE

SVS	NSVS	RATIO
4.7717	1	1
.038612	2	.0080917
.023729	3	.0049729
.018304	4	.0038359
.015671	5	.0032842
.006991	6	.0014657

A		A - BC	
Eigenvalues	Absolute Value	Eigenvalues	Absolute Value
-.304+.38034i	.4869	-.277+.4i	.48654
-.304-.38034i	.4869	-.277-.4i	.48654
.26949-.45524i	.52903	.15904-.56967i	.59146
.26949+.45524i	.52903	.15904+.56967i	.59146
.98431	.98431	.70573	.70573

CPI CYCLE MODEL
IN SAMPLE RESULTS

SERIES	MEANS	MAPE	RMSE	RSQ	AVGERR	MAD	COVCHEKE	HM
CPI	.00594	41.8	.00265	.864	2.41E-4	.00183	1.26	-

LAG	ERROR AUTOCORRELATIONS
1	-.0901
2	-.065414
3	-.06686
4	-.084737
5	.014722
6	.059301
7	.0086804
8	-.11781
9	.16829
10	.11707
11	.022576
12	-.022966

Table 4 (continued): Univariate Consumer Price Index Model

---------- Combined Trend and Cycle Model ----------

CPI COMBINED TREND AND CYCLE MODEL
IN SAMPLE RESULTS

SERIES	MEANS	CONSTANT	MAPE	RMSE	RSQ	AVGERR	MAD	HM
CPI	5.15	5.16	.0352	.00262	1	3.78E-4	.00182	-

LAG	ERROR AUTOCORRELATIONS
1	-.044287
2	-.041054
3	-.06254
4	-.096807
5	.0054606
6	.055931
7	.0067875
8	-.12037
9	.16231
10	.10035
11	.013892
12	-.05526

CPI COMBINED TREND AND CYCLE MODEL
FULL SAMPLE (ALL 177 OBSERVATIONS)

SERIES	MEANS	CONSTANT	MAPE	RMSE	RSQ	AVGERR	MAD	HM
CPI	5.15	5.16	.0363	.00275	1	3.79E-4	.0019	-

Table 5: Univariate Commercial Paper Rate

---------- Univariate RCP Trend and Cycle Model ----------

---------- Trend Model ----------

Np = Nf = 1, A = .95156

RCP TREND MODEL
IN SAMPLE RESULTS

SERIES	MEANS	MAPE	RMSE	RSQ	AVGERR	MAD	COVCHEKE	HM
RCP	2.07	2.66	.0747	.963	.00115	.0553	1.15	1

---------- Cycle Model ----------

Np = Nf = 9, Seven states

A: Eigenvalues	A: Absolute Value	A - BC: Eigenvalues	A - BC: Absolute Value
-.57337+.69438i	.90051	-.46445+.74489i	.87782
-.57337-.69438i	.90051	-.46445-.74489i	.87782
.19895+.78234i	.80724	-.087761	.087761
.19895-.78234i	.80724	.34934-.69539i	.7782
.72374+.31786i	.79046	.34934+.69539i	.7782
.72374-.31786i	.79046	.89172-.21288i	.91678
.84163	.84163	.89172+.21288i	.91678

RCP CYCLE MODEL
IN SAMPLE RESULTS

SERIES	MEANS	MAPE	RMSE	RSQ	AVGERR	MAD	COVCHEKE	HM
RCP	.00115	178	.0701	.118	6.09E-4	.0532	1.01	.626

---------- Combined Trend and Cycle Model ----------

RCP COMBINED TREND AND CYCLE MODEL
IN SAMPLE RESULTS

SERIES	MEANS	CONSTANT	MAPE	RMSE	RSQ	AVGERR	MAD	HM
RCP	2.07	2.07	2.65	.0706	.966	-7.45E-4	.054	.9999

RCP COMBINED TREND AND CYCLE MODEL
FULL SAMPLE (ALL 177 OBSERVATIONS)

SERIES	MEANS	CONSTANT	MAPE	RMSE	RSQ	AVGERR	MAD	HM
RCP	2.07	2.07	2.53	.069	.964	3.21E-4	.052	.997

Table 5 (continued): Univariate Commercial Paper Rate

---------- Stationary Model ----------
Np = Nf = 2, Two States

STATIONARY RCP MODEL

A		A - BC	
Eigenvalues	**Absolute Value**	**Eigenvalues**	**Absolute Value**
-.105	.105	-.275+.256i	.376
.948	.948	-.275-.256i	.376

STATIONARY RCP MODEL
IN SAMPLE RESULTS

SERIES	MEANS	MAPE	RMSE	RSQ	AVGERR	MAD	COVCHEKE	HM
RCP	2.07	2.66	.0756	.962	.00108	.0552	1.16	.999

STATIONARY RCP MODEL
FULL SAMPLE (ALL 177 OBSERVATIONS)

SERIES	MEANS	CONSTANT	MAPE	RMSE	RSQ	AVGERR	MAD	HM
RCP	2.07	2.07	2.5	.0727	.96	.00164	.0524	.999

Table 6: Univariate Money Stock Model

---------- Trend Model ----------

Np = Nf = 1, A = .98008

M1 TREND MODEL
IN SAMPLE RESULTS

SERIES	MEANS	MAPE	RMSE	RSQ	AVGERR	MAD	COVCHEKE	HM
M1	5.73	.111	.00838	1	.00508	.0065	30.3	.877

---------- Cycle Model ----------

Np = N = 12, Seven States

A: Eigenvalues	A: Absolute Value	A - BC: Eigenvalues	A - BC: Absolute Value
-.57642+.77932i	.96933	-.55284+.75755i	.93783
-.57642-.77932i	.96933	-.55284-.75755i	.93783
.17664+.92755i	.94422	.15732+.87916i	.89312
.17664-.92755i	.94422	.15732-.87916i	.89312
.67727+.64605i	.93599	.60286-.59049i	.84387
.67727-.64605i	.93599	.60286+.59049i	.84387
.98282	.98282	.79914	.79914

M1 CYCLE MODEL
IN SAMPLE RESULTS

SERIES	MEANS	MAPE	RMSE	RSQ	AVGERR	MAD	COVCHEKE	HM
M1	.00508	108	.00437	.577	5.68E-4	.00315	1.21	.497

---------- Combined Trend and Cycle Model ----------

M1 TREND AND CYCLE MODEL
IN SAMPLE RESULTS

SERIES	MEANS	CONSTANT	MAPE	RMSE	RSQ	AVGERR	MAD	HM
M1	5.73	5.73	.0561	.0044	1	7.97E-4	.00323	.947

M1 TREND AND CYCLE MODEL
FULL SAMPLE (ALL 177 OBSERVATIONS)

SERIES	MEANS	CONSTANT	MAPE	RMSE	RSQ	AVGERR	MAD	HM
M1	5.73	5.73	.0567	.00448	1	9.22E-4	.00331	.927

Figure 7: Multivariate Commercial Paper Rate Trend
In Logarithms In Sample

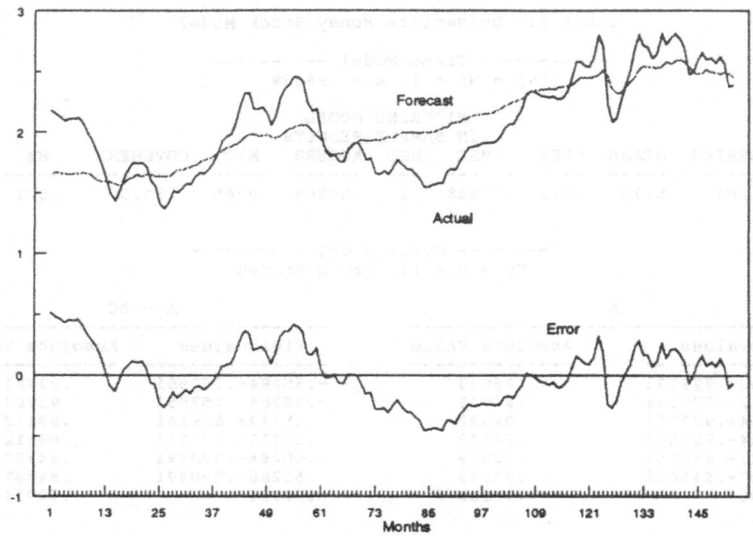

Figure 8: Univariate Commercial Paper Rate Trend
In Logarithms In Sample

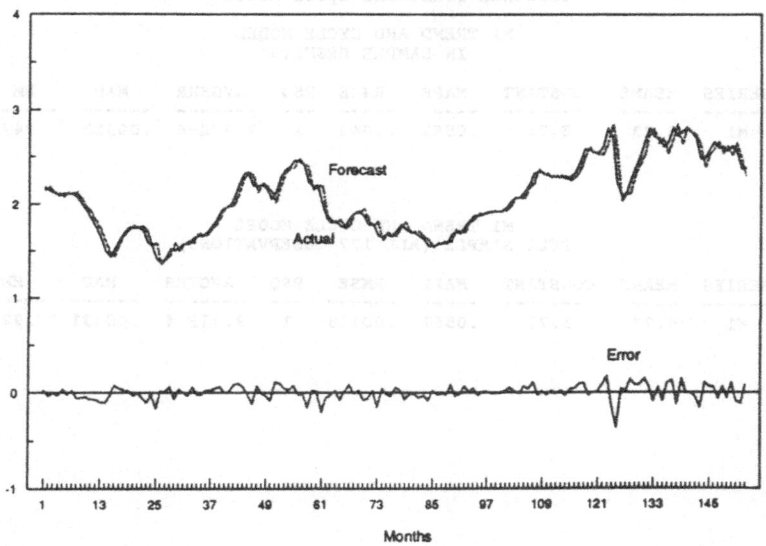

If the trend model reflects true nonstationarities, then the parameter estimates in equations 13 and 14 are believed to be superconsistent, converging at rate T rather than $\sqrt{T}$.[17] In this case, the remaining serial correlation in the errors is of no consequence statistically. If the series in y_t are stationary with one or more eigenvalues substantially greater than the others, then the two step procedure can still prevent the larger eigenvalues from masking the smaller ones, whether or not the larger eigenvalues are of unit modulus. In this case, however, the correlation between the serially correlated states and y_t^* in equations 13 and 14 leads to inconsistent estimates. It is believed that a parallel to the usual GLS iteration to convergence results in consistent estimates, analogous to regression with a lagged dependent variable and a serially correlated error. The iteration proceeds by subtracting the conditional means of the short run dynamics, $C_\eta \eta_{t|t-1}$, from the left hand sides of equations 13 and 14 in order to eliminate the serial correlation from the errors of those equations.

III Specifying the Stationary Model

Suppose the series to be modeled are stationary, either because they were stationary originally or because they have had their deterministic or stochastic nonstationarities removed by the methods described above. Specifying a state space model for stationary series consists of choosing the series to be modeled, the lag parameter, the forecast horizon, and the number of states required, and evaluating the resulting coefficient estimates and forecasts.

1 Setting the lag parameter

The lag parameter N_p and the forecast horizon N_f determine the size of the Hankel matrix of autocovariances (see equation 3 above) to be used for model identification and estimation. For model identification, $N_f \geq N_p$ is required in order not to row constrain the Hankel matrix rank used to determine the number of states needed. When multiperiod forecasts are desired, N_f can be set larger than N_p to cause the coefficient estimates to solve stacked first order conditions over the entire forecast horizon; see Havenner and Aoki (1988b). This would be useful in the current example if N_p was less than three and quarterly forecasts were to be assembled from the monthly data, for example. In most applications N_f is set to N_p, however.

Choosing an appropriate lag parameter can be difficult, especially in multivariate systems. The essential tradeoff is between setting N_p too small, which does not approximate as long an autocovariance sequence (in H) as desirable and in extreme cases can constrain model identification (via the Hankel matrix rank), versus setting N_p too large, which approximates autocovariances with excessive sampling error (at long lags).[18]

17 Havenner and Aoki (1988a,b) show that the systems theory estimators are least squares instrumental variable estimators. See also Havenner and Leng.

18 Choosing the lag parameter is the formal state space analog to deciding which autocorrelations to incorporate into a model and which to ignore in Box-Jenkins time series analysis.

There is often some intuition on the required univariate dynamics. For example, a monthly toy sales model should at least cover the seasonal frequency, with implied dynamics of at least 12. Many agricultural series are also seasonal, of course. If the model is univariate, setting $N_p = 12$ will bring the seasonal autocovariance into the Hankel matrix 12 times (see equation 3). Choosing N_p and N_f such that $N_p + N_f - 1 < 12$ would result in the seasonal autocovariance not entering the Hankel matrix at all, which is clearly undesirable. Other series can be less informative regarding the lag parameter choice, for example the seasonally adjusted macroeconomic data analyzed above. Larger values of the lag parameter permit more cyclic detail, guiding the choice somewhat.

Setting the lag parameter in multivariate systems is considerably harder, due to the richness of their possible dynamic interactions. The size of the lag parameter required is a decreasing function of the number of series in the model. Write the multivariate system as

$$\Phi(L)y_t = \Theta(L)e_t \tag{19}$$

so that

$$y_t = \frac{adj.[\Phi(L)]}{|\Phi(L)|}\Theta(L)e_t \tag{20}$$

or

$$|\Phi(L)|\, y_t = adj.[\Phi(L)]\Theta(L)e_t \tag{21}$$

where *adj.* denotes the adjoint matrix and $\Phi(L)$ and $\Theta(L)$ are matrix polynomials in the lag operator L. Equation 21 gives the stack of implied univariate models extracted from the multivariate model 19. From 21, the upper bound on a given series univariate model order is the number of series times the maximum lag in $\Phi(L)$. Thus, for example, five series in a first order multivariate model imply univariate models of up to fifth order. The implication is that the lag parameter need not and should not be very big for multivariate systems. If the toy sales were embedded in a five series multivariate model, $N_p = 3$ implies univariate dynamics that include the seasonal frequency. Whenever there is a choice, it is desirable to get the seasonal autocovariance into the Hankel matrix, *i.e.*, choose N_p and N_f so that $N_p + N_f - 1$ includes the seasonal frequency. Often these models have excessive sampling error and must be abandoned, however. For monthly toy sales, $N_p = N_f = 7$ will bring the lag 12 autocovariance into the Hankel matrix twice, but with five series a lag parameter of seven would almost surely cause excessive sampling error. When the data are not seasonal -- like the macroeconomic data -- the practical rule is not to forget to try small values of N_p.

For the monthly macroeconomic data, a subjective assessment speculates that the individual series might be expected to require between three and possibly nine lags to provide reasonable forecast accuracy (think of the regression that might be run). For a univariate model this would be the lag parameter range. Multivariate models of three or four series thus require a lag parameter between one and three, unless the best model is at the bound, which would counsel extending the range. The three and four series multivariate cycle models reported in the tables use $N_p = N_f = 3$; $N_p = 4$ was judged inferior based on in-sample statistics. The univariate IPI model uses

$N_p = N_f = 6$; smaller values of the lag parameter were less satisfactory, as expected, and two larger values tried, $N_p = 9$, 12, were about the same. The univariate CPI model also uses $N_p = N_f = 6$, while the univariate RCP model performed better with $N_p = N_f = 9$, and the M1 model required $N_p = N_f = 12$. (The trend models were all based on $N_p = 1$ in order to filter out the short term dynamics, as discussed in section II.2. above).

2 Choosing the number of states

For a given lag parameter, the Hankel matrix can be constructed and its singular values calculated as input into determining its rank which, by the Kronecker Theorem, is the number of states required to model the series. From an ordered display of singular values, it is just necessary to find the point at which they are zero apart from sampling error.

For relatively simple models this can often be done judgmentally, with the advantage that auxiliary information about the problem can sometimes be used as well, *e.g.*, at least two states are required to produce smooth cycles, at least three states are required for smooth cycles around a minor trend, *etc.*. Since the singular values are scale dependent, the ratio of each to the largest is used as a scale free measure analogous to a condition number. When determining the number of states judgmentally it is natural to focus on the largest changes in these condition numbers, choosing as the number of states the point at which these ratios drop precipitously. This is often not a bad strategy in practice. There are two caveats regarding the procedure, however: first, the decision that the singular value is zero apart from sampling error should at least informally incorporate information about the sample size; and second, subtle factors essential to a good forecast can be relatively small compared to the effects of the dominant model.[19]

One entirely *ad hoc* measure that is surprisingly consistent with an informed choice regarding the number of states is an 80% rule. Arranging the singular values from largest to smallest, the rule finds the singular value such that the sum of it and the larger singular values is 80% of the sum of all of the singular values. The number of the singular value at which this is true is the recommended number of states. This choice is based on nothing more substantial than the feeling that analysts like to explain about 80% of the autocovariance sequence.

It is also possible to formally test for the number of states required. In a Monte Carlo investigation by Dorfman and Havenner, a test proposed by Aoki and another test proposed by Dorfman were both found to perform well; see Dorfman and Havenner (1994), where it is also shown that the balanced representation ensures that these tests are free of sequential testing bias. Tests formalize the decision with regard to the number of states, but they are based on a given set of series and lag parameter and

19 For example, strongly trending series usually have one very large singular value followed by near zeros. The detrending procedures discussed in section II allow fitting a cycle model to relatively small errors in order to improve both the direction and the magnitude of the resulting forecasts. In cases where the dominant effect is not due to a trend, the best model often comes from simply going beyond the largest break in the singular value ratios in order to include more detail in the approximation, especially when the sample is large enough to support it.

cannot incorporate the other attributes of the model that are important in model selection, *i.e.*, they ignore the model assessment information described in sections 3 and 4 below. As a result, for all but very large models I personally prefer to go through a pattern-recognition procedure that trades off the multiple criteria described in the sections below.

In general, the number of states increases much less rapidly than the number of series if the modeler is doing a good job, since series that go together should have common cycles. Highly erratic series require more states if the variability reflects changing conditional means rather than just error realizations (which cannot be modeled).

Using the detrended series output from the trend models, calculated singular values and their condition numbers for cycle model specification are given in Tables 1-6.[20] In the four series cyclic model, the singular values drop importantly between the second and third, and the fourth and fifth values (RATIO and SVS, second page of Table 1). With 153 observations (in sample) on monthly data, several states would be expected to be necessary to track the higher frequency components of the series. An initial specification of four states led to an infeasible model due to failure of the symplectic matrix solution, see section III.3.ii below. The maximum number of states for which the calculations were possible was two, which underlie the selected cycle model. The singular values for the three series cycle model are on the second page of Table 2. Here the choice is between two or three states, or as many as five. All of these models are acceptable, with little to distinguish them. The table reports the results for five states.

Finally, statistics for the univariate models are reported in Tables 3-6. The singular values for the univariate IPI model are at the bottom of the first page of Table 3. They suggest that reasonable model choices would have one, two, or four states. A univariate model of high frequency data with a lot of observations often requires and can support models with more states; four states were chosen for the cycle model. Similarly, five states were chosen for the CPI cycle model (see the Cycle Model singular values and ratios in Table 4), and seven states were chosen for both the RCP and M1 cycle models (Tables 5 and 6). Since RCP might be stationary, results based on $N_p = 2$ with two states are reported on the second page of Table 5. The lag parameter value is small for a univariate stationary model of monthly data, but experiments with larger values ($N_p = 6$, 9, 12) were all unsuccessful due to eigenvalues that were too large or symplectic matrix problems, see sections 3.i. and 3.ii. below). Comparing the "stationary" RCP model (second page of Table 5) with the RCP trend model (first page of Table 5), it is clear that second state is not able to add much detail when its effects are estimated jointly with the first state -- the "stationary" model is really an imprecise trend model -- and thus a trend/cycle formulation will be superior.

3 Evaluating the parameter estimates

Since the states are not directly observable, the coefficients of the state space model are not determined without additional constraints. While the state coordinate system can be selected for maximum advantage (see the discussion of the balanced

---

20 Look for the label "-------- Cycle Model ---------" in the tables.

representation in I.6. above, and the references therein), there is no inherent meaning in the resulting coefficients or associated states. Consider the state and observation equations (1 and 2 above), with an arbitrary nonsingular transformation matrix **T** inserted in such a way as to preserve algebraic validity:

$$\mathbf{T}z_{t+1|t} = (\mathbf{T}A\mathbf{T}^{-1})\mathbf{T}z_{t|t-1} + (\mathbf{T})Be_t, \tag{22}$$

$$y_t = \mu + (C\mathbf{T}^{-1})\mathbf{T}z_{t|t-1} + e_t. \tag{23}$$

The equations are equally consistent with the data when the states are $\mathbf{T}z_{t|t-1}$ and the coefficients are $\mathbf{T}A\mathbf{T}^{-1}$, $\mathbf{T}B$, and $C\mathbf{T}^{-1}$ or when the states and coefficients are their original $z_{t|t-1}$ and A, B, and C: any nonsingular transformation of the states is valid with the coefficients adjusted accordingly. Any conclusions that are to be drawn from the model must be invariant with respect to the state coordinate system. For example, it generally does not make sense to test individual coefficients for significance. It can be tempting to forget this, given our training in regression models, and reason from the size of elements in the dynamics matrix A or their pattern, or the correlation of states with observed variables of interest. Except in special cases that guarantee invariance, such analysis is not valid.

Some things are always invariant, however, notably including the forecasts. In addition, the eigenvalues of A are the same as of $\mathbf{T}A\mathbf{T}^{-1}$, so that analysis of the dynamic properties of the system is invariant with respect to state coordinate choice. Similarly, the eigenvalues of A-BC (see the iterative estimator and symplectic matrix discussions below) are invariant. The matrix products $CB = C\mathbf{T}^{-1}\mathbf{T}B$ and $CAB = C\mathbf{T}^{-1}\ \mathbf{T}A\mathbf{T}^{-1}\ \mathbf{T}B$ are also invariant. In a special case, the state space coefficients can be intuitively mapped into ARMA coefficients, which are invariant. When C is square,[21] the observation equation 2 can be solved for the state vector, and the result (at times t and $t+1$) substituted into the state equation 1 to get

$$y_{t+1} = Ay_t + e_{t+1} + (CB - A)e_t \tag{24}$$

which is a vector ARMA form with $\Phi = A$ and $\Theta = CB - A$.

Evaluation of the parameter estimates will center on two areas, the dynamic response implied by the (invariant) eigenvalues, and the numerical stability of the solutions for some of the parameters.

i Eigenvalues

The eigenvalues of the dynamics matrix A must lie inside the unit circle for a stationary stochastic process. A point estimate larger than one in absolute value is logically inconsistent with the stationarity assumption underlying the model and is grounds for rejecting the model. When stationarity is not a maintained hypothesis, as in trend model estimation, larger eigenvalues are logically permissible. The eigenvalues of the trend model dynamics matrix A_τ are usually real and close to one, but they can be complex for some series. Sometimes there are features that show up clearly

21 When the number of states is not equal to the number of series, C is not square and the Cayley-Hamilton Theorem must be used to put the state space model in ARMA form, see Aoki or Aoki and Havenner (1991).

in the roots of A, such as a large negative real eigenvalue in alternate bearing crops, or a complex conjugate pair with near unit modulus in cattle statistics reflecting the long smooth cycle.

The eigenvalues of A in the four series cycle model (Table 1, second page) are complex, which is appropriate for the cycle model. Since the trend run in principle removes the slow dynamics, the absolute values are larger than expected, suggesting the trend/cycle separation was not perfect for this model. The eigenvalues of A in the three series cycle model (Table 2, second page) show the same pattern. The eigenvalues of the univariate IPI cycle model (Table 3) show much better separation of trend and cycle, ranging from .58 to .73 in modulus in contrast to the trend eigenvalue of .98. The dynamics in the remaining series univariate models (Tables 4, 5, 6) are less well separated than those of IPI. There is probably room to improve these models.

ii Stability of the solution for B

Estimation of the moving average matrix B in the state equation using Aoki's original systems theory estimators requires solving a nonlinear matrix Riccati equation for the state covariance matrix. Mathematical details of an analytic solution based on a symplectic[22] matrix can be found section 7.2 and Appendix A.6 of Aoki. The estimates based on solving this Riccati equation are the subject of some debate currently. Leaving technical matters aside, an intuitive overview can be useful.

Define $v_t \equiv Be_t$ in the state equation (equation 1 in the introduction above); then the $n+m \times n+m$ covariance matrix of the state and observation equation errors is

$$F = \mathcal{E}\begin{pmatrix} v_t \\ e_t \end{pmatrix}(v_t{}'\ e_t{}') \tag{25}$$

$$= \begin{pmatrix} \Xi - A\Xi A' & \Omega - A\Xi C' \\ \Omega' - C\Xi A' & \Gamma_0 - C\Xi C' \end{pmatrix} \geq 0 \tag{26}$$

where $\Xi = \mathcal{E}z_{t|t-1}z_{t|t-1}{}'$ (the state covariance matrix), $\Omega = \mathcal{E}z_{t+1|t}y_t{}'$ (the covariance of the states with the observations), and Γ_0 (the observation covariance matrix) is defined in equation 6 above. Since F is a covariance matrix, its eigenvalues must be real and nonnegative. (Because $v_t \equiv Be_t$, all but m should be zero.) This is known as the positive real lemma; see Faurre.

The symplectic matrix algebra provides a matrix Ξ that solves the Riccati equation

$$\Xi = A\Xi A' + (\Omega - A\Xi C')(\Gamma_0 - C\Xi C')^{-1}(\Omega - A\Xi C')'. \tag{27}$$

This is developed in Aoki, or section 3.1.4 of Aoki and Havenner (1991). Real unit eigenvalues in the symplectic matrix indicate that the process has a perfectly predictable component; an example is given in Appendix 1 below under the heading

---

22 A matrix S^* is symplectic when $S^{*\prime}J = JS^{*-1}$, where $J = \begin{pmatrix} 0 & I \\ -I & 0 \end{pmatrix}$. Notice that $-J = J' = J^{-1}$.

Invertibility. Complex eigenvalues with unit modulus in the symplectic matrix[23] indicate that the positivity condition on F is not satisfied, see Heij, Kloek, and Lucas (1992). Since the estimates of A, C, Ω, and Γ_0 are previously computed in the linear systems theory estimators, we tend to think of the Riccati equation estimate of Ξ as causing the problem, although obviously all the parameter estimates together (along with the model specification) create the anomaly.

Thus when the symplectic matrix has complex unit roots the chosen model specification and estimators have produced estimates of A, C, Ω, and Ξ that, with Γ_0, are logically inconsistent with a nonnegative definite state and observation equation error covariance matrix, *i.e.*, the implied state and observation error covariance matrix F has negative and/or complex eigenvalues.

There are several ways to proceed when this happens. They fall into two broad categories: *(i)* change the estimators, or *(ii)* change the model specification. The issue is whether we should attempt to save the model by changing the estimator, or whether the model should be abandoned because the unit roots are conveying useful information indicating an inferior model. In principle, either pure sampling error in the estimators or excessive sampling error produced by model misspecification could be causing the real or complex unit roots in the symplectic matrix.

If we take the model specification as correct, the problem has to lie solely in sampling error in the estimators. In this case, it will disappear asymptotically for consistent estimators, a large class that includes Aoki's original systems theory formulas. In the finite sample we will have to add more information to get better estimates. The work in this area has centered on adjusting the estimators so that the positive real lemma is met, without regard to estimation efficiency in general (and without considering model misspecification), see, *inter alia*, Heij, Kloek, and Lucas' recommendations at the end of their paper, or Vaccaro and Vukina (1993). Alternatively, Havenner and Aoki (1988a,b) interpreted Aoki's original systems theory estimators as instrumental variable estimators and noted that they are in general not statistically efficient due to the choice of instrument.[24] If the model is in fact correctly specified, more efficient estimators can produce estimates that do not have unit eigenvalues in the symplectic matrix. Based on this idea, various alternative estimators have been proposed,including the Havenner and Leng iterative estimator discussed in section IV below, which is both asymptotically efficient in its class and imposes the positivity constraint.

If we want to force model estimation given a model specification, then an alternative estimator is required. However, my usual approach is to change the model specification when the symplectic matrix has unit roots. Intuitively, the criterion is often flatter with respect to the best model choice than it is with respect to the parameter estimates given the model choice. Although the Kronecker Theorem guarantees a consistent estimate of the model order given a sufficiently large Hankel matrix, for finite observation sets there will be sampling error in the choice of the number of

---

23 The symplectic matrix is of size $(2n \times 2n)$, where n is the number of states, and half of its eigenvalues are the reciprocals of the other half, so that unit roots appear in pairs if present. (These are different unit roots than in the dynamics matrix A.)

24 The 1988b paper proposed alternatives, but proved too turgid for publication.

states. Also, the Hankel matrix size may not be suitable, either because of too large a lag parameter with the attendant sampling error in the autocovariances, or too small a lag parameter, rank constraining the Hankel matrix and the resulting model order determination. In addition, inappropriate series may be modeled together as a system. Misspecifying the model in any of these ways can lead to parameter estimates that do not satisfy the positive real lemma. In this case, the appropriate response is not to adjust the parameter estimator but rather to respecify the model. We are usually not wedded to a particular model specification in any case.

Thus if unit roots appear in the symplectic matrix calculations, I usually prefer to reformulate the model -- first with a different number of states, then with a different lag parameter, and finally if necessary with different series being modeled. It has been something of a surprise to me how often we firmly believe multiple series should be modeled together when there are better smaller models.

Models that have serious Riccati equation problems produce eigenvalues of the symplectic matrix that are 1.0000 to five digits.[25] Occasionally the symplectic matrix roots, while not unity to five digits, are sufficiently close to one that the parameter estimates are not robust; estimated covariance matrices may be badly ill-conditioned, for example, and calculations derived from them can be unstable. Two *ad hoc* checks on computational stability are useful in practice, based on the estimated variances of the observation equation errors and of the states. These estimates come from the same calculations that provide the moving average matrix B that is of critical importance in forecasting, and therefore provide some insight into the reliability of B. The symplectic matrix and associated calculations produce estimates of these variances which we will call theoretical for lack of a better label. Once the coefficient matrices A, B, and C are estimated and the state initial condition is set (backcast or set to zero), the state and observation equations can be solved recursively through the sample for empirical estimates of the states and the observation equation residuals. Variances from these series provide "empirical" counterparts to the theoretical estimates from the symplectic matrix. The ratios of the theoretical to empirical variances (for each series) of the observation equation errors should be approximately one for a stable solution of the equations determining them and thus the all-important B. Similarly, the ratios of the theoretical to actual state variances should be approximately one. (These ratios are COVCHEKE and COVCHEKZ, respectively, in the program's terminology).

In the macroeconomic data example, the four series cycle model had serious problems that surfaced with unit roots in the symplectic matrix. Table 1 presents the symplectic matrix eigenvalues (next to the singular values of the cycle model, on the second page of the table) for four states.[26] One pair is unity to five digits indicating a sampling error or model specification problem. With the lag parameter set to three this problem recurred for states from three to eight (the maximum tried). Other values of the lag parameter produced similar problems. As a result, two states were specified for the four series cycle model, even though the desire for cyclic detail would favor more. The COVCHEKE column for the cycle model (in the summary statistics on the bottom of the second page of Table 1) ranges from 1.61 to -1.36, far from one for

---

25 The program described in Appendix 2 customarily reports five digits of accuracy.

26 Since there are four states, there are eight roots in the table with half the reciprocal of the other half, *e.g.*, 1/.25585 = 3.9086.

this many observations. (The sampling distributions of COVCHEKE and COVCHEKZ are completely unknown, but experience shows good models are not this far off with this many observations; compare the univariate cycle model value on the second page of Table 3, for example.) Thus even with only two states the calculations are not promising, and indeed the negative COVCHEKE value stems from a negative eigenvalue in the error covariance matrix. The estimate of *B* (which uses this covariance matrix estimate) is therefore in considerable doubt and cannot be used.[27]

The three series cycle model and the univariate cycle model did not have similar difficulties with unit symplectic matrix eigenvalues. The three series cycle model did have larger COVCHEKE ratios than desirable, however (see the middle of page two of Table 2, under "Based on Aoki's Coefficient Estimates"). The univariate IPI model COVECHEKE was one to five digits (Table 3 page 2), and the other univariate cycle models COVCHEKE values were 1.26, 1.01, and 1.21 for CPI, RCP, and M1 respectively, all suggesting no serious difficulties. The "stationary " RCP model had unit roots in the symplectic matrix for two to six states when the lag parameter was set to 12, and for most of the reasonable selections for the number of states when the lag parameter was set to nine. (The only plausible model without unit symplectic matrix roots had four states, but when the parameters were estimated an eigenvalue of *A* was 1.19, unacceptable in a cycle model.) Stationarity is clearly not a reasonable assumption for this series after all, and models based on it are unstable, a contradiction.

The symplectic matrix problems in the four series model may well simply originate in RCP being forced into the multivariate model when its dynamic properties are a serious mismatch. The problems decrease substantially when RCP is eliminated. It seems unlikely that more efficient estimators will be successful in a model with mismatched dynamics and incorrect stationarity assumptions, and arbitrarily eliminating the problem -- using a procedure such as that proposed by Vaccaro and Vukina -- would not be a good strategy here, if model misspecification is in fact the problem. Based on the overall analysis of these series, I would not be inclined to force the model specification by changing the estimator.

There is a tendency to think like structural modelers when combining series in multivariate time series models; surely industrial production depends on interest rates. Not monthly, and not without a more structured formulation than is inherent in time series analysis. It is better to let the final form algebra (uncoupling individual series) incorporate any interest rate effects in a smaller model -- that is the primary advantage of time series models. Problems of negative and/or complex eigenvalues of the covariance matrix *F* appear to be aggravated by modeling series together that from a purely empirical viewpoint do not have the same characteristics.

4 Forecast Evaluation in Sample

With the coefficients estimated, the state initial condition can be backcast and the state and observation equations solved forward through the sample to provide in-sample predictions and errors. There are a variety of summary statistics based on

27 Sometimes these borderline estimates can be improved by attacking the sampling error using the iterative estimator, see section IV. below. I will argue below that the problem in this macroeconomic case is one of model misspecification, however.

these predictions that can be calculated. Four types of summary information have proven useful in subjective model evaluation: *i*) magnitude information; *ii*) direction information; *iii*) error autocorrelations; and *iv*) graphs.

i Magnitude

The typical forecast evaluation statistics (MAPE, RMSE, RSQ, AVGERR, MAD[28]) provide information on the magnitude of the model errors. Some are more useful than others for specific problems. For example, the MAPE is not very informative for nearly zero mean series (like cycle models) although the RSQ is, while for other problems (such as electricity generation or river flooding) the maximum is the critical statistic. One or more of these measures forms the basic assessment of model success.

Examining Tables 1-6, the immediate conclusion is that the models fit extremely well (as would be hoped for monthly data). With the exception of RCP in the four series model, even the trend models have high RSQ's and low MAPE's, and the RMSE's are all small relative to the means. The statistics for the cycle models (except for the MAPE's) are also very good. The most important statistics are for the combined trend and cycle model (in sample), which are excellent. For example, the lowest RSQ is for RCP, and it is .954 (Table 1, last page). However, the three series model (Table 2, last page) has even smaller combined trend and cycle RMSE's (for IPI, .00769 for the three series model *vs*. .00955 for the four series model) and is more accurate than the univariate IPI model (RMSE .00777). However, apart from this exception for IPI, the univariate models are all more accurate than either of the multivariate models (see the combined trend and cycle statistics in Tables 4-6). Surprisingly (if you think like an economist), the univariate models for three of the four series are better than the multivariate models. If these series are cointegrated, the loss from specifying separate trends is not substantial. Although the four series model is clearly inferior to the other models and the univariate models are generally better, these models all forecast exceptionally accurately one month ahead in sample.

ii Direction

In some problems (such as in finance) the size of the error is not as important as the direction so that the MAPE, RMSE, RSQ, *etc*., are all less relevant than usual. Even when the magnitude measures are important, the Henriksson-Merton nonparametric test of direction provides information that is independent of the magnitude measures. There are very few cases in which we are truly indifferent regarding the sign of the prediction error as long as it is small.

The Henriksson-Merton test ignores information about the sizes of the errors and simply considers whether or not the forecasts predict the correct direction of change. By eliminating magnitude information and using a nonparametric test it is possible to obtain a test statistic with a known finite sample distribution under plausible and weak assumptions. The null hypothesis of the test is

$$H_0:\quad P(y_{it}-y_{it-1}>0\mid \hat{y}_{it}-y_{it-1}>0)+P(y_{it}-y_{it-1}\le 0\mid \hat{y}_{it}-y_{it-1}\le 0)=1 \quad (28)$$

---

28 Recall from the explanation of Tables 1-4 that MAPE is the mean absolute percentage error, RMSE is the root mean squared error, RSQ is the correlation of each series' actual and forecast values, AVGERR is the average error, and MAD is the mean absolute deviation.

for each of the $i = 1, 2, \ldots, m$ series in turn. To carry out the test, define

N_1 = the number of observations where y_{it} actually rose,

N_2 = the number of observations where y_{it} actually did not rise,

$N = N_1 + N_2$ = the total number of observations,

n_1 = the number of successful predictions given that y_{it} rose,

n_2 = the number of unsuccessful predictions given that y_{it} did not rise,

$n = n_1 + n_2$ = the number of times y_{it} was forecasted to rise.

Then the confidence level, c, is given by

$$c = 1 - \frac{\sum_{x=n_1}^{\min(N_1,n)} \binom{N_1}{x}\binom{N_2}{n-x}}{\binom{N}{n}} \tag{29}$$

The Henriksson-Merton confidence levels are in the columns under HM in Tables 1-6. (Where a hyphen occurs a logarithm of zero was encountered and the confidence was not calculated.) We are very sure that the univariate model is useful in determining the direction of IPI: the trend, cycle, and combined model confidence levels are all one (Table 3). The direction of M1 is difficult for the trend components of the four and three series multivariate models to predict (confidence levels of .091 and .477), but the combined trend/cycle models do very well, with confidence levels of .997 and .801 for the four series and three series models respectively. The separate trends of the univariate models (rather than the shared single state trends in the multivariate models) are much more effective in determining the directions (Tables 3-6), although the advantage declines when combined models are considered. Overall, the models do very well in predicting direction one month ahead.

iii Error autocorrelations

The error autocorrelations tell us whether or not everything systematic has been extracted from the series; a good model will have error autocorrelations that are zero apart from sampling error. Two warnings are in order, however. First, the tests, if used, are asymptotic against the general autocorrelation found in time series model errors. Knowing this inexperienced analysts tend to overfit models in pursuit of very small autocorrelations. Second, autocorrelations by construction do not depend on the size of the errors. It is important to balance autocorrelation information against other information when selecting a model; autocorrelated errors that are tiny are certainly preferred to uncorrelated but large errors. The best model will have uncorrelated errors that are also the smallest possible and predict the direction of change accurately.

The combined trend/cycle error autocorrelations for the macroeconomic data models are reported in Tables 1-4. They are much higher than desired in the four series model, signaling the possibility of a better model. In an ARMA world, more terms would be added to the model (not necessarily improving it if the symplectic matrix calculations are informative). In the state space context, adding states to either the trend or cycle model is ruled out by the symplectic matrix calculations in this case (see the discussion in section III.3.ii above). The stochastic structure of RCP is a problem, as we have seen, and it led to the three series model. The autocorrelations of the three series combined trend/cycle model are much lower than for the four series model. The only exceptional value is .597 at lag one for CPI, surely not just sampling error.[29] Since CPI tested trend stationary the anomaly may be related to CPI itself, suggesting that the CPI autocorrelation be disregarded in this model to avoid adversely affecting the performance of other series while trying to eliminate this autocorrelation. On the other hand, with the exception of IPI the univariate models are more accurate and their error autocorrelations (not all reported) are satisfactory.

iv Graphs

Finally, a visual assessment is immensely helpful and is usually the single most important evidence of model performance. When the forecast values are less volatile than the actual values, for instance, a model with more states should be investigated; the opposite is also true. The graph scaling can importantly influence what is perceived. Plotting the errors on the same graph as the forecast and actual values supplies a frame of reference for the errors that is otherwise absent. It is true, however, that nearly mean-zero series like the cycle model can become jumbled (*e.g.*, Figure 9). Also, for series such as the macroeconomic data analyzed here, the smallness of the errors relative to the actual values conceals visual clues to model failures like autocorrelation or special events. Nevertheless, the graphs do make the overall forecasting success or failure clear at a glance. For example, the surprising accuracy of the trend model plots indicates that most of the forecast precision originates here, and not in the cycle model. Also, despite the clutter in Figure 9, it is apparent that the trend model left a systematic component in the cycle (the rise on the right side of the plot). In addition, if the increasing magnitudes of the cycle errors persisted through a trend model re-specification, heteroskedasticity could be an issue.

IV Iterative Estimation

Aoki's conventional state space estimators can be interpreted as instrumental variables estimators that use y_{t-1}^{-} as the instrument, see Havenner and Aoki (1988a,b) and Aoki (1990). Havenner and Leng proposed iterating the estimators using the states as instruments and formally proved that these estimators are asymptotically efficient in their class.

On average there is typically a small but persistent improvement in the accuracy of estimates based on the iterative formulas. The efficiency proof is asymptotic; for any finite sample the conventional estimator can be superior, and sometimes it is. Empirical experience with a variety of large and small models confirms the conclusions based on the Monte Carlo experiment reported in Havenner and Leng. The conventional estimator is the first step in the iteration, so if the later steps are not making headway it is obvious from the statistics.

---

29 The Durbin-Watson statistic provides a test that is valid in the small sample for this lag. Since $d \approx 2(1 - .597) = .806$, the hypothesis of no serial correlation is overwhelmingly rejected.

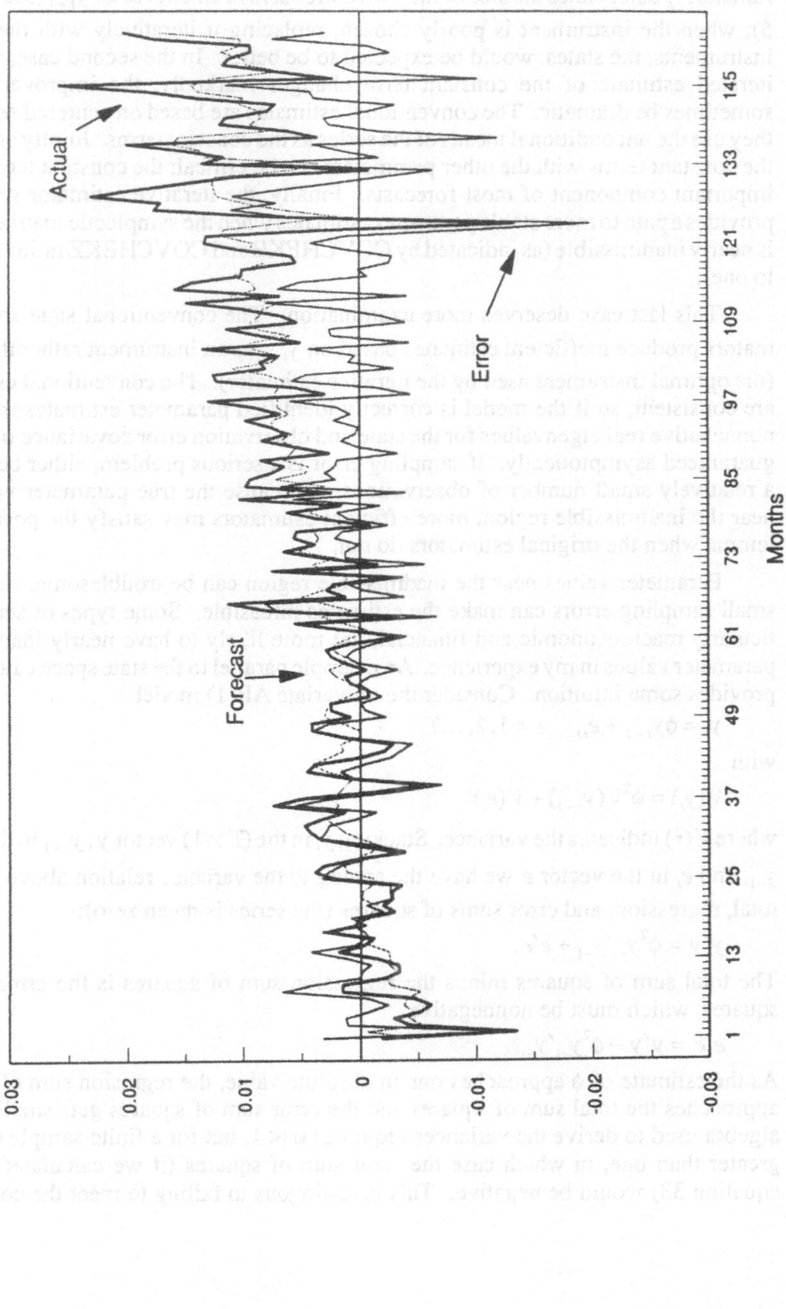

Figure 9: Univariate Money Stock Cycle
In Logarithms In Sample

In some cases the iterative estimator can be a great improvement. These cases tend to occur when the symplectic matrix calculations were nearly unstable, or when the unconditional means of the series are not good estimates of the constants, or when the lag parameter is badly specified. Examining them in reverse order, the lag parameter N_p determines the size of the conventional instrument vector y_{t-1}^- (see equation 5); when the instrument is poorly chosen, replacing it iteratively with the optimal instruments, the states, would be expected to be better. In the second case, when the iterated estimate of the constant term changes markedly, the improvement can sometimes be dramatic. The conventional estimates are based on centered series, *i.e.*, they use the unconditional means of the series as the constant terms. Jointly estimating the constant terms with the other parameters can be critical; the constant terms are an important component of most forecasts. Finally, the iterative estimator sometimes provides a path to more stable parameter estimates when the symplectic matrix solution is nearly inadmissible (as indicated by COVCHEKE and COVCHEKZ ratios not equal to one).

This last case deserves more examination. The conventional state space estimators produce inefficient estimates based on y_{t-1}^- as an instrument rather than $z_{t|t-1}$ (the optimal instrument used by the iterative estimator). The conventional estimators are consistent, so if the model is correctly identified parameter estimates that imply nonnegative real eigenvalues for the state and observation error covariance matrix are guaranteed asymptotically. If sampling error is a serious problem, either because of a relatively small number of observations or because the true parameter values are near the inadmissible region, more efficient estimators may satisfy the positive real lemma when the original estimators do not.

Parameter values near the inadmissible region can be troublesome, since even small sampling errors can make the estimates infeasible. Some types of series, particularly macroeconomic and financial, are more likely to have nearly inadmissible parameter values in my experience. An example parallel to the state space calculations provides some intuition. Consider the univariate AR(1) model

$$y_t = \phi y_{t-1} + e_t, \quad t = 1,2,\ldots T \tag{30}$$

with

$$V(y_t) = \phi^2 V(y_{t-1}) + V(e_t) \tag{31}$$

where $V(\bullet)$ indicates the variance. Stacking y_t in the $(T \times 1)$ vector y, y_{t-1} in the vector y_{-1}, and e_t in the vector e we have the analog to the variance relation above defining total, regression, and error sums of squares (the series is mean zero):

$$y'y = \phi^2 y_{-1}'y_{-1} + e'e. \tag{32}$$

The total sum of squares minus the regression sum of squares is the error sum of squares, which must be nonnegative:

$$e'e = y'y - \phi^2 y_{-1}'y_{-1}. \tag{33}$$

As the estimate of ϕ approaches one in absolute value, the regresion sum of squares approaches the total sum of squares and the error sum of squares gets smaller. The algebra used to derive the variances requires $|\phi| < 1$, but for a finite sample $\hat{\phi}$ can be greater than one, in which case the error sum of squares (if we calculated it from equation 33) would be negative. This is analogous to failing to meet the conditions of the positive real lemma (extending to the matrix case requires nonnegative, real eigenvalues of the matrix analog to $e'e$). The closer the true parameter ϕ is to unity, the smaller is the sampling error needed to cause a logical inconsistency. In the state space model, the anomaly signals its presence in the moving average calculations rather than in the autoregressive parameter calculations used as an example above. The principle is the same, however. For example, the lower right block in the matrix F in equation 26 above, $\Gamma_0 - C\Xi C'$ can be interpreted as the total sum of squares minus the regression sum of squares for the observation equation. Sometimes it is not possible to find a matrix Ξ such that F has nonnegative real eigenvalues, given the estimated values of A, C, Ω, and Γ_0, just as it is not possible to find a positive error sum of squares $e'e$ in the AR(1) example. When this occurs, the symplectic matrix will have unit roots. To examine a specific example in terms of the state space moel, consider the bivariate ARMA(1,1) analyzed in Appendix 1. The symplectic matrix will have a real unit root when the estimate of the moving average parameter θ_2 is equal to 1 or -1. If the true value is large in absolute value minor sampling error can cause this to happen, even if the model is correctly specified.[30] In these cases we have to estimate right against the border, so an efficient estimator is required; it may have to be constrained as well.

In the four series cycle model reported in Table 1, Aoki's conventional estimates produced COVCHEKE ratios ranging from -1.36 to 1.61 indicating possible problems. A negative eigenvalue in the observation equation error covariance matrix was a more pronounced sign. The problem was resolved in the iterated estimates. The constant terms and coefficients adjusted accordingly. However, the model is not much better (if any at all), as evidenced by the summary statistics and error autocorrelations. Here the problems come from a poor model specification; iterating the estimator is not a substitute for eliminating the mismatched RCP series. In the three series cycle model of Table 2, the conventional estimates again have COVCHEKE ratios that suggest instability (1.56, 3.69, 2.99). In this case, the iterated estimator does slightly improve the model, however.

Once the step is taken from the conventional estimator to the iterated estimator the symplectic matrix calculations[31] are not used, so there are no theoretical and empirical variances to form the COVCHEKE and COVCHEKZ ratios. The program described in Appendix 2 reports the ratio of the variances from iteration to iteration as COVCHEKE and COVCHEKZ (in an altered display). These are not measures of computational instability, however; among other things, they are a function of step size.

---

30 This is the case analyzed in the simulation study of Heij *et al.* where they compare the performance of the inefficient systems theory estimators to the efficient maximum likelihood estimators. They call it a "failure" when the systems theory estimators signal trouble due to a parameter estimate in the infeasible range, but it could equally well be interpreted as a valid warning. The conventional maximum likelihood computational formulas do not check for feasibility, leaving the analyst unaware of a contradiction in assumptions. Although the more efficient estimator would have less sampling error, it should produce infeasible estimates as well.

31 The empirical relation reported in Appendix 1 provides a link between the eigenvalues of A-BC and the symplectic matrix.

The iteration of the estimator is constrained by two side conditions. If the series are stationary, the eigenvalues of A must lie inside the unit circle. In addition, the iterated estimator must run through the sample to calculate the states it uses as instruments. If the eigenvalues[32] of $A\text{-}BC$ are outside the unit circle, these calculations will explode and the states cannot be obtained. (It is this constraint that forces the positive real lemma to hold.) As a result, the step sizes used in iteration must be closely monitored in order to not fall off of either of these eigenvalue "cliffs" into infeasibility. These are strongly nonlinear constraints, imposed by a naive step size procedure that takes a linear combination of the proposed new estimates and the prior iteration estimates, with the proportion of the prior estimates interpreted as a damper on the steps taken. For models with eigenvalues of A nearly equal to one, it is prudent to take very small steps in each iteration in order to give the algorithm time to veer away from the infeasibility cliff over several iterations rather than march right up to it and stop. The eigenvalues of $A\text{-}BC$ appear less systematic (in that they are a function of all three estimated matrices and do not seem to repeat as persistently from iteration to iteration as the eigenvalues of A), and here the task is simply to get past the problem point in the given iteration.

On balance, iterating will improve a good model (at least asymptotically, see the proof in Havenner and Leng), and can occasionally dramatically improve a poor model through more efficient constant and other parameter estimates. Since the sampling error that causes the symplectic matrix unit roots is a function of the model specification, it is advisable to examine other models to ensure that a poor specification is not the primary cause of the unit roots. When the true parameter estimates are near the infeasible region (as indicated by unit symplectic matrix roots in all of the plausible trial model specifications), the constrained iterative estimator can sometimes be used to find efficient estimates inside the feasible region that produce a good model.

V Conclusion

The point of the model construction methods described above is to forecast the series out of sample.[33] Forecasts can be made for a single period ahead or for multiple periods ahead, for the levels of the variables or for the detrended values (analogous to differences), and for the actual values or some transformation of them (such as logarithms).

Multiple period forecasts in a state space model use the moving average matrix B for the first period, but from then on the moving average term $B e_t$ is zero since e_t is set to its unconditional expected value of zero. Any higher order moving average terms are automatically incorporated as the included error moves through the state difference equation. See Aoki, or Aoki and Havenner (1991), for the augmentation details. If the forecast period of interest exceeds N_p, setting N_f to the forecast horizon expands the Hankel matrix and the stack of first order conditions to formalize the tradeoff between errors in different forecast periods, see Havenner and Aoki (1988b).

32 See Appendix 1 for the relation between these eigenvalues and those of the symplectic matrix.

33 It is good practice to reserve some observations for model validation before putting the model to its real use. Good models do not have large differences in the summary statistics for in sample versus out of sample. In the macroeconomic data example, 24 of the 177 observations were held back to test the models, as reported in the tables.

For the macroeconomic data examined here, the coefficient estimates would not change for up to quarterly forecasts since the cycle model lag parameters are all at least three. Forecasts beyond a quarter would be based on a different set of coefficients if the analyst elected to trade off errors in multiple periods by increasing N_f.

Forecasts of levels or differences are facilitated by the stochastic nonstationarity trend/cycle model. If the series all have separate unit roots the model includes it as a subcase with $A_\tau = B_\tau = C_\tau = I$[34] (there must be as many states as series in this case). Then the forecasts from the cycle model are forecasts of the differences, and the forecasts from the trend model are forecasts of the levels.

The effects of data transformations are not specific to state space procedures, of course, but their implications are often ignored in practice whatever the method employed. If the series have been transformed into logarithms, for instance, the simple exponential of the forecast of the logarithm of the series will be biased downward since $exp(e_t) = \sigma^2/2$ for $e_t \sim iid\ N(0, \sigma^2)$. The relation $\hat{y}_t = \exp(\widehat{\ln y} + \hat{\sigma}^2/2)$ can be used to correct univariate series. This has a larger effect with larger error variances, of course. The multivariate case requires that the covariances be included by factoring the error covariance matrix.

Figures 10-13 present the univariate combined trend and cycle models for each of the four series, back in their original units (rather than in logarithms). The last 24 periods are out of sample one month ahead predictions. None of the models show any deterioration in accuracy out of sample. The autocorrelations of the errors are small and both the direction and magnitude statistics are excellent. Because the data were transformed into logarithms for analysis, the forecasts have been exponentiated and multiplied by $\exp(\sigma^2/2)$ as discussed above. The variances are so small that the adjustment factor made no difference in the foecasts for any series except RCP; the RCP factor was 1.0025, still not very large. With larger error variances the adjustment can be more substantial, however.

Given stationary series, the state space model construction procedure is essentially a three or four dimensional search over selected ranges of options. Specifically, decisions must be made about the series to be modeled together, the lag parameter values to use in constructing the Hankel matrices, the numbers of states in the models, and, possibly, whether or not to iterate on the estimates. Aoki's state space method simplifies the search in several ways (see the introduction). Building accurate models still requires careful analysis and good decisions based on the correct interpretation of the information provided by the algorithm, however. The sections above have attempted to offer guidance in these decisions in general and by examining four macroeconomic series in detail. The series studied turned out not to exploit a major feature of Aoki's method, *i.e.*, the ability to model multivariate systems. (The myriad of interactions possible in a multivariate system represents a far more difficult test of a procedure than univariate modeling.) Nevertheless, they make an excellent example because they incorporate so many of the issues that are confronted in practice. For instance, the treatment of nonstationarity, the meaning of unstable computations, and

---

34 A nonsingular transformation of $A_\tau = B_\tau = C_\tau = I$ is equivalent, see equations 22 and 23.

Figure 10: Industrial Production Index

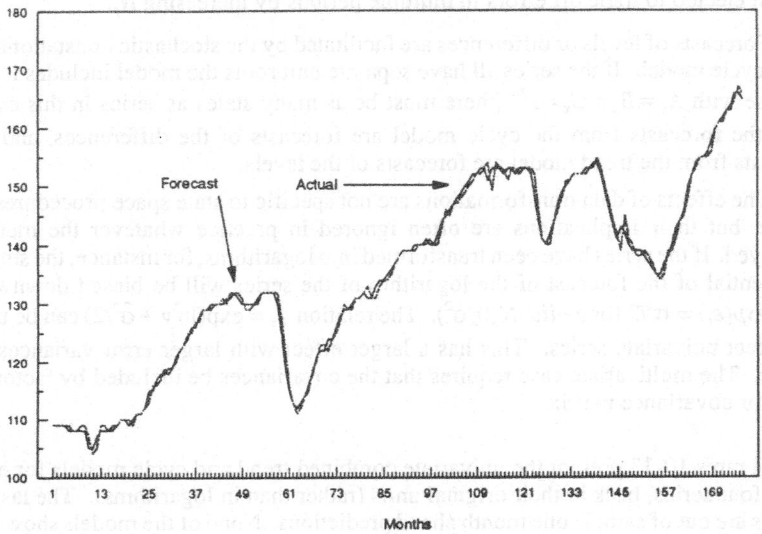

Figure 11: Consumer Price Index

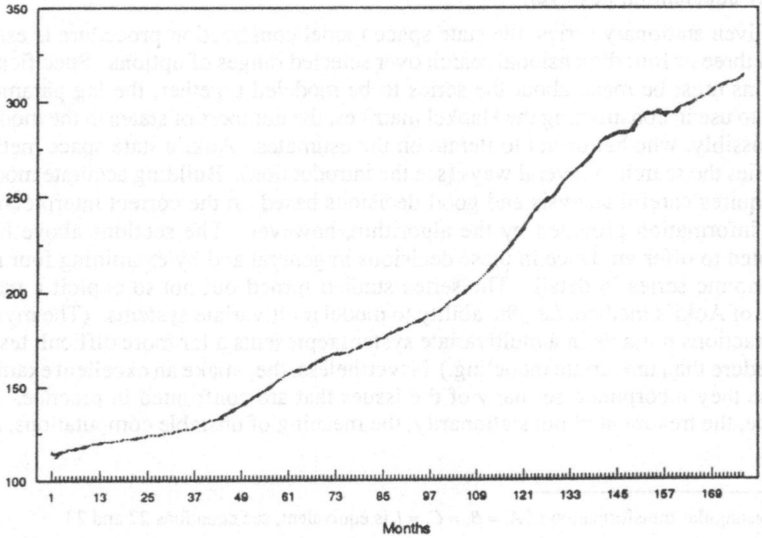

Figure 12: Commercial Paper Rate

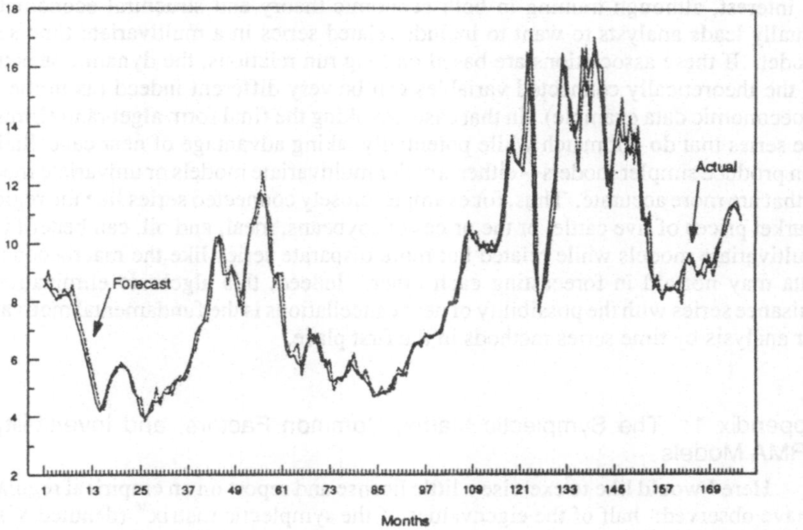

Figure 13: Money Stock

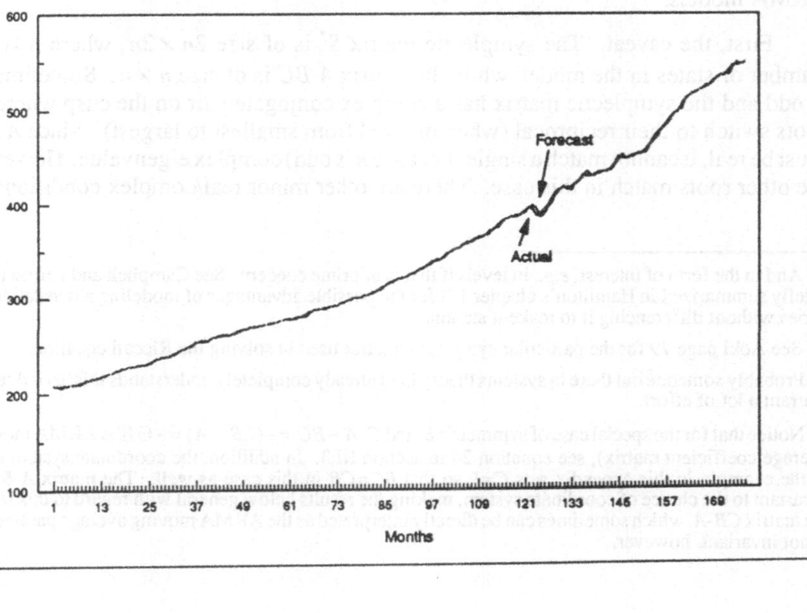

the choice of series to model together are all questions that arise routinely in state space modeling. On the latter issue, the macroeconomic example is unquestionably representative. My standard advice is to model what is of interest[35] and only what is of interest, although training in both economic theory and structural econometrics usually leads analysts to want to include related series in a multivariate time series model. If these associations are based on long run relations, the dynamic structures of the theoretically connected variables can be very different indeed (as in the macroeconomic data example). In that case, invoking the final form algebra to eliminate the series that do not match while potentially taking advantage of near cancellations can produce simpler models -- either smaller multivariate models or univariate models -- that are more accurate. Thus, for example, closely connected series like the regional market prices of live cattle, or the prices of soybeans, meal, and oil, can benefit from multivariate models while related but more disparate series like the macroeconomic data may not aid in forecasting each other. Indeed, this algebraic elimination of nuisance series with the possibility of near cancellations is the fundamental motivation for analysis by time series methods in the first place.

Appendix 1: The Symplectic Matrix, Common Factors, and Invertibility in ARMA Models

Here I would like to exercise a little license and report on an empirical regularity I have observed: half of the eigenvalues of the symplectic matrix[36] (denoted S^*) are generally the same as the eigenvalues of the matrix *A-BC*. I have not seen a previous reference to this relation, and have been unsuccessful in limited attempts[37] to establish it algebraically, but it has held true (subject to the caveat below) in every one of a large number of empirical runs since I initially noticed it. This relation is tantalizing because it allows a direct link (and thus some intuition) between the symplectic matrix and the coefficients of the model, and, under some restrictive conditions, corresponding ARMA models.[38]

First, the caveat. The symplectic matrix S^* is of size $2n \times 2n$, where n is the number of states in the model, while the matrix *A-BC* is of size $n \times n$. Sometimes n is odd and the symplectic matrix has a complex conjugate pair on the cusp where its roots switch to their reciprocal (when ordered from smallest to largest). Since *A-BC* must be real, it cannot match a single (because n is odd) complex eigenvalue. However, the other roots match in this case. There are other minor real/complex conditions.

---

35 And in the form of interest, *e.g.*, in levels if that is of prime concern. See Campbell and Perron (also briefly summarized in Hamilton's chapter 17) for the possible advantages of modeling a nonstationary series without differencing it to make it stationary.

36 See Aoki page 79 for the particular symplectic matrix used in solving the Riccati equation.

37 Probably someone out there in systems theory land already completely understands this, so it doesn't warrant a lot of effort.

38 Notice that for the special case of symmetric B and C, $A - BC = -(CB - A) = -\Theta$ (the ARMA moving average coefficient matrix); see equation 24 in section III.3. In addition, the coordinate system used in the example in this appendix sets *C=I*, so that *BC=CB* in this case as well. The matrix *A-BC* is invariant to the choice of coordinate system, making the results below general with regard to this issue; the matrix *CB-A*, which sometimes can be directly interpreted as the ARMA moving average parameter, is not invariant, however.

The matrix A-BC is interesting in its own right. Solve the observation equation 2 for $e_t = y_t - Cz_{t|t-1}$ and substitute it into the state equation 1 to get

$$z_{t+1|t} = Az_{t|t-1} + B(y_t - Cz_{t|t-1})$$
$$= (A - BC)z_{t|t-1} + By_t$$

the relation between the states $z_{t|t-1}$ and the observations y_t. Because the form of the equation above is an update based on the innovation, the matrix A-BC is sometimes called a Kalman matrix. Its eigenvalues must be inside the unit circle for the states not to explode, a condition required by the Havenner-Leng iterative estimator.

The symplectic matrix signals trouble when its roots are equal to one, not greater than one as is the case for A-BC. (Since half the symplectic matrix roots are the reciprocals of the other half, greater than or less than one is irrelevant.) Under what conditions will there be unit eigenvalues of A-BC?

Let us examine two issues, common factors and invertibility of the moving average process, in terms of a simple bivariate ARMA(1,1) with a triangular structure to facilitate obtaining the requisite eigenvalues:

$$(y_{1t} - e_{1t}) = \phi_1(y_{1t-1} - e_{1t-1}) + \phi_{12}(y_{2t-1} - e_{2t-1}) + \phi_1 e_{1t-1} + \phi_{12} e_{2t-1}$$

$$(y_{2t} - e_{2t}) = \phi_2(y_{2t-1} - e_{2t-1}) + (\phi_2 + \theta_2)e_{2t-1}$$

or

$$\begin{pmatrix} y_{1t} - e_{1t} \\ y_{2t} - e_{2t} \end{pmatrix} = \begin{pmatrix} \phi_1 & \phi_{12} \\ 0 & \phi_2 \end{pmatrix} \begin{pmatrix} y_{1t-1} - e_{1t-1} \\ y_{2t-1} - e_{2t-1} \end{pmatrix} + \begin{pmatrix} \phi_1 & \phi_{12} \\ 0 & \phi_2 + \theta_2 \end{pmatrix} \begin{pmatrix} e_{1t-1} \\ e_{2t-1} \end{pmatrix}$$

$$\begin{pmatrix} y_{1t} \\ y_{2t} \end{pmatrix} = \begin{pmatrix} 1 & 0 \\ 0 & 1 \end{pmatrix} \begin{pmatrix} y_{1t} - e_{1t} \\ y_{2t} - e_{2t} \end{pmatrix} + \begin{pmatrix} e_{1t} \\ e_{2t} \end{pmatrix}$$

in state space format. The state space coefficient matrices A, B, and C are

$$A = \begin{pmatrix} \phi_1 & \phi_{12} \\ 0 & \phi_2 \end{pmatrix},\ B = \begin{pmatrix} \phi_1 & \phi_{12} \\ 0 & \phi_2 + \theta_2 \end{pmatrix},\ C = \begin{pmatrix} 1 & 0 \\ 0 & 1 \end{pmatrix}.$$

Since A is a triangular matrix, its eigenvalues are ϕ_1 and ϕ_2.

Common factors:

Write the bivariate ARMA(1,1) model above in terms of the lag operator L

$$\begin{pmatrix} 1 - \phi_1 L & -\phi_{12} L \\ 0 & 1 - \phi_2 L \end{pmatrix} \begin{pmatrix} y_{1t} \\ y_{2t} \end{pmatrix} = \begin{pmatrix} 1 & 0 \\ 0 & 1 + \theta_2 L \end{pmatrix} \begin{pmatrix} e_{1t} \\ e_{2t} \end{pmatrix}.$$

Using the fact that the inverse is the adjoint divided by the determinant,

$$(1 - \phi_1 L)(1 - \phi_2 L) \begin{pmatrix} y_{1t} \\ y_{2t} \end{pmatrix} = \begin{pmatrix} 1 - \phi_2 L & \phi_{12} L \\ 0 & 1 - \phi_1 L \end{pmatrix} \begin{pmatrix} 1 & 0 \\ 0 & 1 + \theta_2 L \end{pmatrix} \begin{pmatrix} e_{1t} \\ e_{2t} \end{pmatrix}$$

$$= \begin{pmatrix} 1 - \phi_2 L & \phi_{12} L(1 + \theta_2)L \\ 0 & (1 - \phi_1 L)(1 + \theta_2 L) \end{pmatrix} \begin{pmatrix} e_{1t} \\ e_{2t} \end{pmatrix}$$

so that

$$(1 - \phi_1 L)(1 - \phi_2 L)y_{1t} = (1 - \phi_2 L)e_{1t} + (1 + \theta_2 L)\phi_{12} L e_{2t}$$

and

$$(1-\phi_1 L)(1-\phi_2 L)y_{2t} = (1-\phi_1 L)(1+\theta_2 L)e_{2t} \Rightarrow (1-\phi_2)y_{2t} = (1+\theta_2 L)e_{2t}.$$

If $\theta_2 = -\phi_2$, then there are common factors that can be cancelled to simplify the model.

Evaluation of *A-BC* using the definitions of *A*, *B*, and *C* above gives

$$A-BC = \begin{pmatrix} 0 & 0 \\ 0 & -\theta_2 \end{pmatrix}$$

which, using the common factor relation $\theta_2 = -\phi_2$ implies that the eigenvalues of *A-BC* are ϕ_2 and 0.

Using the empirical relation between the eigenvalues of S^* and *A-BC*, we conclude that common ARMA factors do not necessarily produce unit roots in the symplectic matrix S^*.

Invertibility:

Now suppose $\theta_2 = -1$, *i.e.*, the moving average process is not invertible. In this case, the eigenvalues of *A-BC* are 1 and 0; thus noninvertible moving average processes can induce unit roots in *A-BC* and, based on the empirical relation between *A-BC* and S^*, in the symplectic matrix used to solve the Riccati equation.

Appendix 2: Using the Speakeasy Version of the Program

SSATS is a Speakeasy[39] program to do univariate and multivariate time series analysis using the techniques described in Aoki, *State Space Modeling of Time Series* (1990), and in Aoki and Havenner, "State Space Modeling of Multiple Time Series," *Econometric Reviews*, 10 (1), 1991, *pp.* 1-59.

Model:

$$\underset{(n^*\times 1)}{z_{t+1|t}} = \underset{(n^*\times n^*)(n^*\times 1)}{A z_{t|t-1}} + \underset{(n^*\times K)(K\times 1)}{B e_t} \qquad \textit{(state equation)}$$

$$\underset{(K\times 1)}{y_t} = \underset{(K\times 1)}{\mu}\ \underset{(K\times n^*)(n^*\times 1)}{C z_{t|t-1}} + \underset{(K\times 1)}{e_t} \qquad \textit{(observation equation)}$$

Notation:

The notation used here is that of the program. As the procedure evolved, the notation matured in the papers but remained as initially set in the program. A table of equivalences between Aoki's notation, the program notation, and the notation used above (and in the *Econometric Reviews* paper) is presented below:

39 Speakeasy Computing Corporation, 224 S. Michigan Avenue, Chicago, Il 60604, (312)427-2400. The program runs on personal computers equipped with Speakeasy IV eta or above.

	Econometric Reviews	Aoki	Program
Number of series	m	-	K
Input data	y_t	y_t	$Y \sim (T \times K)$
Number of observations	T	T	T
Estimated number of states	$\hat{n}$	n	n^* *(NSTAR)*
Hankel matrix size parameters	N_p, N_f	J, K	N, NP, NF
Coefficient matrices	A, B, C	A, B, C	A, B, C
State covariance matrix	Ξ	Π	Ω *(OMEGA)*
Covariance of states and data	Ω	M	M
Error covariance matrix	Ψ	Δ	Σ *(SIGMA)*
Data covariance matrix	Γ_0	Λ_0	Γ_0 *(GAMMA0)*

Data Preparation Program:

A Speakeasy program to link the data to the SSATS program is required, with all information specific to each data set contained in it. A typical data preparation program is provided below. The statements numbered (1) - (10) would be sufficient after the initial dataset is read and kept on disk in Speakeasy format.

```
PROGRAM DEMODATA                                              (1)
$ (Dollar signs start comment fields)
$ REQUIRED:     K, number of series
$               T, number of observations
$               Y, data matrix, T by K
$               YSAVE, complete data matrix, K columns, possibly more
$                    than T rows for out of sample forecast validation
$ The next four statements were used to input the original data (on a PC)
$   and keep it on disk in Speakez format:
$     ASSIGN(11, "C:\SIMS.DAT") $ Master dataset in file SIMS.DAT
$     SIMDAT=MAT(177,4:) $ 177 observations on 4 variables
$     READ("(4F7.1)",11:SIMDAT) $ Fortran format for data
$     KEEP SIMDAT $ Writes the data to disk in Speakeasy format under the
$     (arbitrary) name SIMDAT
$
      GET SIMDAT $ Reads Speakez data matrix after initial run  (2)
      SPACE 2
      "          M A C R O E C O N O M I C   D A T A"
      "             (Differenced Logarithms)"
      "             Order: IPI, CPI, RCP46, M1"
      $ Industrial production index, consumer price index,
      $   commercial paper rate (4-6 months), money stock
      SPACE 2
      Y=LOG(SIMDAT)
      IY=INTEGERS(2,177)
      Y=Y(IY) - Y(IY-1) $ First differences
      YSAVE=Y
      Y=Y(INTEGERS(1,177-24), ) $ Save 24 observations for forecasts
      SERIES='IPI','CPI','RCP46','M1' $                         (3)
$     Note name labels are 8 characters maximum, in single quotes
$     An ampersand (&) as the first character continues a Speakeasy line
      IC=1,2,3,4 $ Defaults to all series; or, omit this to use prior IC
```

```
        "Select the model series by setting IC to the column numbers"$        (4)
$       Select the series at the PAUSE, e.g., IC=1, 2, 4  to get IPI, CPI, M1
        PAUSE to select the series to be modeled                              (5)
        Y=Y( ,IC) $ Use only the columns in IC in the model                   (6)
$       Use only columns IC for the out of sample forecast data set YSAVE:
        YSAVE=YSAVE( ,IC) $                                                   (7)
        SERIES=SERIES(IC) $                                                   (8)
        K=NOCOLS(Y); T=NOROWS(Y) $ Rember to set K and T                      (9)
        INPUT any final checks on data $ (debug, delete when data ok)
        END $                                                                 (10)
```

As noted, this program must supply (case sensitive):

K	the number of series jointly modeled
T	the number of observations in the estimation sample
Y	a K column matrix of the data (with means in)
YSAVE	a K column master matrix of the data (with means in), starting at the first row in the sample, possibly with rows beyond the sample for out of sample forecast evaluation

Running SSATS:

To use the program, invoke Speakeasy (usually by entering SPEAK at the system prompt). A record of the session can be kept for printing later by typing JOURNAL ON at any Speakeasy prompt, see HELP JOURNAL. At the Speakeasy :_prompt enter

```
:_GET SSATS;SSATS $ Gets the program from disk; executes it
```

Characters beyond $ are comments, while the semicolon is equivalent to entering what follows on a new line.

Required Parameters:

There are three required parameters, the NAME of your Speakeasy program providing the data, an identifying TITLE of up to 50 characters, and the number of lags back and periods forward to be included in the Hankel matrix, *NP*, *NF*, or, if both are the same, just *N*. Suppose your data program is DEMODATA in the following example. (A program DEMODATA similar to but not exactly the same as listed above is supplied with the program to make a representative run, so you can try this and list DEMODATA to model your own program after it or the shorter version above.) The SSATS program calls other programs, including the main driver program STTS (System Theoretic Time Series). At the STTS> prompt, enter

```
STTS>NAME=ARGUMENT(DEMODATA)
STTS>TITLE="D E M O N S T A T I O N  R U N"
STTS>N=6 $ Automatically sets NF=NP=N=6
```

If any of these parameters are not given, the program will recycle for the remaining unset values. Previously set values remain in force, so if nothing changes after the initial input you need not reenter the values. (This is true even between SSATS runs, although it is a good idea to CLEARDATA between major runs). Note the use of ARGUMENT to pass the name of the program that handles all details associated with the particular dataset.

Options:

Defining a variable OPTIONS, *e.g.*, :_OPTIONS=1 followed by a null carriage return will provide limited on-line help in the form of a list of options available, if entered in the required parameter loop (above).

One option that is particularly noteworthy is MODELS, an array that you can TABULATE at any Speakeasy prompt that provides a record of the key summary statistics for each of the models you have tried in this session. This array can get very large in some sessions, and you may have to list it (just type MODELS) with the JOURNAL ON for printing and then FREE MODELS if you run into space problems. STTS offers a NOSAVE option each time you enter, asking if you want to union the summary statistics from the models so far with the new ones you are about to try. Setting NOSAVE=1 at the STTS question starts a new MODELS summary, overwriting the records of the models previously tried, so you may want to list MODELS with the JOURNAL ON if you are selecting NOSAVE=1. (On PC's, you can scan back through the output file using page up and page down, if you turn REVIEW ON.)

Another useful option is to define a variable named SERIES loaded with the names (up to eight characters) of the variables in your dataset. If you define SERIES as a name literal or character literal [using ' ' (preferred), or " "], the labels will print on the summary statistics and graphs automatically. See the sample program above.

UPDATE should probably be ignored. Recursive updating of the coefficient estimates becomes pretty complicated in its interactions with your data preparation program until you gain sophistication. Similarly, USEBIGH and AUTCOR should be ignored.

Defining a variable BRIEF=1 causes long forecast and actual prints to not come to the screen but to go into the log if JOURNAL is ON. BRIEF now defaults to 1, so if you want the long version on the screen, you must FREE BRIEF.

ABCLIM is the iteration limit on the new Havenner-Leng iterative estimator (*Journal of Economic Dynamics and Control*, May 1996). It defaults to 1, but can be changed interactively depending on the summary statistics from each iteration.

BFLAG is used for multiple period forecasts, which do not have moving average updates beyond the first period.

UNDO provides an opportunity to insert a program to reverse any transformations on the data required to induce stationarity. For example, if the data were differences of logarithms, the summed exponentials of the changes with variance adjustment can be recovered if a program is written to do this. The program is special in each case, and so must be supplied by the user. I have an example named INTDEMO that can serve as a guide. To use it, the program must be fetched from disk before the UNDO statement, *e.g.*, GET INTDEMO. Then the statement UNDO=ARGUMENT(INTDEMO) tells the SSATS program to use INTDEMO to recover the original data. Watch out: if you give the UNDO statement before the program is fetched from disk you have told the processor to use **no** program to reverse the transformations (because the name you gave to ARGUMENT doesn't exit), in which case no error occurs but the UNDO does nothing.

Final Print:

If JOURNAL was set to ON, you can print a log of the session after leaving Speakeasy by printing the file SPEAKEZ.LOG. The file DEMODATA.LOG is a renamed SPEAKEZ.LOG file from a run on DEMODATA. You can avoid large output files by running with JOURNAL OFF and then turning it on only to list MODELS. For example, you might run several models, and then type

:_JOURNAL ON
:_MODELS
:_JOURNAL OFF

and then print the SPEAKEZ.LOG file that will have a list of MODELS. Changing the TITLE for importantly different models allows easier identification of the results.

On PC's, graphics can come to the screen or go into a file. SSATS will request information for graphing, or else type RETURN if you do not want these graphs. If you elected to write the graphics output into a file, the default name is MYGRAPH.PS unless you specified an alternative. The default graphics use postscript, so this is a postscript file to print for hardcopy plots.

References

Akaike, H. (1974), "Markovian Representation of Stochastic Processes and Its Application to the Analysis of Autoregressive Moving Average Processes," *Annals of the Institute of Statistical Mathematics* 26, *363-87*.

Akaike, H. (1975), "Markovian Representation of Stochastic Processes by Canonical Variables," *SIAM Journal on Control* 13, *162-73*.

Akaike, H. (1976), Canonical Correlation Analysis of Time Series and the Use of an Information Criterion. In R. Mehra and K. Lainiotis, eds., *System Identification: Advances and Case Studies*. Academic Press, Inc., New York.

Aoki, M. (1983), *Notes on Economic Time Series Analysis: System Theoretic Approach*, Springer-Verlag, Heidelberg.

Aoki, M. (1987), *State Space Modeling of Time Series*, Springer-Verlag, Berlin.

Aoki, M. (1990), *State Space Modeling of Time Series: Second, Revised and Enlarged Edition*, Springer-Verlag, Berlin.

Aoki, M., and A. Havenner (1991), State Space Modeling of Multiple Time Series. *Econometric Reviews* Vol. 10 (1), *1-59*.

Arndt, C., and K. Foster (1996), "A State Space Model of Monthly Wheat Prices," *Applications of Computer Aided Time Series Modeling* (this volume).

Berwald, D., and A. Havenner (1996), "Evaluating State Space Forecasts of Soybean Complex Prices," *Applications of Computer Aided Time Series Modeling* (this volume).

Campbell, J. Y., and P. Perron (1991), "Pitfalls and Opportunities: What Macroeconomists Should Know About Unit Roots," *NBER Macroeconomics Annual*, MIT Press, Cambridge.

Cerchi, M., and A. Havenner (1988), "Cointegration and Stock Prices: The Random Walk on Wall Street Revisited," *Journal of Economic Dynamics and Control,* Vol. 12, 2/3, *333-346.*

Criddle, K., and A. Havenner (1991), "An Encompassing Approach to Modeling Fishery Dynamics: Modeling Dynamic Nonlinear Systems," *Natural Resource Modeling,* Vol. 5, 1, *55-90.*

Dickey, D. A., and W. A. Fuller (1981), "Likelihood Ratio Statistics for Autoregressive Time Series with a Unit Root," *Econometrica,* Vol. 49, *1063.*

Dorfman, J., and A. Havenner (1991), "State Space Modeling of Cyclical Supply, Seasonal Demand, and Agricultural Inventories," *American Journal of Agricultural Economics,* Vol. 73, 3, *829-840.*

Faurre, P. (1976), "Stochastic Realization Algorithms," *System Identification: Advances and Case Studies,* R. K. Mehra and D. G. Lainiotis, editors, Mathematics in Science and Engineering, Vol. 126, Academic Press, New York.

Dorfman, J., and A. Havenner (1994), "Model Specification Tests for Balanced Representation State Space Models," *Communications in Statistics: Theory and Methods,* Vol. 24, 1, *97-119.*

Dorfman, J., and A. Havenner (1992), "A Bayesian Approach to State Space Multivariate Time Series Modeling," *Journal of Econometrics,* Vol. 52 (3), *315-346.*

Egan, L., and and A. Havenner (1996), "Managing the Herd: Price Forecasts for California Cattle Production," *Applications of Computer Aided Time Series Modeling* (this volume).

Foster, K., A. Havenner, and A. Walburger (1995), "System Theoretic Forecasts of Weekly Live Cattle Prices," *American Journal of Agricultural Economics,* Vol. 77 (4), *1012-1023.*

Hamilton, J. D. (1994), *Time Series Analysis.* Princeton University Press, Princeton.

Hannan, E. J., and M. Deistler (1988), *The Statistical Theory of Linear Systems.* John Wiley and Sons, Inc.

Harvey, A. C. (1989), *Forecasting, Structural Time Series Models and the Kalman Filter.* Cambridge University Press, Cambridge.

Havenner, A., and M. Aoki (1988a), "An Instrumental Variable Interpretation of Linear Systems Theory Estimation," *Journal of Economic Dynamics and Control,* Vol. 12, 1, *49-54.*

Havenner, A., and M. Aoki (1988b), "Econometrics and Linear Systems Theory in Multivariate Time Series Analysis," University of California, Agricultural Economics Department Working Paper 88-6.

Havenner, A., and M. Aoki (1990), "Deterministic and Stochastic Trends in State Space Models of Nonstationary Time Series," University of California, Working Paper 90-9.

Havenner, A., and K. Criddle (1988), System Theoretic Time Series: An Application to Inventories and Prices of California Range Cattle. *Computers and Mathematics with Applications* Vol. 17, *1177-1187.*

Havenner, A., and Z. Leng (1996), "Improved Estimates the Parameters of State Space Time Series Models," *Journal of Economic Dynamics and Control,* Vol. 20 (5), *767-789.*

Havenner, A., and J. Tracy (1992), "Flooding on the Eel River: Systems Theoretic Time Series Versus Structural Model Forecasts," *Natural Resource Modeling*, Vol. 16 (2), *171-190*.

Heij, C., T. Kloek, and A. Lucas, (1992). "Positivity Conditions for Stochastic State Space Modeling of Time Series," *Econometric Reviews*, Vol. 11, *379-396*.

Henriksson, R. D., and R. C. Merton, (1981). "On Market Timing and Investment Performance, II. Statistical Procedures for Evaluating Forecasting Skills," *Journal of Business* Vol. 54, *513-533*.

Kailath, T. (1980), *Linear Systems*, Prentice-Hall, Englewood Cliffs.

Litterman, R. (1986), "Forecasting with Bayesian Vector Autoregressions -- Five Years of Experience," *Journal of Business and Economic Statistics*, Vol. 4, *25-38*.

Nelson, C. R., and C. I. Plosser, (1982), "Trends and Random Walks in Macroeconomic Time Series: Some Evidence and Implications," *Journal of Monetary Economics*, Vol. 10, *139-162*.

Popov, V. M. (1969), "Some Properties of Control Systems with Matrix Transfer Functions," *Lecture Notes in Mathematics*, Vol. 144, Springer, Berlin, 1969, *169-180*.

Rosenbrock, H. H. (1970), *State Space and Multivariable Theory*, Wiley, New York.

Shumway, R. H. (1988), *Applied Statistical Time Series Analysis*. Prentice Hall, N.J.

Vaccaro, R. J., and T. Vukina (1993), "A Solution to the Positivity Problem in the State-Space Approach to Modeling Vector-Valued Time Series," *Journal of Economic Dynamics and Control*, Vol. 17, *401-421*.

Wei, W. S. (1990), *Time Series Analysis: Univariate and Multivariate Methods*. Addision-Wesley Publishing Company, Inc.

Part II

Applications of State Space Algorithm

Evaluating State Space Forecasts of Soybean Complex Prices

Derek Berwald and Arthur Havenner
University of California
Davis, California 95616

Abstract

A multivariate state space time series model is fitted to monthly data on six series, soybeans, meal, and oil prices in both the spot and futures markets. These are highly volatile series that are difficult to forecast accurately. The usual statistical criteria are applied to evaluate model performance on the 155 months in sample, and on 47 months out of sample. The model's root mean squared errors and other summary statistics (including graphs of the actual, forecast, and error) indicate that the model fits very well, both in sample and out. However, a small root mean squared error is not enough to guarantee even a profit, let alone a gain in risk-adjusted utility, from using the model forecasts in actual trading. For example, small errors with random signs can make even a relatively accurate model a financial disaster, as can a few ill-timed large errors. It is very difficult to evaluate the usefulness of model forecasts for financial decisionmaking, where both the direction and magnitude of risky strategies combine to produce success or failure for investors whose well-being not only depends on the mean returns, but also the associated variances. We examine a number of nonstandard methods for evaluating forecasts. For example, a nonparametric test is applied to examine the null hypothesis of randomness in the price direction forecasts, with the result that the randomness hypothesis is rejected in general. Even if the direction forecasts are right on average, however, there is no assurance that a risk averse investor would choose to employ the model given the risk levels involved. To investigate this issue, a particular utility function, logarithmic, is applied to the model moments to calculate the expected utility from a naive speculation rule utilizing the model forecasts, and then realized utility is computed for the 47 months out of sample. We find that the model performs well, earning risk-adjusted utility for a hypothetical logarithmic utility investor. In the course of evaluating the forecasts from the soybean complex model, several important issues in financial model evaluation are raised.

1 Introduction

Futures markets exist to exchange risk, at a price, between those willing to bear it and those who are not: market participants form diverse expectations about future prices, contracts are entered and risk is transferred. Many commodity prices show enormous variation with subtle seasonal and other patterns, suggesting the possibility of using mathematically sophisticated modeling procedures such as state space time series in forming expectations of future prices.

Soy products -- soybeans, meal, and oil -- are characterized by highly volatile prices. Besides the natural variation from agricultural production, they are traded internationally, introducing another large element of demand (and thus price) uncertainty through exchange rate variation. Soybeans are grown primarily as inputs to other agricultural products. They are usually crushed and turned into soybean meal and soybean oil. The meal is used for high protein livestock feed, especially for hogs and chickens. When meal prices are low, soybean meal can be substituted for corn in cattle feed. The oil is a residual from meal production, and is used for cooking and salad oil.

The soybean complex is heavily traded and important to United States agriculture.[1] There would be a high return to accurate price forecasts. Efficiently forecasting these prices is technically difficult, since there are potential arbitrage opportunities if the spot and futures prices of beans, meal, and oil spread too far apart, requiring the joint modeling of all six series if all information is to be utilized.

The state space time series procedure due to Masanao Aoki (1983, 1987, 1990) is particularly well suited to a problem like soybean complex price forecasting. Since it is based on states, *i.e.*, dynamic factors, it is possible to discover the underlying common dynamics in interconnected series. If the series do in fact have common cyclical components, the reduced parameter estimation burden from utilizing shared dynamics (fewer states) allows the modeling of a larger number of series than is possible with other methods. In addition, the difficulty of modeling a large number of very complicated time paths is reduced by the orderly inclusion of sample information implicit in Aoki's procedure, since the method is robust with respect to the choice of the number of states.[2]

In the sections below, we will describe the data and the reasons for its choice, estimate the parameters of a state space model of the six soybean complex prices, and examine the results of an admittedly naive trading rule based on these price forecasts. The evaluation of the forecast performance in trading differs from the usual model summary statistics analysis by more tightly connecting the model forecasts to their final use, shifting the emphasis from overall closeness of fit to accurate direction forecasting and utility compensation for the risk assumed.

2 The Data

Daily closing prices on monthly intervals for soybean, meal, and oil spot and futures markets for the period from 1974 through 1991 were drawn from the Dunn & Hargitt Commodity Data Bank, and spot checked by published data from the Chicago Board of Trade Statistical Annual. Although data were available from 1959 to 1991, we discarded the earlier years and used data beginning with 1974 contracts because soybean markets underwent a major structural change after the collapse of the Bretton Woods exchange rate agreement in 1973. This data set contained 202 monthly observations for six series.[3] The last 47 observations were withheld to see how well the state space model would predict out of sample, leaving 155 observations for estimation.

---

1 Although soybeans are subject to price supports, the market prices are high enough that supports do not directly affect the observed price, so that changes in prices are determined by market forces rather than agricultural policy.

2 Each successive state added to the model must be of less value in approximating the autocorrelation sequence than the previously included states, see Aoki and Havenner (1991).

3 There was one missing data point, for soybean meal on July 1, 1985. This was filled in with the previous price, an optimal forecast under a null hypothesis of a random walk.

The prices used are the closing price on the first day of the month, not monthly averages. The use of monthly data is appropriate since we are interested in forecasting longer, low frequency cycles rather than short-run day to day fluctuations. Anecdotal evidence suggests that market professionals generally hold only short-term positions. If traders are unwilling to hold long-term positions, then perhaps soybean futures markets are efficient at driving out short-term profit potential, but there may be opportunities available from using accurate forecasts for profitable longer strategies.

The data on spot prices for the three commodities were already in continuous format. The monthly series on futures prices are observations on the prices of July contracts maturing less than a year in the future, but more than a month away. That is, on the first trading day in July of every year, the contract rolls over to the contract for the following July. This ignores prices from more than a year before maturity, which may be subject to light trading, and prices from the month of delivery, which may have different characteristics due to the impending termination of the contract. By organizing the futures data in this way, we establish a fixed contract that is being priced in any given year (as opposed to a rolling one year ahead contract, for example January soybeans next year, February soybeans next year, *etc.*). Thus the changes in market futures prices are market price revisions for this fixed contract, and accurate forecasts translate into profit (after being adjusted for transactions costs). The revisions in these futures series reflect the risk reallocation between agents in the market.

3 The State Space Time Series Model

Arrange the soybean complex prices in a (6×1) vector y_t, $t=1,2,3,\ldots$. Then the state space model[4] is

$$z_{t+1|t}=Az_{t|t-1}+Be_t \tag{1}$$

$$y_t=Cz_{t|t-1}+e_t \tag{2}$$

where $z_{t|t-1}$ is the $(n\times1)$ vector of state variables at time t, e_t is a vector of serially uncorrelated random errors, and A, B and C are matrices of coefficients to be estimated. The n states $z_{t|t-1}$ are minimum sufficient statistics for y_t using all the information available at time t-1. The number of states, n, can be greater than, equal to, or less than the number of series. If the series have shared low frequency common dynamics, n will be less than the number of series, while if they have separate and/or shared high frequency dynamics, the number of states will increase. A strength of the state space method is its ability to allow the data to determine the optimal number of states, while being robust with respect to errors in this number.

To specify the number of states and estimate the parameters of this model, the (6×6) matrices of cross-autocorrelations at lags $1, 2$, and 3 were used, arranged in the Hankel matrix

$$\underset{(12\times12)}{\hat{H}}=\begin{pmatrix}\hat{\Gamma}_1 & \hat{\Gamma}_2\\ \hat{\Gamma}_2 & \hat{\Gamma}_3\end{pmatrix}. \tag{3}$$

4 The basic references for this section are Aoki (1983, 1987, 1990). See also Aoki and Havenner (1991), and, for an introduction of minimal complexity, Havenner and Aoki (1988).

Every time series procedure implicitly or explicitly determines how far back in time a lagged autocorrelation is important. In this case, we have set the lag parameter to two (the maximum lag in the first block row of the Hankel matrix) because we are interested in finding longer-term patterns in soybean complex prices.[5] With six series and a lag parameter of two, the possibility of dynamic interactions implies that each series can have a univariate model of up to order 12, not coincidentally including the seasonal frequency.

The rank of this Hankel matrix can be shown to be the number of states required to characterize the dynamics of the system. The sample autocorrelation estimates were calculated and arranged in a sample estimate of the Hankel matrix H and its singular values were computed. All of the singular values beyond the third were judged to be zero apart from sampling error, and a state space model with three states was formulated and its parameters estimated. These parameter estimates are given in Table 1 for completeness; given these estimates, the model can be forecasted through the period studied and the results replicated. On the other hand, the estimates are not invariant to the choice of coordinate system for the latent states, and so have no intrinsic value of their own. Since there are orthogonality conditions between functions of them, they are not even the elementary free parameters of the system; see Aoki and Havenner (1991).

Table 1. **Estimated Soybean Complex State Space Model**
Order: Futures Prices of Beans, Oil, Meal, Spot Prices of Beans, Oil, Meal

A

.81602	.1117	-.047806
-.06073	.92162	.046127
.058345	-.046757	.69567

B'			*C*		
-1.43E-4	-5.30E-4	-.00232	-1133.8	-25.145	-126.38
-3.86E-4	7.28E-4	6.57E-4	-381.13	269.01	-45.275
-2.20E-4	1.42E-5	2.27E-5	-2959.3	-1394.14	-402.06
-5.58E-4	-2.35E-4	.0023	-1049.9	34.492	53.903
3.08E-4	.00139	-1.67E-5	-362.62	442.96	21.366
1.01E-4	-4.35E-5	1.22E-4	-2485.3	-1833.1	243.9

Given the estimated coefficients, the initial condition of the state vector can be backcast and the state space model equations (1) and (2) solved forward through the sample period. The summary statistics for this operation are recorded in Table 2, which gives the in-sample root mean squared error (RMSE), average error, mean absolute deviation (MAD), and R^2, defined as the correlation of y and $\hat{y}$ for each series. The mean of each series is given for purposes of comparison. The estimation used the data in the original form from the tape, so the scale of the error statistics is not

5 Intuitively, if we plot a univariate series y_t against time and imagine the regression coefficient that results from regressing y_t on just y_{t-1}, it is obvious that just the long term trend will be obtained. Alternatively, if we imagine the results that would come from regressing y_t on a series of regressors $y_{t-1}, y_{t-2}, y_{t-3}, \ldots$ more short-term detail will result. Thus the length of the lag in the Hankel matrix is inversely related to the maximum detail of the dynamics of the model, and acts as a filter to separate out short-term dynamics. See especially Aoki and Havenner (1991).

obvious. For soybeans the figures should be interpreted as tenths of a cent per bushel, while for soybean oil the numbers are in cents per hundredweight, and for soybean meal, cents per ton.

Table 2. **In-Sample Summary Statistics**

Series	*Mean*	*RMSE*	*Average Error*	*MAD*	R^2
Soybean Futures	6,654	512	-12.3	386	.82
Soy Oil Futures	2,326	216	-10.2	157	.81
Soy Meal Futures	18,166	1,450	-4.54	1,084	.84
Soybean Spot	6,281	470	-9.12	363	.83
Soy Oil Spot	2,351	244	-12.5	183	.84
Soy Meal Spot	17,084	1,460	22.5	1,075	.82

The correlations of the one month ahead predictions, R^2, are good for these extremely volatile series. The root mean squared errors, average errors, and mean absolute deviations are all small relative to the means of the series.

The model also appears to predict well out of sample. The data reserved for out-of-sample validation were kept completely clean: the out-of-sample predictions and comparisons were made only after competing model specifications were tested and a final decision reached based on the in-sample data. The summary statistics are given in Table 3. The correlation of the actual and predicted values (R^2) falls a little in all cases, and the average errors tend to be larger, except for oil futures. However, the root mean squared error declined for bean and oil futures and oil spots. Overall, the summary statistics are perhaps not quite as good but essentially not a lot different out of sample than in sample.

Table 3. **Out-of-Sample Summary Statistics**

Series	*RMSE*	*Average Error*	*MAD*	R^2
Soybean Futures	490	-98.7	374	.75
Soy Oil Futures	149	-1.46	114	.73
Soy Meal Futures	1,667	717	1,145	.73
Soybean Spot	497	-107	394	.76
Soy Oil Spot	204	-65.4	152	.65
Soy Meal Spot	2,306	1,333	1,678	.72

Further evidence regarding the quality of the model is presented in Figures 1 through 6, graphs of the actual values (dots), forecasts (solid line), and errors (solid line around zero) against time; the errors are plotted on the same scale so that it is immediately apparent how large they are relative to typical values. The in-sample period is the 155 months to the left of the vertical bar, while the 47 months to the right of the vertical bar are out of sample. Overall, the model appears to track the swings in the prices reasonably well, although there are indeed some misses that would give a speculator pause.

We find this to be strong evidence of a correctly functioning model when the statistics and plots are this good for six such volatile series over a nearly four year period out of sample, with no model respecification (or even coefficient revision).

Soybean Futures: Actual, Forecast and Error

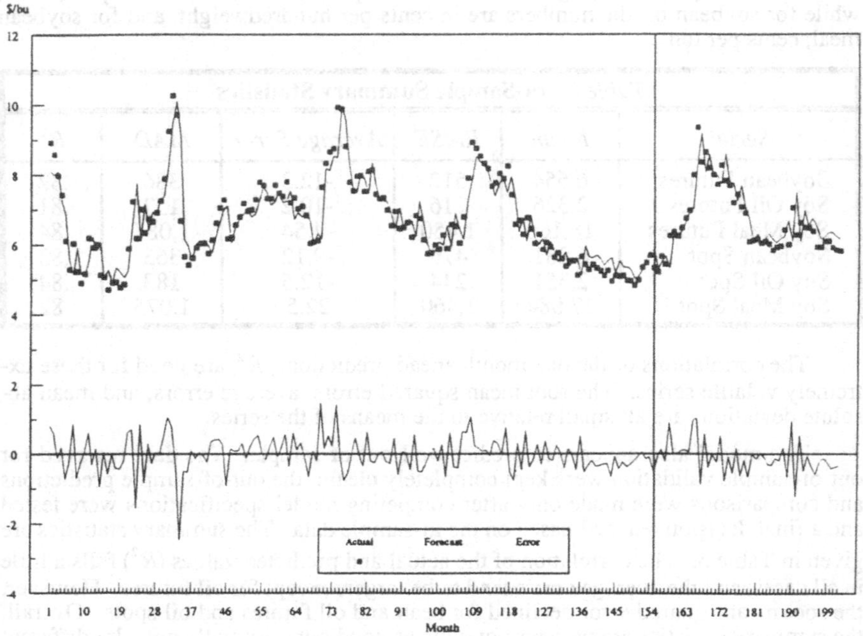

Soybean Oil Futures: Actual, Forecast and Error

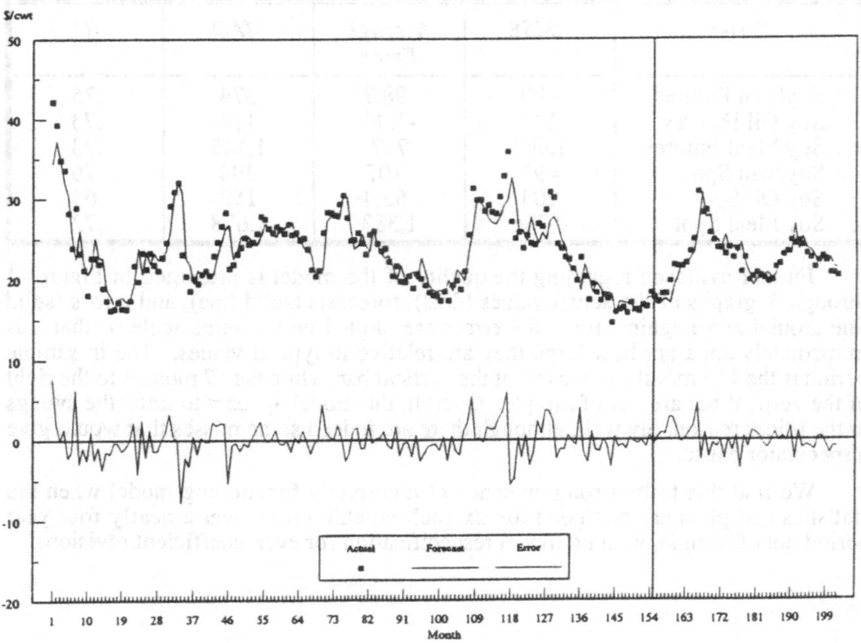

Soybean Meal Futures: Actual, Forecast and Error

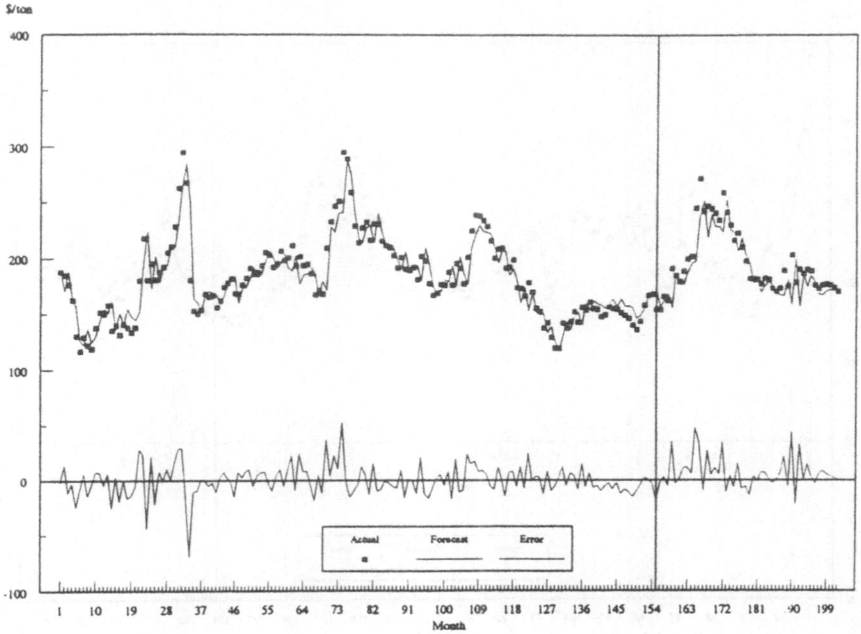

Soybean Spots: Actual, Forecast and Error

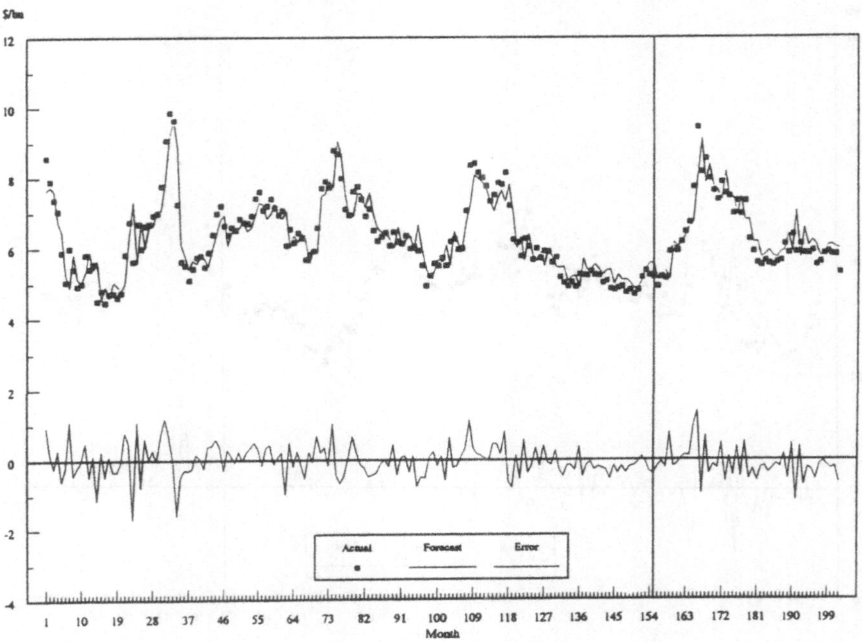

Soybean Oil Spots: Actual, Forecast and Error

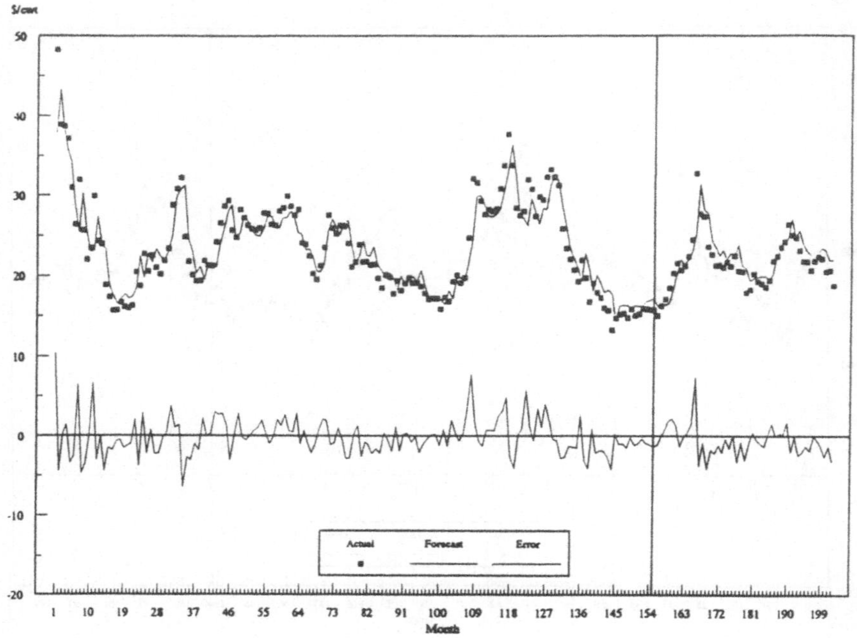

Soybean Meal Spots: Actual, Forecast and Error

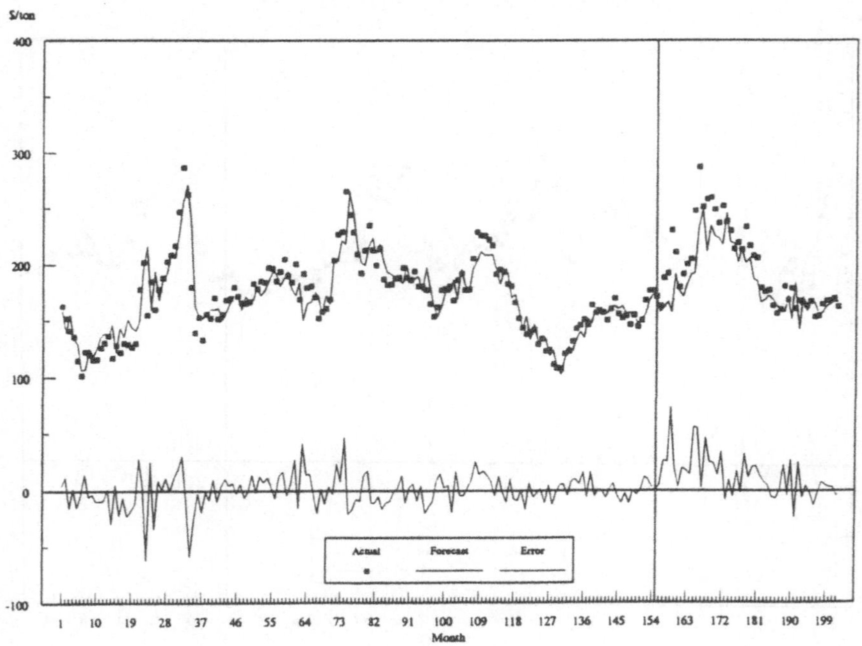

4 Speculation

In this section we will use some less traditional analysis to examine the usefulness of the state space soybean complex model. By constructing forecast evaluation measures based on finance rather than statistics, we hope to elucidate issues that are encountered in actual trading. In particular, we will analyze the reliability of the forecast direction, and use a particular utility function to examine the effects of risk and uncertainty. In addition, we will calculate the certainty equivalent rate of return to be indifferent to the model forecasts, and finally we will examine the whole path of returns for critical liquidity considerations.

Tests of Direction

The closeness of the forecasts to the actual values does not guarantee a profit to a speculator. If the speculator's decision is not based on magnitudes, but only directions, a near forecast with a completely random wrong sign can be disastrous. We calculated a non-parametric test of direction proposed by Henriksson and Merton to evaluate the model's ability to forecast the directions of price movements. The test computes how often predicted price directions (not magnitudes) match actual price directions. This test has a known finite sample distribution (important for small numbers of forecasts, although much less so for our 47 observations per series), and is not dependent on an assumption about the distribution of prices. For a complete description of the test, see Henriksson and Merton (1981). The resulting confidence levels are presented in Table 4, for the overall model forecast and individually by series.

Table 4. **In-Sample Henriksson-Merton Direction Test Results**

Series	*Confidence In Sample*	*Confidence Out of Sample*
Overall	.929	.839
Soybean Futures	.262	.882
Soy Oil Futures	.933	.277
Soy Meal Futures	.572	.149
Soybean Spot	.786	.515
Soy Oil Spot	.755	.723
Soy Meal Spot	.559	.727

This is sobering news. In the futures market, the reliability of the forecast of direction is heavily dependent on the particular sample draw, as evidenced by the changing confidence levels in and out of sample.[6] The spot price forecasts have both in-sample and out-of-sample confidence in excess of one half; if this pattern were to continue, in principle a speculator could make money on the directional information in these forecasts on an individual series basis. The model predictions are generally correct in direction over the six series, on average: the overall confidence levels of .929 in sample and .839 out of sample are high for financial market data of this sort. But the only way to get this level of confidence is to take six positions simultaneously -- a speculator cannot just operate in one or two markets and expect to succeed.

6 We remind the reader that the confidence levels are not the proportion of correct directions, but instead one minus the probability of rejecting the null hypothesis of random directions when it is true.

To make this more concrete, we evaluate a (non-optimal) speculation rule based on the forecast directions to see if it generates a positive profit. This test goes beyond direction, using the direction to determine the position held, but using the actual prices to determine the resulting profit or loss. The rule is simple: if the forecast of the price next month is higher than the actual current price, take a long position in that commodity, and if the forecast price is below the actual current price, take a short position. The total positions are limited to one contract[7] for each commodity at a time. If the position is long, and the price is expected to rise, then no buying or selling takes place, which is also true for a short position and a falling forecast. If the rule says to reverse a position, then two contracts are bought or sold: one to close out the position, and then another to take the opposite position. The returns from following this strategy are reported in Table 5 for all series, for out-of-sample observations only.[8]

Table 5. **Out-of-Sample Speculation**

Series	*Profit*	*Number of Round Turns*	*Break-Even Transaction Costs*
Soybean Futures	$21,180	14	$1,513
Soy Oil Futures	$6,840	16	$428
Soy Meal Futures	$60	1	$60
Soybean Spot	$18,000	12	$1,500
Soy Oil Spot	$4,452	8	$557
Soy Meal Spot	-$150	10	-
Total	$50,382	61	$826

If nothing from the model forecasts except the expected direction of the price change was used, holding six contracts for 46 months[9] out of sample would produce a gross profit of $50,382. The model prescribed 61 position reversals in total, so that breakeven transaction costs are $826 per contract. (Actual transaction costs would be more in the neighborhood of $50 to $100.) Soybean futures and spots did very well, with the model correctly reversing position frequently and generating large returns. The model did poorly in the meal markets, both futures and spot, generating a small loss in the spot meal market. The Henriksson-Merton test results above suggest that it would be a mistake to try to pick the most profitable series (from a prior sample) and trade only them. While the model gets the directions right on average, the six-series average is more stable than the individual series. Soybeans were profitable for these four years, but the meal and oil contracts might well be the big winners in the next time interval.

There is, of course, no assurance that profits will be positive, so this seems to us like a pretty good performance based on very little information from the model. Magnitudes of price predictions and their associated variances, both available from the model, have been ignored, and, given the forecasts, an actual speculator would surely write a more sophisticated set of contracts, perhaps involving options. It is hard to determine the percentage return on these trades since only the margin is invested if all goes well, but the downside risk is potentially very large. The futures contract

7 A contract is 5,000 bushels for beans, 60,000 pounds (one tank car) for oil, and 100 tons for meal. At the sample mean prices, the contract values are $33,270, $13,956, and $18,166 respectively.

8 Realized profit over the 12 years and 11 months in sample was $502,254, again on a margin of $3,000. For this period, current wealth always exceeded initial wealth.

9 The first month was lost in startup.

margins were $1,500 for soybeans, $800 for oil, and $700 for meal in 1984, so approximately $3,000 would have been invested for four years, along with $14,000 to cover some initial losses, to earn the $50,382 *apart from downside liability.*

Utility Function Analysis

Despite this high *ex post* return, a speculator might or might not have been willing to participate in the markets using the forecasts, depending on the attitude toward risk and the uncertainty in the prices. We turn now to a utility function analysis to take the effects of price uncertainty into account in judging the value of the state space model forecasts. Since investors are generally not risk neutral, the variance of returns must be considered in addition to the expected value to see if a risk averse investor could make utility from following the simple rule we have proposed. We must assume a utility function from among the family of concave utility functions. We chose the logarithmic. The logarithmic utility function has been described as "everyman's utility function" by Jan Mossin (1977). It has several properties that make it useful in this context.[10] It satisfies the von Neumann-Morgenstern expected utility property, is time-separable, and numeric results can easily be calculated to compare the cardinal utility levels derived from different investment paths. Log utility functions display decreasing absolute risk aversion, so an initial wealth level must be specified to compare investment strategies. We chose an initial wealth level of $500,000, perhaps a little low for a commodity market speculator. (A low wealth position implies higher risk aversion and thus favors no investment over speculation; this in turn makes the model forecasts less useful.)

Given this particular form of utility function, the utility of a hedging or speculation strategy can be calculated and compared to other strategies. If the utility of the investment is higher than the utility of the initial wealth, then speculating increases the welfare of an investor even after the effects of uncertainty are taken into account.

The log utility function at a point in time can be written

$$U_{t+1} = \ln(\tilde{W}_{t+1}) \tag{4}$$

where $\tilde{W}_{t+1}$ is random wealth. A moving second order Taylor series expansion[11] taking the uncertainty into account yields an approximation in terms of expected wealth and the variance:

$$U_{t+1} \approx \ln E(\tilde{W}_{t+1}) - 1/2\frac{V(\tilde{W}_{t+1})}{E(\tilde{W}_{t+1})^2}. \tag{5}$$

This can be written in terms of initial wealth W_t, expected returns $E(\tilde{W}_{t+1} - W_t)$ and the variance of returns $V(\tilde{W}_{t+1} - W_t)$ as:

$$U_{t+1} \approx \ln[W_t + E(\tilde{W}_{t+1} - W_t)] - \frac{1}{2}\frac{V(\tilde{W}_{t+1} - W_t)}{[W_t + E(\tilde{W}_{t+1} - W_t)]^2} \tag{6}$$

see, *e.g.*, Anderson, Dillon and Hardaker (1977).

---

10 Other utility functions that could be used for this purpose include negative exponential or quadratic. The relative weights of the risk/return tradeoff must be supplied in these other utility functions. These weights are implicit in the log utility function, although it does still require specification of an initial wealth level.

11 The moving approximation can be very accurate, see Levy and Markowitz (1979).

At every time period, the (random) returns from the portfolio are

$$\tilde{W}_{t+1} - W_t \equiv \Delta\tilde{W}_{t+1} = q_t'(\tilde{p}_{t+1} - p_t), \tag{7}$$

where $\tilde{p}_{t+1}$ is the unknown price next period, p_t is the known price this period, and q_t is a vector of the positions in each commodity at time t, with elements either plus or minus one. (The convention is that q_t is the vector of contracts held between period t and period $t+1$.) For the state space model, the expected returns are

$$\Delta\hat{W}_{t+1} = q_t'(\hat{p}_{t+1} - p_t), \tag{8}$$

where $\hat{p}_{t+1}$ is the state space prediction of next period's prices. Because the only random variable in equation (8) is $\hat{p}_{t+1}$, the variance of expected returns is the q-weighted variance of the state space price forecast, or

$$V(\Delta\hat{W}_{t+1}) = q_t'C\Xi C'q_t, \tag{9}$$

where C is the coefficient matrix from the observation equation (2) above, and $\Xi \equiv E z_{t|t-1} z_{t|t-1}'$, the covariance matrix of the states. Substituting equations (8) and (9) into the moment expansion of the utility function (6) provides the expected utility from each transaction based on the state space model forecasts.

Of course, the expected utility is just a prediction based on the model forecasts and estimated variances. The realized utility in each period is the logarithm of wealth based on the realized prices and the positions determined from the model forecasts. Table 6 gives the average expected and realized utility for the in-sample and out-of-sample periods for an investor with $500,000 in initial wealth, and the percentage of the total monthly utility due to the soybean complex positions.[12]

Table 6. **Average Monthly Utility from Soybean Complex Speculation**

	Expected	*Realized*
In Sample Average		
Utility	13.623	13.617
Soybean Percentage	3.82%	3.77%
Out of Sample Average		
Utility	13.205	13.195
Soybean Percentage	.63%	.56%

When interpreting the utility figures, recall that they apply to positions in six contracts for one month; the return from this ought not to be a large proportion of total utility. In sample, there were expectations of great opportunities, on average, so that these six contracts alone represented a relatively large portion of total utility (3.82%). The reality was not quite as good as expected, but the trades still provided 3.77% of all utility each month, on average. It seems remarkable to us that the risk adjusted utility of just six contracts is such a large proportion of the utility of an investor with half a million dollars in wealth. In sample, the risk adjusted value of the state space forecasts is substantial, significantly more than might be expected. Out of sample, the opportunities were more limited. The expected utility was much lower, and the realized utility matched, so that the percent of total utility due to these trades on average was .63% expected and .56% realized, values more in keeping with six contracts and

12 The initial wealth of $500,000 has as a logarithm 13.122; the percentage change in utility is defined as the change in utility from this value divided by this value.

$500,000 initial wealth. (In both cases, the expected and realized utility are close, validating the accuracy of the model forecasts.) We conclude that for the risk aversion implicit in the logarithmic utility the model forecasts are of value.

Risk Free Rates of Return

An alternative evaluation of forecast accuracy is based on calculating the risk-free rate of return necessary to produce the same total utility as that derived from the trading rule using the state space model forecasts. Again, we can calculate *ex ante* and *ex post* values for the in-sample and out-of-sample periods.

In sample, an annual risk free rate of 7.73% produces monthly utility that sums to the soybean speculation expected utility, while an annual rate of 7.63% is required to match the in-sample realized utility. Out of sample, the required rate to match the expected soybean utility is 4.23%, while the corresponding realized number is 3.71%. Note that these values are utility based, *i.e.*, they are not the rates that produce the same wealth, but rather are the rates that produce the certainty equivalent wealth (without the variance penalty). Annual rates of return in these ranges are sometimes available, but the entire $500,000 must be invested at these rates to match the risk adjusted soybean complex speculation return.

The Path of Returns

Finally, it is useful to examine the entire out-of-sample[13] path of returns, wealth, and utility implied by the model. There is a lot of detail in this out-of-sample output, but it is really the only way an individual who is not sure they have a logarithmic utility function (and who is sure?) can assess the model forecasts. Table 7 provides information on monthly returns and utility. Extrema are in bold italic.

The third column, realized wealth, is of immediate interest. From an initial wealth of $500,000 the model makes money for two months before sustaining the largest single period loss of all, -$16,414. At this point, wealth is at the minimum ($485,546), although the investor does not know it at the time. It takes six more months to get back to initial wealth. But the 10th and 11th months make up for the wait, with returns of $17,358 and $22,438 respectively. From that point on, wealth stays in the black, with returns ranging from -$4,932 to $12,754. In the 34th month, the maximum wealth of $568,954 is attained, although again the investor does not know it and lets it slide steadily downward to the final value of $550,382.

(In sample, wealth never went below the initial value, and ended up $502,254 greater than the initial wealth after 12 years and 11 months. The maximum gain in wealth over the period was $507,640.)

The essential question for each investor is whether or not they would be willing to sustain this kind of variation for these returns. *Expost*, this was obviously a profitable set of trades: $50,382 was made on the $3,000 margin investment plus the short term investment of about $14,000 to cover early losses. The potential investment is very large, however, for short positions in exceptionally high price months.

13 Since the 155 months in sample are too large to conveniently table, only summary statistics are provided for this period.

Table 7. Monthly Returns and Utility Out of Sample

Period	*Expected Returns*	*Realized Returns*	*Realized Wealth*	*Expected Utility*	*Realized Utility*
1	4,875	3,540	503,540	13.139	13.129
2	3,773	-1,580	501,960	13.134	13.126
3	6,019	*-16,414*	*485,546*	*13.105*	*13.093*
4	6,615	1,120	486,666	13.109	13.095
5	6,059	3,420	490,086	13.115	13.102
6	3,243	-168	489,918	13.109	13.102
7	3,104	1,348	491,266	13.111	13.105
8	3,619	3,182	494,448	13.118	13.111
9	*2,781*	1,916	496,364	13.121	13.115
10	6,863	17,358	513,722	13.163	13.149
11	8,769	*22,438*	536,160	13.208	13.192
12	*10,114*	-4,932	531,228	13.202	13.183
13	4,886	4,634	535,862	13.201	13.192
14	7,914	4,228	540,090	13.214	13.199
15	4,500	1,730	541,820	13.211	13.203
16	3,790	-3,728	538,092	13.203	13.196
17	6,542	-1,336	536,756	13.205	13.193
18	5,706	4,042	540,798	13.211	13.201
19	3,025	1,802	542,600	13.210	13.204
20	3,695	-3,20	542,280	13.210	13.204
21	4,099	-4,660	537,620	13.202	13.195
22	3,963	-758	536,862	13.201	13.193
23	7,655	12,754	548,616	13.231	13.217
24	5,491	2,770	552,386	13.232	13.222
25	5,659	-330	552,056	13.232	13.221
26	7,441	4,152	556,208	13.242	13.229
27	4,026	360	556,568	13.237	13.230
28	4,091	-1,376	555,192	13.234	13.227
29	3,835	1,792	556,984	13.237	13.230
30	3,374	4,306	561,290	13.244	13.238
31	3,862	1,056	562,346	13.247	13.240
32	4,249	2,202	564,548	13.251	13.244
33	5,145	-72	564,476	13.253	13.244
34	4,527	4,478	*568,954*	*13.259*	*13.252*
35	8,258	-6,658	562,296	13.254	13.240
36	6,600	-4,788	557,508	13.243	13.231
37	5,689	-3,120	554,388	13.236	13.226
38	3,809	-406	553,982	13.232	13.225
39	5,279	78	554,060	13.235	13.225
40	3,275	-2,804	551,256	13.226	13.220
41	4,301	1,920	553,176	13.231	13.223
42	4,364	3,348	556,524	13.237	13.229
43	3,813	168	556,692	13.237	13.230
44	3,747	-3,178	553,514	13.231	13.224
45	4,349	-308	553,206	13.231	13.223
46	4,399	-2,824	550,382	13.226	13.218

The expected and realized returns are distressingly far apart. The correlation of in-sample expected and realized returns is .42, while the out of sample correlation is .22. It is hard to take the effects of variance into account in an *ad hoc* way when examining these series. (This is the essential problem of risky forecast evaluation addressed in this section.) We know these returns are highly variable, but there is not much intuition from looking at the returns series about whether or not the outcomes are reasonable, *i.e.*, something on which to base speculation and expect to earn risk-adjusted profits.

The utility values formally take the variance into account, if the form of the utility function and the initial wealth match the investor. Not surprisingly, it is considerably more stable than returns, or even wealth. Expected and realized utility both have their minimum in the third month and their maximum in the 34th month. Incidentally, every value of expected utility exceeds realized utility, possibly due to truncation in the second moment approximation of the logarithmic utility function. (The average difference of expected and realized utility is .006 in sample, and .009 out of sample.)

5 Conclusion

We have constructed and evaluated a model of soybean, meal, and oil spot and futures prices using the state space time series methods of Masanao Aoki. The model was fitted from 155 monthly observations on the six series, withholding 47 months to test the model's ability to produce a profit, and, more appropriately, a risk-adjusted profit, out of sample.

The model tests well both in sample and out of sample using the normal statistical criteria of root mean squared error, correlation of actual and forecast, mean absolute deviation, and graphical analysis. Because its main purpose is commodity market speculation, a further set of tests was applied. These included the Henriksson-Merton nonparametric test of direction, and a number of utility based statistics designed to incorporate the variance effects that are fundamental to these types of series. For an investor with an initial wealth of $500,000 and a logarithmic utility function, the model earns risk-adjusted returns in and out of sample.

References

Anderson, Jock R., John L. Dillon and Brian Hardaker (1977), *Agricultural Decision Analysis*. Ames, Iowa: Iowa State University Press.

Aoki, Masanao (1983), *Notes on Time Series Modeling*. Heidelberg: Springer-Verlag.

Aoki, Masanao (1987, 1990 2nd *ed.*), *State Space Modeling of Time Series*. New York, New York: Springer-Verlag.

Aoki, Masanao and Arthur Havenner (1991), "State Space Modeling of Multiple Time Series," *Econometric Reviews*, 10(1), pp. 1-59.

Cerchi, Marlene and Arthur Havenner (1988), "Cointegration and Stock Prices: The Random Walk on Wall Street Revisited," *Journal of Economic Dynamics and Control*, 12, pp. 333-346.

Havenner, Arthur and Masanao Aoki (1988), "State Space Modeling of Stationary Time Series: Theory and Applications," American Statistical Association: *1988 Proceedings of the Business and Economic Statistics Section*, pp. 51-58.

Henriksson, R.D and R.C. Merton, (1981), "On Market Timing and Investment Performance, II: Statistical Procedures for Evaluating Forecasting Skills," *Journal of Business*, 54, pp. 513- 533.

Levy, H., and H. M. Markowitz (1979), "Approximating Expected Utility by a Function of Mean and Variance," *American Economic Review*, 69(3), pp.308-317.

Mossin, Jan (1977), *The Economic Efficiency of Financial Markets*. Lexington, Massachusetts: D.C. Heath and Company.

Forecasts of Monthly U.S. Wheat Prices: A Spatial Market Analysis

Channing Arndt and Kenneth Foster
Purdue University
West Lafayette, Indiana 47907

1. Introduction

The value of timely and accurate forecasts can hardly be overstated. Because remuneration for accurate forecasting is so great, an army of economists, traders, and speculators expend considerable effort to gain insight into the future path of asset prices. Theoretically, the efforts of these agents drives markets to efficiency. In a frictionless world, efficiency implies that asset values would tend to follow a random walk. Under this sort of stochastic process, the best predictor of price in period $t+1$ is the price at time t.

In real markets, the random walk hypothesis runs afoul of other theoretically logical propositions. For asset prices that follow a random walk, expected returns to research into the probable price movements are zero. Yet, as mentioned above, tracking and predicting asset prices such as stock prices, exchanges rates, or prices of agricultural commodities is a veritable industry. Why would rational agents expend time and resources to predict asset prices when the best predictor is the current price? Furthermore, if agents did not actively research and speculate on these markets, how could they be efficient?

The unit root hypothesis also implies that asset prices may infinitely diverge. Yet, economic theory postulates relationships between assets. For example, the stock prices of electric utilities should respond in similar fashion to a variety of shocks and hence be related over time. Occasionally, a vector of economic time series which appears to embody individual random walks can be shown to contain a special relationship such that they form a stationary linear combination. This is the concept which Engle and Granger call cointegration. However, as Granger points out, cointegration implies that both current period prices are of value in predicting future price of either asset. That is, there exists an error correction specification that can be used to predict asset prices. If these predictions yield positive utility for traders then cointegration violates the efficient market assumption embedded in the random walk hypothesis (Granger, 1986).

Another quandary which exists between speculative efficiency and spatial market efficiency (The Law of One Price). Speculative efficiency appears to imply no cointegration, at first glance, while the Law of One Price demands that spatial market prices must be cointegrated. However, Walburger points out that speculative efficiency simply demands that, if prices are cointegrated, then no profit or utility can be gained from the model forecasts.

Other empirical regularities tend to militate against the random walk hypothesis. For instance, the variance of a random walk tends to increases without bound over time. Yet, asset price series in existence for literally thousands of years, such as the price of wheat, exhibit relatively stable variance. Finally, simple versions of the efficient market hypothesis ignore transactions costs, risk aversion, and the impacts of government policy on asset valuation.

For these reasons, one might postulate that asset values tend toward a random walk. If the series only tend towards a random walk, there exists systematic information about the direction of price movements which remains unexploited. As a result, sophisticated time series analysis might generate forecasts which could be used to produce positive trading profits.

In this paper, a state-space time series model is used to forecast spot prices of wheat in nine markets across the United States. A betting scheme uses the directional forecasts from the model to determine whether to go long or short in each market. In the following sections, we introduce the state-space time series representation, discuss data and estimation, and present results and conclusions. Results from out of sample speculation tentatively indicate above market returns to trading based upon the state-space predictions. Hence, the results tend to support the hypothesis of inefficient spot markets for wheat.

2. The Model

The primary advantage of state-space formulations, for modelling spatial market relationships, stems from a method to determine model specification based upon the data and the ability to handle a relatively large number of series. Walburger successfully modeled 19 series using Aoki's state space approach. The problem inherent in the specification of vector valued time series is that as the number of series increases the difficulty for the modeler to specify the appropriate across series lags intensifies.

Proper specification is important for two primary reasons. First, as Granger pointed out relative to cointegration, values of related series can supply valuable information for forecasting. Second, as Aoki and Havenner point out, because of the dynamic properties of systems of equations, "small changes in coefficients can dramatically alter the dynamic properties of the model as series feed into one another at lags". Given the importance and difficulty of proper specification, the existence of a reliable procedure for specifying multivariate time series models lends great appeal to the state-space form for spatial market analysis.

It is well known that, when there are unit or near unit roots in the dynamics of the system, improvements in modeling can be made by modeling the slow and fast dynamics separately (see De Jong and Whiteman). The empirical model used in this paper follows the two step approach suggested by Aoki and Havenner. Borrowing notation from Aoki and Havenner, the state and observation equation pair for the trend (slow dynamics) model specification is:

$$\tau_{t+1|t} = A_\tau \tau_{t|t-1} + B_\tau y_t^* \tag{1a}$$

$$y_t = C_\tau \tau_{t|t-1} + y_t^* \tag{1b}$$

Any nonstationarity or apparent nonstationarity will be removed by the implied generalized differencing in (1b). The stationary residual, y_t^*, becomes the dependent variable in the state space cycle (fast dynamics) model below.

$$\eta_{t+1|t} = A_\eta \eta_{t|t-1} + B_\eta e_t \tag{2a}$$

$$y_t^* = C_\eta \eta_{t|t-1} + e_t \tag{2b}$$

In the above system, $\tau_{t|t-1}$ and $\eta_{t|t-1}$ are the conditional means of the trend and cycle state variables, A_τ , A_η , B_τ , B_η , C_τ , and C_η are parameters matrices, and e_t is a serially uncorrelated disturbance vector. The *(mx1)* vector y_t is, for this application, the vector of prices being modeled in period *t*. With the above system, the model specification question is condensed into questions concerning the number of states to incorporate in the model and how to associate them with the long and short run dynamics of the system.

Methods for estimation are well documented in Aoki and Havenner. First, the data were centered and scaled such that the terms in the Hankel matrix correspond to autocorrelations. Second, a state-space model, equations (1a) and (1b), was estimated with the lag parameter set to one.[1] This procedure might be viewed as a generalization of differencing or alternatively as a method of trend removal. However, in the multivariate series context, differencing each individual series may be unnecessary and incurs a substantial loss of information if cointegration exists. Namely, the common trend is lost if the series are each differenced and spurious autocorrelations are introduced into y_t^*. Since the series share trend components, the trend components may be removed by calculating the residuals, y_t^*, from the estimate of equation (1b). Third, another state-space model, equations (2a) and (2b), was estimated on these residuals in order to capture the faster dynamics. In the second stage model, the lag parameter was set to two. Using a larger lag parameter in the second stage model allows higher order lags to enter the model in an effort to completely capture the faster dynamics. With monthly data allowing at least 12 lags is desirable to account for any seasonality. Combining nine series and a lag parameter equal to two allows the implied univariate models for each series to contain up to 18 lags.

Joining the trend and cycle models is accomplished by substituting (2b) into (1a) and (1b), rearranging, and stacking with (1a) and (2a) to give

$$\begin{pmatrix} \tau_{t+1|t} \\ \eta_{t+1|t} \end{pmatrix} = \begin{pmatrix} A_\tau & B_\tau C_\eta \\ 0 & A_\eta \end{pmatrix} \begin{pmatrix} \tau_{t|t-1} \\ \eta_{t|t-1} \end{pmatrix} + \begin{pmatrix} B_\tau \\ B_\eta \end{pmatrix} e_t \tag{3a}$$

$$y_t = (C_\tau, C_\eta) \begin{pmatrix} \tau_{t|t-1} \\ \eta_{t|t-1} \end{pmatrix} + e_t, \tag{3b}$$

which is the form of the model used for forecasting and model validation. A more compact and standard notation for equations (3a) and (3b) is given below.

$$z_{t+1|t} = A^* z_{t|t-1} + B^* e_t \tag{4a}$$

$$y_t = C^* z_{t|t-1} + e_t. \tag{4b}$$

Despite the property of strict nesting, this procedure yields different parameter estimates and forecasts than estimating the trend and cycle components jointly in a single step procedure. Specifically, with the procedure employed, parameter estimates in the lower left-hand block of the matrix A^* are restricted to zero. In the case of the single step state-space estimation, these values would not necessarily be zero. Previous experience indicates that the two step procedure produces more robust forecasts (see Cerchi and Havenner and Foster and Havenner).

1 Even with the lag parameter set to one, some cycle dynamics are likely fall into the trend model. This is especially true in this application with nine series.

3. The Markets for Wheat

There are a number of different types of wheat grown and consumed in the United States. Both the consumption and production of these "varieties" are geographical as well as biologically diverse. As a result, the dynamics in the prices reflect substitutability (in production and processing) between different varieties of wheat as well as potential spatial and speculative arbitrage.

While all types of wheat are members of the same species, they are not perfect substitutes in either production or consumption. Protein content represents the major differentiating factor between the wheats. Hard wheats, such as hard red winter and dark spring, contain higher levels of protein. Soft wheats have a lower protein content. Genetic differences have the greatest effect on protein content, but climatic factors and soil type also play a role. Regions with a mild, humid climate, especially during the ripening period, tend to produce wheats lower in protein content. In contrast, regions with a hot dry climate during the ripening period, such as the Great Plains, tend to produce higher protein wheats. Soft wheat tends to be produced in regions east of the Great Plains. Thus, soft wheat varieties predominate in states such as Ohio, Indiana, and Illinois while farmers in the Great Plains region tend to plant hard wheats.

Winter wheats are planted in the fall. They grow rapidly up to the onset of winter then lie dormant during the winter before resuming growth in the spring. Due to their fall growth, winter wheats tend to have higher yields and earlier harvests than spring wheats. However, severe winter weather can decimate winter wheat stands. Consequently, farmers in the Northern plains and Canada tend to plant spring wheat while farmers in the Southern plains tend to plant winter wheat.

Soft white wheat is morphologically similar to soft red wheat. White wheat is grown predominantly in Washington state and Michigan. Of all wheats, exports constitute the highest fraction of production in white wheat. Strong demand for white wheat from Southeast Asia augurs for production of white wheat in Washington state. In addition, certain specialty products require white wheat flour. Hence, some production, historically centered in Michigan, occurs nearer to the eastern seaboard.

At the margin, production of different varieties of wheat is strongly sensitive to price. In most regions, alternative crops exist allowing wheat acreage to expand or contract according to market needs. Furthermore, while some regions are clearly better suited for certain types of wheat, large swathes of territory may choose between different varieties of wheat without substantial impact on expected yield. Thus, for example, high premiums for high protein hard red winter and spring wheat tend to expand wheat acreage in the Northern plains and extend high protein wheat acreage to more Southern and Eastern regions previously devoted to lower protein wheats.

Substitution possibilities exist on the consumption side as well. High protein wheats yield a dough high in gluten content. Gluten lends an elastic structure to the dough allowing it to capture gases released by leavening agents such as yeast. Thus, high protein wheats yield a flour well suited for baking bread. Cakes, pastries, and crackers are produced from low protein wheats. The lower gluten content allows a less fluffy more crumbly structure.

Just as production is precluded in certain regions by climatic considerations, certain processing needs preclude the use of some flours. For example, one cannot bake pastries using flour from high protein dark spring wheat. However, a continuum of substitution possibilities exist. Within some technical margins, flour from low protein hard red winter wheat may be blended with soft wheat flour for end uses such as crackers, cookies, and biscuits. Flour from lower protein hard red winter wheat may also be blended with flour from high protein dark spring wheat to create flour for bread. Due to the presence of bleaching equipment in the overwhelming majority of U.S. mills, U.S. millers are basically indifferent between soft red and soft white wheats on technical grounds for many end uses. In general, strong substitution possibilities exist (i.e. nearly linear isoquants) within a limited technical range. Beyond the limits, substitution possibilities drop dramatically. The width of the technical range depends upon the final product.

Trade in wheat can stabilize or destabilize relative wheat prices. Some large wheat importers, such as China, tend to purchase the cheapest variety of wheat available thus attenuating price variability within wheat classes. Other importers have marked preferences for certain types of wheat. India, for example, exhibits a strong preference for white wheat. In years of poor production in Australia, a major white wheat exporter, and poor production in India, white wheat prices can spike upwards dramatically.

In summary, one should expect wheat prices within the various categories listed above to be strongly linked in the long run due to substitution possibilities. The series thus lend themselves well to time series analysis through the cointegrated state-space representation developed in Aoki and Havenner.

4. Estimation and Results

Monthly prices for nine wheat markets were obtained from various issues of USDA's *Wheat Situation and Outlook* between June 1953 and December 1993. The total number of observations per series is 488. In order to validate model forecasting performance, the most recent 98 observations, corresponding to the period January 1986 to December 1993, were withheld during estimation. Consequently, parameter estimates were obtained based upon the first 390 observations per series. The series modeled are described in Table 1.

Table 1. Description of the Nine Wheat Price Series.

Series	*WHEAT TYPE AND MARKET LOCATION*
1	KANSAS CITY, NO. 1 HARD RED WINTER (ORDINARY PROTEIN)
2	KANSAS CITY, NO. 1 HARD RED WINTER (13 % PROTEIN)
3	CHICAGO, NO. 2 SOFT RED WINTER
4	ST. LOUIS, NO. 2 SOFT RED WINTER
5	TOLEDO, NO. 2 SOFT RED WINTER
6	TOLEDO, NO. 2 SOFT WHITE
7	PORTLAND, NO. 1 SOFT WHITE
8	MINNEAPOLIS, DARK NO. 1 SPRING (13% PROTEIN)
9	MINNEAPOLIS, DARK NO. 1 SPRING (15% PROTEIN)

Various numbers of states were evaluated in both the trend and cycle models based on the first 390 observations of each series. The model chosen has one trend and two cycle states (three total states)[2]. The single trend state was based on a hypothesis of at most one apparent nonstationarity. The ratio of successive cycle model singular values of the Hankel matrix were 1, 0.26, 0.15, 0.13, 0.09, 0.04, 0.03, 0.02, 0.01, 0.01, 0.01, 0.0, 0.0, 0.0, 0.0, 0.0, 0.0, and 0.0. The rule suggested by Aoki and Havenner is to choose the number of states such that the ratio $\lambda_{n+1}/\lambda_1 < \frac{1}{\sqrt{T}}$, where

2 Estimates of initial states were obtained by backcasting.

$\hat{n}$ is the number of chosen states, λ_i is the ith singular value (ranked from largest to smallest) of the Hankel matrix, and T is the number of time periods in the in-sample data. This criterion suggests a model with five cycle states. However, such a model does not have a stable solution as evidenced by unit eigenvalues in the simplectic matrix. In fact, it was necessary to reduce the number of cycle states to two in order to find a model which did not suffer from this problem. Table 2 contains the summary in-sample statics for the model with one trend and two cycle states. It is worth noting that adjusting the lag parameter up or down in the cycle model did not alter the difficulty with the simplectic matrix.

Table 2. Summary Results for In-Sample Model Performance

Series	Squared Correlation[a]	Average Error	RMSE
1	0.98	0.003	0.17
2	0.98	0.003	0.15
3	0.97	0.003	0.17
4	0.97	0.003	0.17
5	0.97	0.003	0.17
6	0.97	0.002	0.18
7	0.98	0.002	0.15
8	0.98	0.003	0.15
9	0.98	0.003	0.17

[a]Squared Correlations between actual and predicted values are used as an analog to R^2.

Unlike classical applications in econometrics, parameter values in the state-space model are of little interest. The model aims to produce quality forecasts. Consequently, we will focus on out-of-sample forecast results in this section. First, we examine plots of the actual series, the forecasts, and the error.

Consider figures 1 through 9 which contain the actual, state space forecasts, and model errors for the entire sample. The massive price increase in 1972 represents the salient feature of all of the figures. Prior to 1972, price variability was light. Following 1972, prices shifted up to a higher average level and variability increased substantially. One could hypothesize a structural shift in the wheat market circa 1972. Alternatively, one might hypothesize a general upward trend in nominal prices and a consistent structure to the faster dynamics throughout the time frame. Under this hypothesis, the greater variance in the post '72 period stems from a series of larger shocks hitting the wheat markets as opposed to a structural change in the dynamics of the system. In view of the quantity of shocks impacting wheat markets in the post 1972 period (oil shocks, exchange rate changes, large fluctuations in interest rates, etc.), this second view of the system seems plausible and represents the approach relied upon here.

The figures indicate that the forecasts fit the data quite well even out of sample. Recall that the forecast horizon depicted in the graphs is one period ahead. Table 3 details out-of-sample performance for each series.

The out-of-sample squared correlations between actual and predicted values indicate some, though not much, deterioration in model fit. In terms of the average error, t-tests reject, at $\alpha=0.10$, the hypothesis that the average error equals zero in all instances except the first, third and fifth series. The other series tend to consistently overestimate or underestimate next period prices. Nevertheless, the average errors are small relative to the average price of about \$3.00 per bushel.

Figure 1. Kansas City Hard Red Wheat (Ordinary Protein), State Space Model Performance.

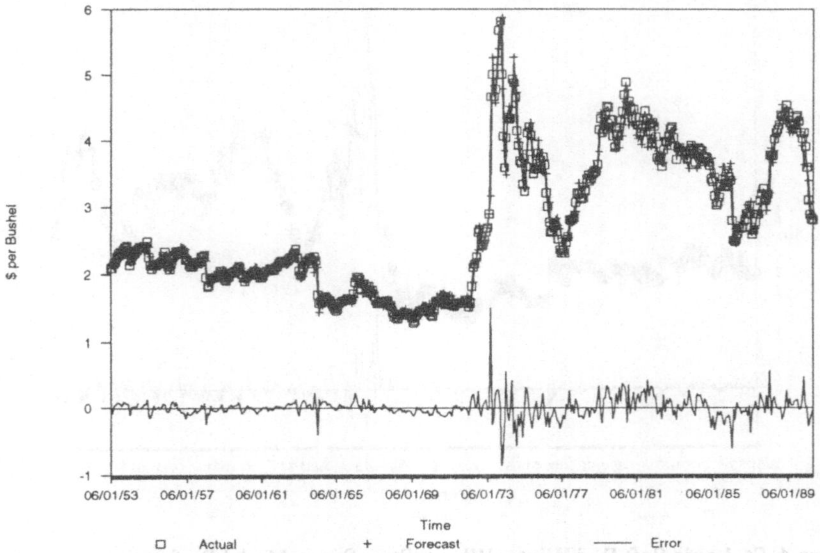

Figure 2. Kansas City Hard Red Wheat (13% Protein), State Space Model Performance.

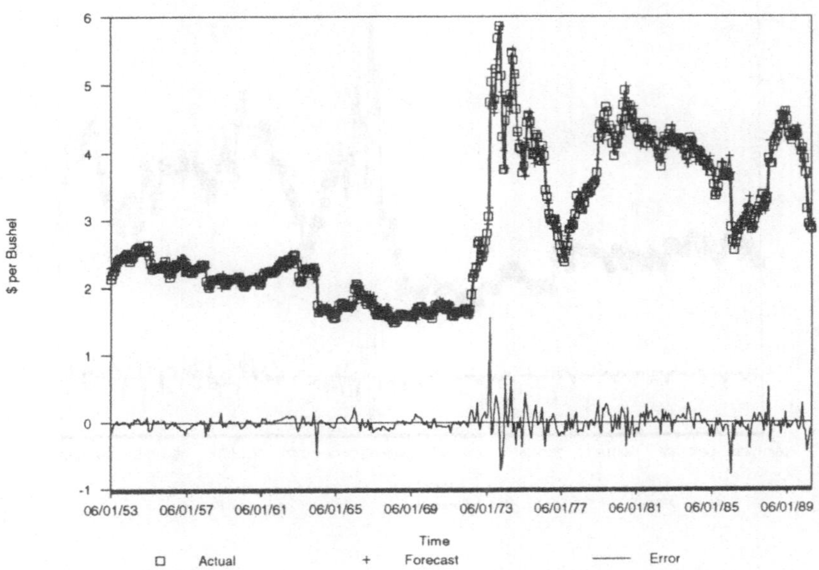

Figure 3. Chicago Soft Red Winter Wheat, State Space Model Performance.

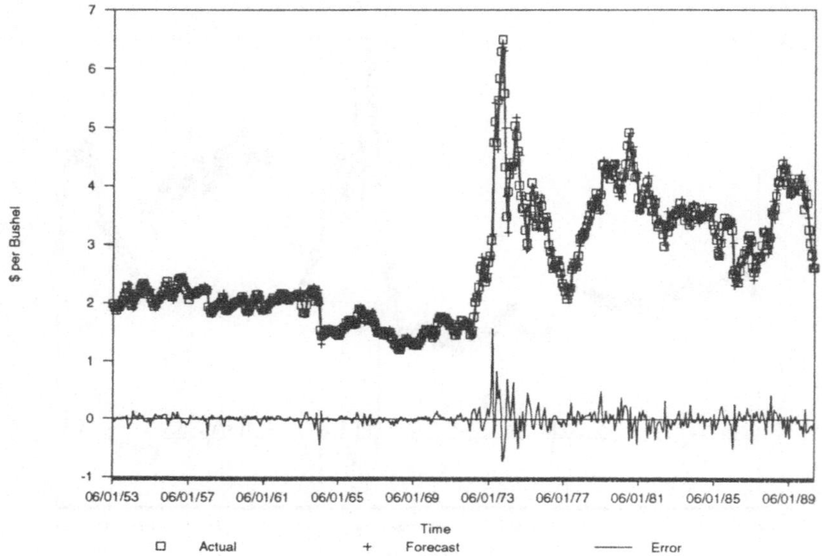

Figure 4. St. Louis Soft Red Winter Wheat, State Space Model Performance

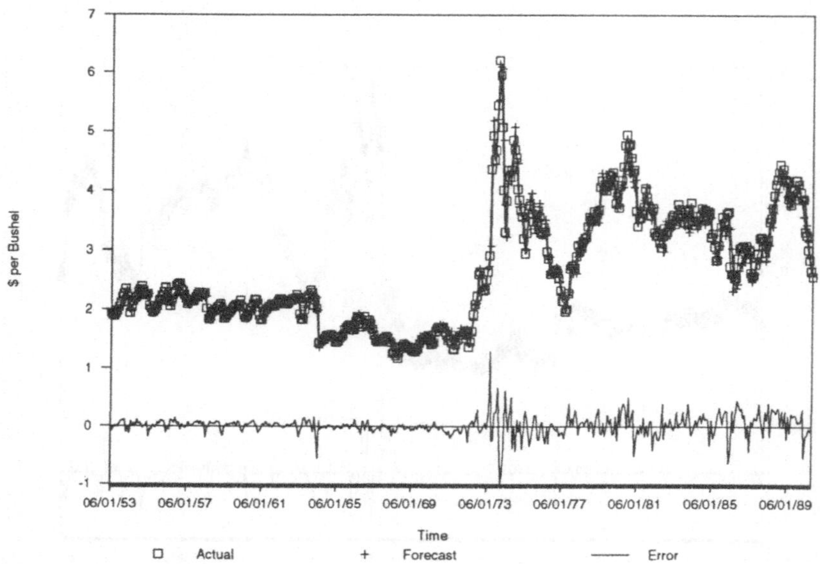

Figure 5. Toledo Soft Red Winter Wheat, State Space Model Performance.

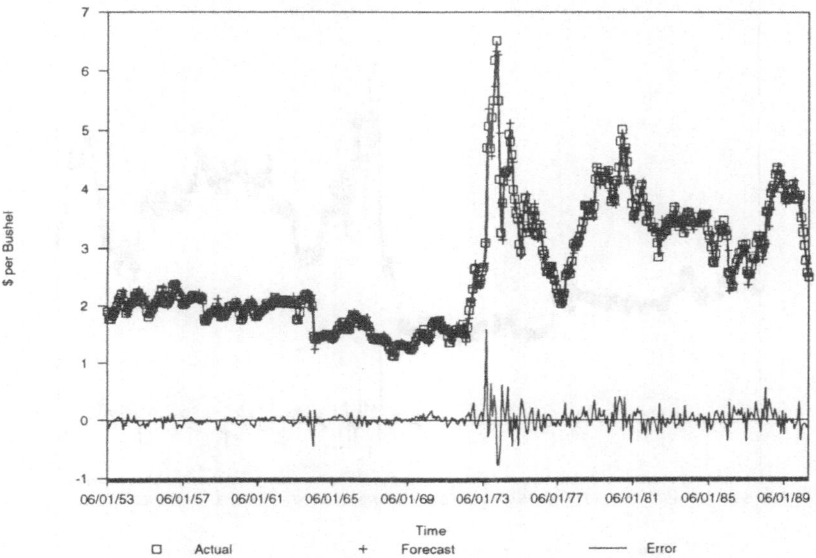

Figure 6. Toledo Soft White Wheat, State Space Model Performance

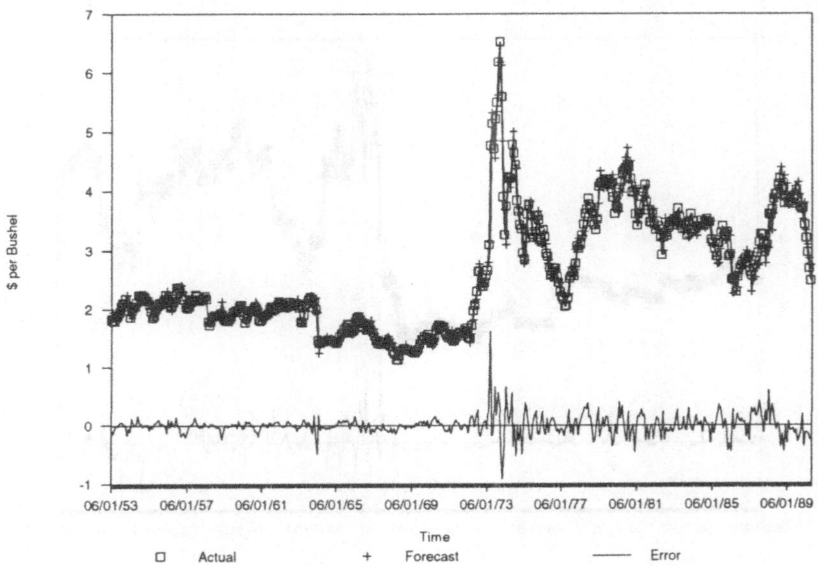

Figure 7. Portland Soft White Wheat, State Space Model Performance.

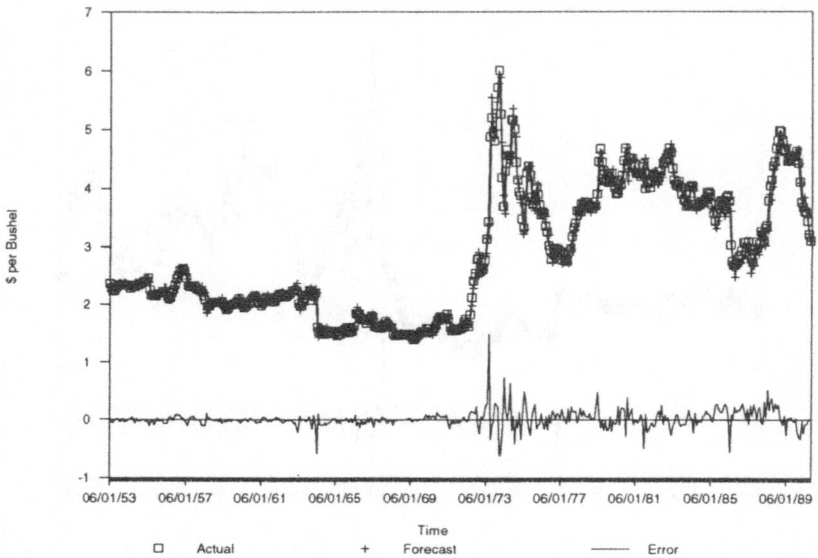

Figure 8. Minneapolis Dark Spring Wheat (13% Protein), State Space Model Performance.

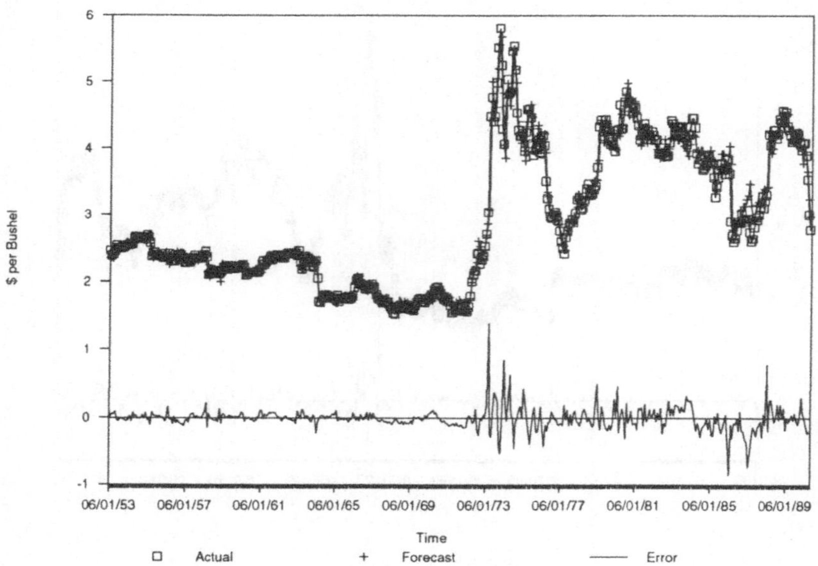

Figure 9. Minneapolis Dark Spring Wheat (15% Protein), State Space Model Performance.

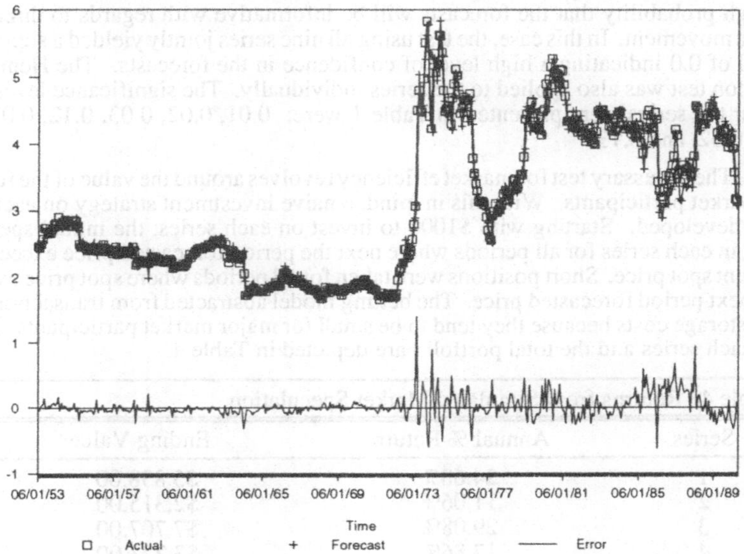

Table 3. Summary Results for Single Period Out-of-Sample Forecasts.

Series	Squared Correlation[a]	Avg. Error	RMSE
1	0.91	-0.02	0.19
2	0.87	-0.06	0.23
3	0.90	-0.02	0.17
4	0.82	0.04	0.23
5	0.87	0.02	0.20
6	0.81	-0.04	0.24
7	0.92	0.08	0.21
8	0.86	-0.09	0.26
9	0.78	0.22	0.45

[a]**Squared Correlations between actual and predicted values are used as an analog to R^2.**

The Henriksson-Merton test was applied to the out-of-sample forecasts in order to determine if the model is informative in terms of predicting direction of price movement. The test yields a significance level between 0 and 1. A low value indicates a high probability that the forecasts will be informative with regards to direction of price movement. In this case, the test using all nine series jointly yielded a significance level of 0.0 indicating a high level of confidence in the forecasts. The Henriksson-Merton test was also applied to the series individually. The significance levels in the order the series were presented in Table 1 were: 0.01, 0.02, 0.03, 0.12, 0.07, 0.05, 0.0, 0.52, and 0.13.

The necessary test for market efficiency revolves around the value of the forecasts to market participants. With this in mind, a naive investment strategy on each series was developed. Starting with $1000 to invest on each series, the model speculated long in each series for all periods where next the period forecasted price exceeded the current spot price. Short positions were taken for all periods where spot price exceeded the next period forecasted price. The betting model abstracted from transactions costs and storage costs because they tend to be small for major market participants. Results for each series and the total portfolio are depicted in Table 4.

Table 4. Returns from Simulated Market Speculation.

Series	Annual % Return	Ending Value
1	24.68%	$5,838.00
2	11.06%	$2,315.00
3	29.08%	$7,707.00
4	17.86%	$3,723.00
5	10.85%	$2,280.00
6	-1.92%	$856.00
7	6.33%	$1,634.00
8	23.58%	$5,439.00
9	-0.09%	$993.00
Total	16.62%	$30,786.00

Clearly, some series performed much better than others. However, in total, the naive portfolio returned a highly respectable 16.62% annual return over the 8 year out-of-sample period. Figures 10 and 11 give some idea of the variability of returns. The figures indicate that, while strong positive returns are possible, the portfolio managed to avoid losses in every year except 1989. Even in 1989, losses were relatively small. At this level, the portfolio seems to offer high returns with relatively low risk.

5. Multiperiod Forecasts

Sequential multiperiod forecasts were developed in an effort to extend the state space model's usefulness as a marketing tool for wheat marketers. We first attempted to forecast four months ahead into the post-sample data. The squared correlations between actual and predicted dropped off dramatically. None of the squared correlations exceeded 0.70 but none fell below 0.40. The Henriksson and Merton test on all nine series jointly returned a significance level of 0.34. This is probably insufficient success at predicting directional movement to generate profit for traders.

In contrast, the two month ahead forecasts (again into the post-sample data) are very good. The Henriksson and Merton significance level for all nine series jointly is 0.04. The significance levels for the individual series were also calculated. In the order in which the series were presented in Table 1, these significance levels were: 0.47, 0.54, 0.26, 0.05, 0.14, 0.07, 0.30, 0.85, and 0.75. Clearly, the soft wheat markets (series 3 through 7) are where one would expect to profit from the two month ahead forecasts based on these results. The model statistics for the two month ahead out-of-sample forecasts are presented in Table 5.

Table 5. Results of Two Month Ahead Forecasts Into Out-of-Sample Data.

Series	Squared Correlation[a]	Average Error	RMSE
1	0.78	0.10	0.28
2	0.71	-0.03	0.33
3	0.73	-0.01	0.30
4	0.69	0.06	0.31
5	0.70	0.03	0.32
6	0.63	-0.03	0.35
7	0.81	-0.10	0.32
8	0.75	-0.06	0.32
9	0.58	0.27	0.60

[a]Squared Correlations between actual and predicted values are used as an analog to R^2.

While the results tentatively indicate that attempting to forecast the paths of prices of agricultural commodities may not be a pointless pursuit, a number of caveats must be highlighted especially with regard to the naive betting strategies described above. First, the data used are monthly average prices. Consequently, the model predictions are based on average prices, but trades are unlikely to occur at average prices. This fact may alter the utility of the model in a practical trading situation and mitigate any claims of market inefficiency. Second, transactions costs and storage costs would certainly drive down average returns. Third, it may be practically difficult to take a short position in some of these markets during some months.

Figure 10. Annual Rate of Return from State Space Motivated Trading Simulation.

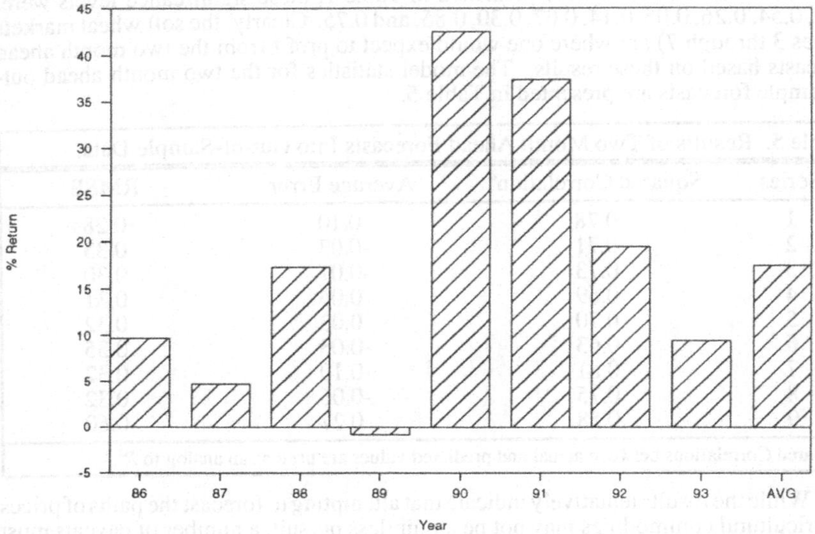

Figure 11. The Value of the Equally Weighted Portfolio Over Time.

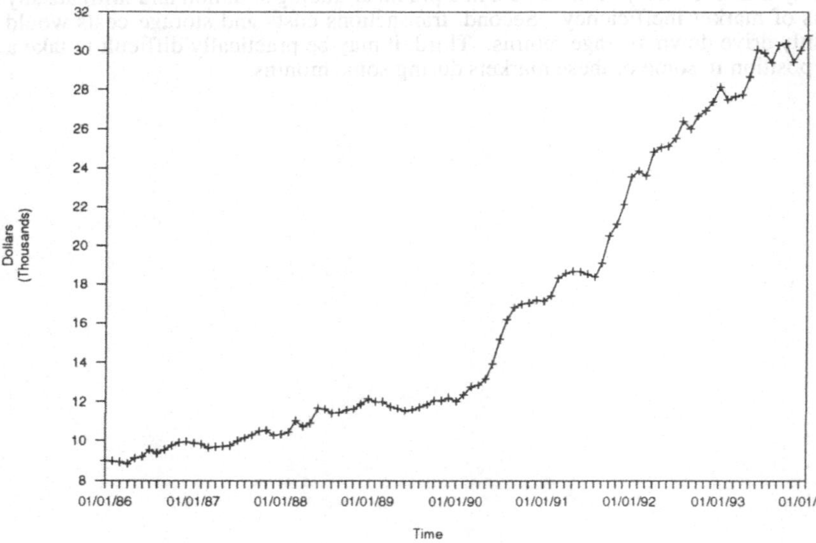

6. References

Aoki, Masanao and Arthur M. Havenner. "State Space Modeling of Multiple Time Series." *Econometric Reviews.* Vol. 10, (1991):1-59.

Aoki, Masanao. *State Space Modelling of Time Series.* 2nd edition. Springer-Verlag, 1990.

Cerchi, Marlene and Arthur M. Havenner. "Cointegration and Stock Prices: The Random Walk on Wall Street Revisited." *Journal of Economic Dynamics and Control.* Vol. 12 (1988): 333-346.

De Jong, David, N. and Charles H. Whiteman. "The Forecasting Attributes of Trend- and Difference-Stationary Representations for Macroeconomics Time Series." *Journal of Forecasting.* Vol 13 (1994): 279-88.

Engle, R. G. and C. W. J. Granger. "Cointegration and Error Correction: Representation, Estimation, and Testing." *Econometrica.* Vol. 55 (1987): 251-76.

Foster, Kenneth A., and Arthur M. Havenner. "Cointegration and Settlement of Commodity Futures Contracts." University of California, Davis, Department of Agricultural Economics, Working Paper No. 92-09, 1992.

Granger, C.W.J. "Developments in the Study of Cointegrated Economic Variables." *Oxford Bulletin of Economics and Statistics.* Vol. 48 (1986): 213-228.

Henriksson, R. D. and R. C. Merton. "On Market Timing and Investment Performance. II. Statistical Procedures for Evaluating Forecasting Skills." *Journal of Business.* Vol. 54 (1981): 513-53.

Walburger, Allan. *Modeling and Testing the Efficiency of Price Discovery in Live Cattle Markets: A Linear Systems State Space Approach.* Ph.D. Dissertation, Purdue University, 1994.

Managing the Herd: Price Forecasts for California Cattle Production

Lorraine M. Egan and Arthur M. Havenner
University of California
Davis, California 95616

Abstract

A state space model of vertically integrated California cattle prices is constructed and tested in and out of sample. The state space model provides more accurate forecasts than a vector autoregressive model or futures market prices. Since the weight gain characteristics of holding cattle over the relevant production periods are well known, the primary uncertainty lies in the prices. Dependable price forecasts permit market participants to optimize their purchases and sales.

1 Introduction

Beef cattle are sold in various forms--slaughter steers/heifers, cow-calf pairs, feeder steers, and slaughter cows, to name a few. While all types of cattle can be converted to hamburger, and eventually most, if not all, will come to such an end, the various types are generally used as inputs to produce slaughter steers. Although the cattle are sold as different types, the market prices are highly related with exogenous impacts extending forward through time. Cow-calf pairs eventually become slaughter cows and slaughter steers, so an event affecting the price of a cow-calf pair in the current period will have an effect on the slaughter steer price several periods ahead.

Cattle ranchers are faced with the decision to sell cattle today or at some future date. Although there is an aspect of perishability to beef cattle, they can be stored over a period of time and sold later if the expected increase in value is sufficiently large to balance the costs of holding the cattle an extra month. The daily expected weight gain, broken down by breed classification, is well known (Feuz et al 1992, Sims 1993); what is unknown is the increase or decrease in the sale price of the cattle in some future period.

This paper will examine the relative efficiency of three forecasting methods: Aoki's State Space Time Series, Vector Auto Regression, and the futures market in an application to forecasting slaughter steer and heifer prices, feeder steer prices and cow-calf prices. Comparisons of the accuracy of the forecast prices in terms of both magnitude and direction are made in and out of the model specification and estimation sample.

2 Industry Background

The initial phase of beef cattle production occurs on cow-calf operations where a ranch will have an existing herd of breed cows that produces one calf per bred cow per year. The cow-calf ranch operator faces several decisions related to maintaining, expanding or contracting the mother herd. The process works as follows: cows have calves and these calves are raised with their mothers until weaning at approximately six months of age. At this point, the calves can be 1) sold to a feedlot, 2) retained as replacement breeding stock, or 3) retained and fed out to be sold later. Beef cattle can be sold at almost any age or weight classification. The Federal State Market News Service publishes prices for cows with small calves, cows with larger calves, slaughter steers and heifers of various weights, slaughter cows, and feeder steers and heifers of various weights. The cattle rancher starts with a cow-calf herd and when the calves reach a certain age or weight threshold, the decision is made whether to sell those calves as feeders (animals mature enough to be placed in a feedlot), or hold back the calves to be used as potential breed stock replacements, or plan to sell the calves later. Since steers are castrated males, they can only become slaughter steers; they have no value as breeding stock. However, the feeder heifers can become either slaughter heifers or replacement heifers--heifers that are retained to maintain or improve the original herd. Slaughter steers and slaughter heifers are likely to share some common dynamics as they are apparently the same product with only a gender difference. However, slaughter heifers can also be used as breeding stock and so they will tend to have additional dynamics somewhat different from their brothers.

Figure 1. Possible Uses of the Types of Cattle

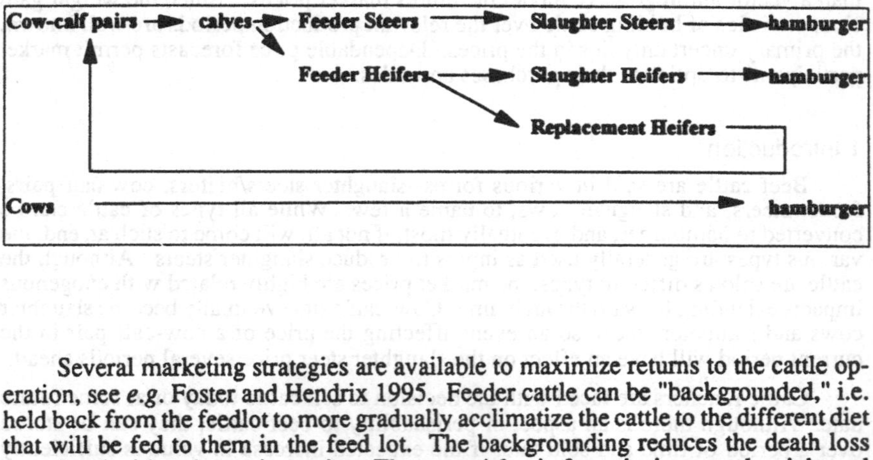

Several marketing strategies are available to maximize returns to the cattle operation, see *e.g.* Foster and Hendrix 1995. Feeder cattle can be "backgrounded," i.e. held back from the feedlot to more gradually acclimatize the cattle to the different diet that will be fed to them in the feed lot. The backgrounding reduces the death loss associated with a change in routine. The potential gain from the increased weight and the reduced death rate will be balanced with the costs of feeding and holding the cattle. Retained ownership is also a viable marketing strategy encompassing such tactics as maintaining ownership of calves placed in a custom feedlot or retaining calves on the ranch. Cow-calf pairs can be sold as cows with small calves (up to 300 pounds) or cows with calves over 300 pounds. As before, the decision is whether to hold the cow-calf pair until the calf is over 300 pounds or to hold the pair until the calf is ready to sell as a feeder. The determination of whether it is profitable to hold the cattle or to sell them is based on the forecast price.

At any stage in the production process, the cattle can be marketed through a livestock auction or through private sales. The profit maximizing cattle ranch would prefer to have a reliable prediction of future cattle prices when faced with the sell now or sell later decision. The decision of when to sell is based on many factors including availability of range or feed supplies, tax liabilities, and capital considerations, in addition to the relative prices of the various types of cattle. Thus reliable forecasts of future prices of feeder and slaughter animals are a valuable input into the decision process. We will concern ourselves with four representative classifications, slaughter steers, slaughter heifers, feeder steers, and cow-calf pairs.

The dynamic and integrated nature of the system implies that events in the past will have an extended impact on the prices of the various types of cattle in the future, *e.g.* if prices rise, the quantity of slaughter animals may actually decrease as more slaughter heifers are withheld to build up cow-calf pairs in anticipation of future needs. The dynamic interactions are complex and subtle--critical to accurate forecasting, but difficult to model without the insights of time series analysis.

3 The Data

Data for the monthly prices of slaughter steers, slaughter heifers, feeder steers, and cow-calf pairs from January 1980 to December 1992 were taken from the Livestock and Meat Price Receipts published by the Federal State Market News Service. The steer and heifer prices are denominated in dollars per hundredweight while the cow-calf pairs are denominated in dollars per pair. The steer and heifer prices are a simple average of California feedlot and range sales by grade and month. We used the slaughter steer select 2-3 grade, 900-1100 pounds, slaughter heifer select and choice 2-4 grade, 900-1100 pounds, feeder steers medium frame 700-800 pounds. The cow-calf pair price is a simple average by grade and month at Cottonwood, California (a major Northern California auction yard). The cow-calf pairs are mixed age cows with small to 300 pound calves at side. The futures prices are from the *Chicago Mercantile Exchange Yearbook*, generously provided by Zhiqiang Leng. Each of the forecasting techniques was implemented using the first 117 observations. Twenty five percent of the data (39 observations) were reserved for out of sample validation.

Calves are not the only input into slaughter steers; the animals need to eat, and so we originally considered using hay prices as an indicator of range conditions. An initial model was estimated using five data series: the cattle prices described above and the monthly price quotation for good quality hay on a delivered basis from the Escalon-Modesto-Turlock region. The hay series added almost no information to the model. In California, hay is exported, used as a feed supply for horses, and used as an input into the dairy industry. These influences on the hay prices are sufficiently large to effectively uncouple the hay series from the cattle prices so that adding the hay series to the model did not generate useful information.

4 State Space Time Series Model

Aoki's (1983, 1987, 1990) State Space Time Series Analysis method is appropriate for this system of equations with dynamic interrelations. The state space procedure can be broken into two phases: model identification and model estimation. The first phase, determining the correct model, involves choosing three things: the series to be modeled together (described above), a lag parameter (N_p) such that the lags to N_p periods back are sufficient for the history of the series, and the number of states (n) necessary to characterize the series dynamics.

There are several considerations in choosing the lag parameter N_p. Cattle prices are believed to follow a repeating long cycle over several years (Gilliam 1984) as the breeding stock is adjusted by withholding heifers from the slaughter market or culling the breeding stock (sending breeders to the slaughter market), depending on price expectations. Graphical analysis of the data -- plotting the prices against time -- appeared to show both long and short cycles, leading to an initial analysis based on a trend and cycle model. For this model, the lag parameter is used to separate long run and short run dynamics: a small lag parameter value averages over all the high frequency dynamics and thus produces a trend model, while a large lag parameter value has the opposite effect (see Aoki 1990, or Aoki and Havenner 1991). For the data examined here, the dynamics were not sufficiently distinct to benefit from the separation into trends and cycles, however, and so the trend-cycle model was abandoned in favor of a combined model. In this combined model, the lag parameter must include enough autocorrelations to capture short run effects essential to accurate forecasts without introducing excessive sampling error in the Hankel matrix. Accordingly, the lag parameter was set to three. With the four series in the model, the implied univariate lag polynomials are up to order twelve, which is appropriate for monthly data.

The final decision in the model specification phase is the determination of n, the number of states necessary to characterize the dynamics of the system. By the Kronecker Theorem, n is equal to the rank of the Hankel matrix formed from the cross autocorrelations of the future values of the vector of series prices with their past values. The singular value decomposition of the Hankel matrix gives a computationally robust determination of its rank. A sample estimate of the Hankel matrix and the (scale free) ratios of each successive singular value to the first singular value were constructed; these initial calculations suggested a maximum of three states were required.

The sympletic matrix solution for the stochastic realization model (see Aoki 1990 for a detailed description) provides an initial filter to determine plausible models. If the sympletic matrix factorization gives rise to a nonpositive spectrum, then the eigenvalues of the sympletic matrix map into one in pairs. Thus, unit eigenvalues of the sympletic matrix are an indication of an unsuccessful solution to the Riccati equation. This can result from an inappropriate choice of the number of states, or of the lag parameter of the series being modeled together. It can also be a warning of nonstationarity in the data.

After the preliminary model specification, the model parameters were estimated and the model was subjected to a number of performance checks (using only the in sample data). These included a number of measures of calculation stability, as well as the more typical forecast evaluation measures. We used the singular value decomposition of the Hankel matrix combined with graphical analysis, the Henriksson Merton nonparametric test of direction, and summary statistics such as the Root Mean Squared Error (RMSE) and R^2 to determine the final model specification.

Two *ad hoc* diagnostic checks to determine the stability of the solution were also used. These tests were calculated using the state space parameter estimates along with an estimate of the state vector initial condition to recursively evaluate the state and observation equations, producing estimates of the states and errors. We first used these estimates to generate an empirical estimate of the variances of the errors of each series, and compared these estimates to the theoretical variances of the errors obtained by solving the matrix Riccati equation. The second *ad hoc* test was constructed in a similar manner, using the empirical and theoretical estimates of the variances of the states (rather than the errors). Ratios close to one are desired, indicating the realized variances closely match the theoretical variances.

Based on the in sample data, the model specification chosen was for the four series slaughter steers, slaughter heifers, feeder steers, and cow-calf pairs (excluding hay), with the number of states and the lag parameter set to three. The eigenvalues of the state dynamics matrix were inside the unit circle, consistent with a stationary data generating process. Analysis of the graphs of forecasts, actual values and errors showed that the model captured most of the changes in magnitude and direction of the steer and heifer prices. The prices appeared to lag one period behind the actual values on the cow-calf pairs, and the error autocorrelations for this series were higher than desired suggesting systematic omissions from the model. Other models were inferior, however.

The model was then re-estimated using the Havenner-Leng iterative estimator, which converged after three iterations. The in sample summary statistics are shown in Table 1. The confidence level of the Henriksson-Merton test ranges from 0.999 to 0.712. The Root Mean Squared Error (RMSE) ranges from 2.97% to 6.05% of the series mean, evidencing considerable model accuracy.

Table 1. In Sample Summary Statistics for the State Space Model

	Slaughter Steers	Slaughter Heifers	Feeder Steers	Cow-calf Pairs
%RMSE	2.97	3.05	4.00	6.05
%AE	0.16	0.02	0.27	0.06
%MAD	2.40	2.40	3.06	4.78
R^2	0.852	0.864	0.899	0.884
H-M	0.978	0.999	0.717	0.712

The root mean squared errors (%RMSE), the average errors (%AE), and the mean absolute deviations (%MAD) are all expressed as percentages of the sample means. R^2 is the correlation of the actual and forecast values, and HM is the Henriksson Merton confidence level.

The state dynamics matrix has a pair of complex eigenvalues, $0.70317 \pm 0.37595i$ (modulus 0.795), and a real eigenvalue of 0.94931. The dominant real eigenvalue reflects a secular downturn in the price of beef cattle during the sample period. There are several explanations for the decline in the price of cattle ranging from changes in demand for beef resulting from changes in tastes, to changes in the price of beef relative to other meats, to an increase in supply resulting from relaxed trade restrictions. This decline is evident in the sale price of cattle over this period and is captured in the model.

5 Vector Autoregessive Model

Vector Autoregressive models are often used to forecast systems of interrelated time series variables; see, *inter alia,* Hamilton. While the coefficients of the VAR model are difficult to interpret and are often unstable due to inherent multicollinearity, the system can be useful for forecasting.

Let y_t denote a stationary (4 x 1) vector containing the cattle prices at time t. The dynamics of y_t are presumed to be governed by a pth order Gaussian vector autoregression of the form:

$$y_t = \delta + \Phi_1 y_{t-1} + \Phi_2 y_{t-2} + \dots + \Phi_p y_{t-p} + v_t, \qquad t = 1,2,\dots,T$$

with $v_t \sim i.i.d.(0,\Omega)$.

Several lag lengths were tested to determine the correct model structure. Using the Akaike information criterion and the Schwartz criterion the correct lag was determined to be four. This implies that sixteen parameters will be estimated to determine the forecast of each of the prices. A VAR (4) model structure was estimated and used to forecast the cattle prices over the same time period as before. The VAR model can be viewed as a seemingly unrelated regression with the same set of regressors, thus ordinary least squares estimates are efficient. Overall, the VAR model fits fairly well. The in sample statistics are listed in Table 2. The regressors explain between 78% and 84% of the variation in the dependent variable. The Root Mean Squared Errors range from 3.6% to 7.18% of the means of the series. All the eigenvalues were less than one in absolute value, indicating a stable model solution; see Table 3. As in the state space case, the dominant eigenvalue is real, consistent with a secular decline in beef cattle prices.

Table 2. In Sample Summary Statistics for the VAR Model

	Slaughter Steers	Slaughter Heifers	Feeder Steers	Cow-calf Pairs
%RMSE	3.60	3.87	5.03	7.18
%AE	0.03	0.03	0.02	0.02
%MAD	2.91	3.06	4.03	5.60
R^2	0.784	0.782	0.842	0.837
H-M	0.979	0.970	0.867	0.285

The root mean squared errors (%RMSE), the average errors (%AE), and the mean absolute deviations (%MAD) are all expressed as percentages of the sample means. R^2 is the correlation of the actual and forecast values, and HM is the Henriksson Merton confidence level.

Table 3. VAR Model Eigenvalues

0.9820	$0.2218 \pm 0.7720i$
0.7893	0.0265
$0.7032 \pm 0.4300i$	$-0.1940 \pm 0.4533i$
$0.5522 \pm 0.3493i$	-0.5641
$0.3756 \pm 0.5959i$	$-0.6881 \pm 0.1561i$

The VAR model requires four lags of the four series to capture the series dynamics. Put into state space form, this would require sixteen states, in substantial disagreement with the three states found necessary in the state space model. Either this model tracks more complicated short run movements of prices, or the original series have important moving average components (automatically incorporated in the state space states) that require an excessively large model when inverted to autoregressive form.

6 The Futures Market

Commodity futures markets involve trading a promise to deliver a commodity at a specific time rather than delivery of the actual commodity. Either a producer or investor can sell the promise to make delivery of a certain quality and amount of the commodity in question. This contract can be fulfilled either by delivering the physical commodity at the designated delivery spot or by offsetting by buying back a similar contract. Delivery is physically difficult and involves significant costs and organization, and so will occur only when the market and futures prices are sufficiently divergent to recover both the delivery and transactions costs. Offsetting the contract is the usual choice.

Because of transportation costs and market differences, the cash and futures prices will seldom be identical, but the opportunity for delivery always exists. The futures price is the rationally based expectation formed through the aggregate mental process of the market (Rausser and Just) when agents are risk neutral and information is costless. If the difference between the futures price and the cash price is sufficiently large, arbitrage opportunities exist and will force convergence between the cash and futures price. Thus, although the futures price is not generated by an econometric model, the futures price is a dollar weighted prediction of the cash price.

There are futures markets for fat cattle--animals ready for slaughter--but not for the feeder steers or the cow-calf pairs. Thus we can only make model comparisons with the slaughter steers and heifers. As the futures market price is not segregated by gender, there is no separate classification for slaughter steers and slaughter heifers. We compared the futures price for fat cattle with the prices for slaughter steers and slaughter heifers.

Table 4. In Sample Summary Statistics for the Futures Prices

	Slaughter Steers	Slaughter Heifers
%RMSE	10.57	13.02
%AE	-8.07	-10.89
%MAD	8.59	11.17
R^2	0.623	0.595
H-M	1	1

The root mean squared errors (%RMSE), the average errors (%AE), and the mean absolute deviations (%MAD) are all expressed as percentages of the sample means. R^2 is the correlation of the actual and forecast values, and HM is the Henriksson Merton confidence level.

Surprisingly, the futures market fared the worst in predicting in sample in terms of magnitude. As shown in Table 4, the R^2 implies that the futures prices explain approximately 60% of the variation in the cash prices and the RMSE is more than ten percent of the mean of the series. Both the average error and the mean absolute deviation are high as percentages of the series means. However, the confidence level for the Henrikkson-Merton test is 1 for both series, indicating that the futures market price is an extremely accurate predictor of direction for these two price series.

7 Conclusions

The results of the out of sample forecasts for the three models are summarized in Table 5. Examining the out of sample predictive performance of the futures market, VAR, and State Space models, we find that the VAR model was the best predictor for cow-calf pairs (for cow-calf pair prices, the state space model has a RMSE of 8.92% of the mean, compared with the VAR model's marginally better RMSE of 8.41% of the mean), but the state space model was the best predictor for the other three series: slaughter steers, slaughter heifers, and feeder steers. This pattern is especially apparent in the summary statistics based on magnitudes, *i.e.*, RMSE, AVG. ERR., MAD, and R^2, where the state space model is substantially superior to the alternatives.

The nonparametric Henriksson Merton direction test is more mixed, with the futures market best for slaughter steers, the state space model best for slaughter heifers and (by a small edge) feeder steers, and the VAR model best for cow-calf pairs.

Table 5: Out of Sample Summary Statistics for the Three Models

	Slaughter Steers			Slaughter Heifers			Feeder Steers		Cow-Calf Pairs	
	SS	VAR	F	SS	VAR	F	SS	VAR	SS	VAR
%RMSE	3.57	4.26	11.97	3.05	4.45	13.47	3.22	4.56	8.92	8.41
%AE	1.60	1.82	-10.67	1.89	2.01	-12.24	11.1	1.27	3.17	2.01
%MAD	2.49	3.68	10.77	2.51	3.89	12.31	2.69	3.64	7.28	6.88
R^2	0.59	0.12	0.07	0.64	0.10	0.06	0.59	0.26	0.05	0.12
HM	0.70	0.78	0.84	0.80	0.62	0.74	0.17	0.17	0.03	0.90

SS indicates the state space model, VAR the VAR model, and F the futures price. The root mean squared errors (%RMSE), the average errors (%AE), and the mean absolute deviations (%MAD) are all expressed as percentages of the sample means. R^2 is the correlation of the actual and forecast values, and HM is the Henriksson Merton confidence level.

As shown on the graphs, the model captures most of the in sample and out of sample variation for both the slaughter steers and slaughter heifers. The forecast values (plotted with a solid line) track the actual values through most of the in sample and out of sample period. (On all graphs, the actual values are plotted using a dotted line and the forecast values with a solid line, and the out of sample period is delineated with a solid vertical line on all the graphs.) The errors are relatively small and do not appear to have a systematic component. Although the forecast for the feeder steer prices appear to lag behind the actual values by one period, the errors are small relative to the magnitude of the actual series and appear to be random. The model predicts well out of sample.

The predictions for the cow-calf pairs are the worst of all the series for both the VAR and State Space models (futures prices are not available). The forecasts fit the actual data in the middle of the sample, but miss on both the magnitudes and directions at the beginning and out of sample. The variation of the errors increases sharply out of sample. This could be a function of a thin market, since the data came from a single auction yard, rather than an average of all statewide sales.

When using the forecasts, both the magnitude and the direction of the price movement is important. The rancher must use the price forecast to balance the marginal gain in value from the increasing weight of cattle against the marginal costs of holding the animals another month. Included in the decision process is the knowledge of the weight gain the livestock will undergo during the time period. For example, suppose a rancher has several 900 pound steers to be sold this month or next month[1]. The cattle are expected to weigh 1000 pounds in one month's time (with a negligibly small variance). If the price today is fifty dollars per hundredweight, next month's forecast price is forty-eight dollars per hundredweight, and the cost of holding the cattle is twenty dollars per month, by waiting one month to sell the cattle each animal is worth ten dollars more than the previous month even if the price falls as expected. Whether or not waiting is preferred depends on the discount rate and the rancher's risk aversion as well as the price forecasts, but accurate estimates of the final worth of the cattle, including the more precise magnitude information available from the state space model, are essential to rational decisionmaking.

---

1 Although cattle auction yards operate on a daily basis, the labor and transportation requirements inherent in working and moving cattle require that sale decisions be made in advance.

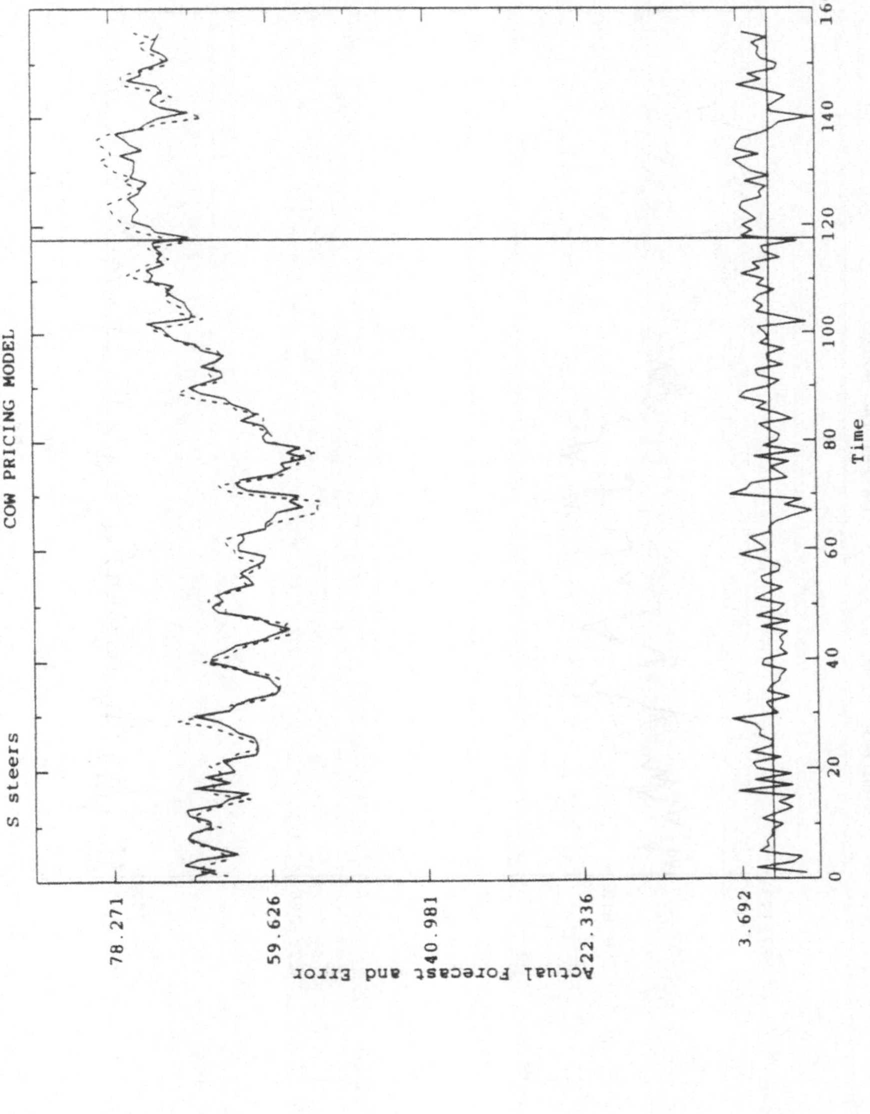
S steers
COW PRICING MODEL
78.271
59.626
40.981
22.336
3.692
Actual Forecast and Error
0
20
40
60
80
100
120
140
160
Time

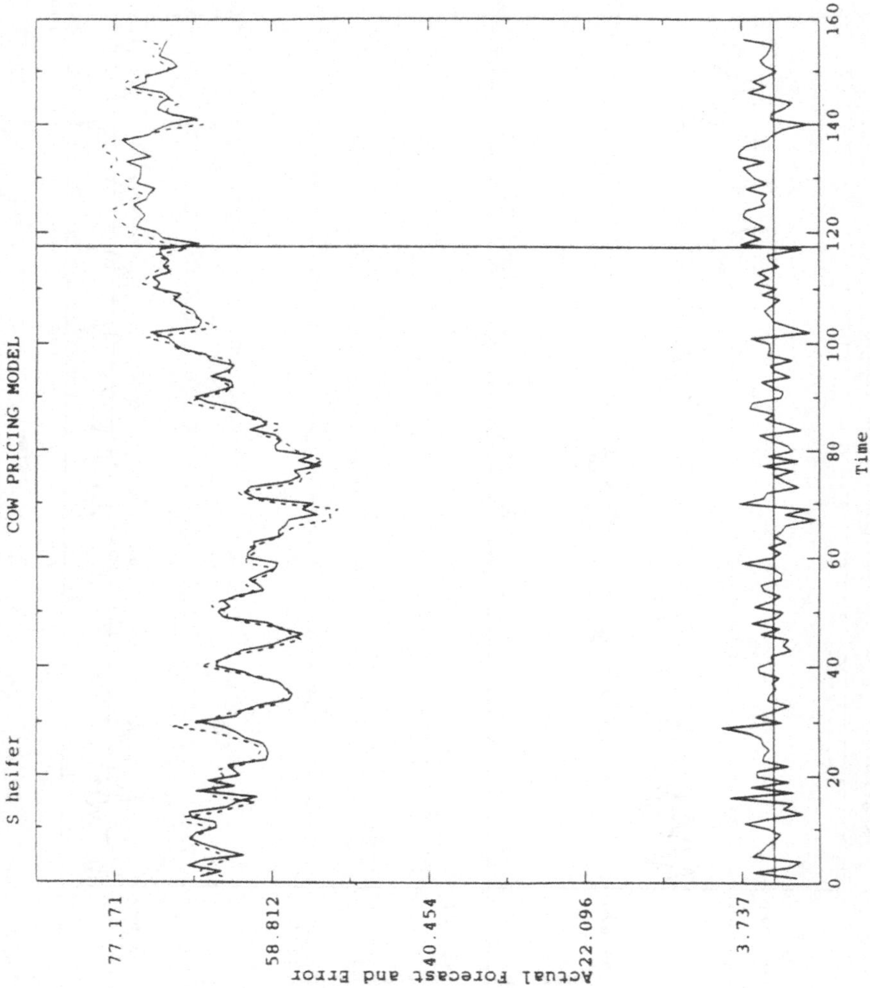
S heifer
COW PRICING MODEL
Actual Forecast and Error
77.171
58.812
40.454
22.096
3.737
0
20
40
60
80
100
120
140
160
Time

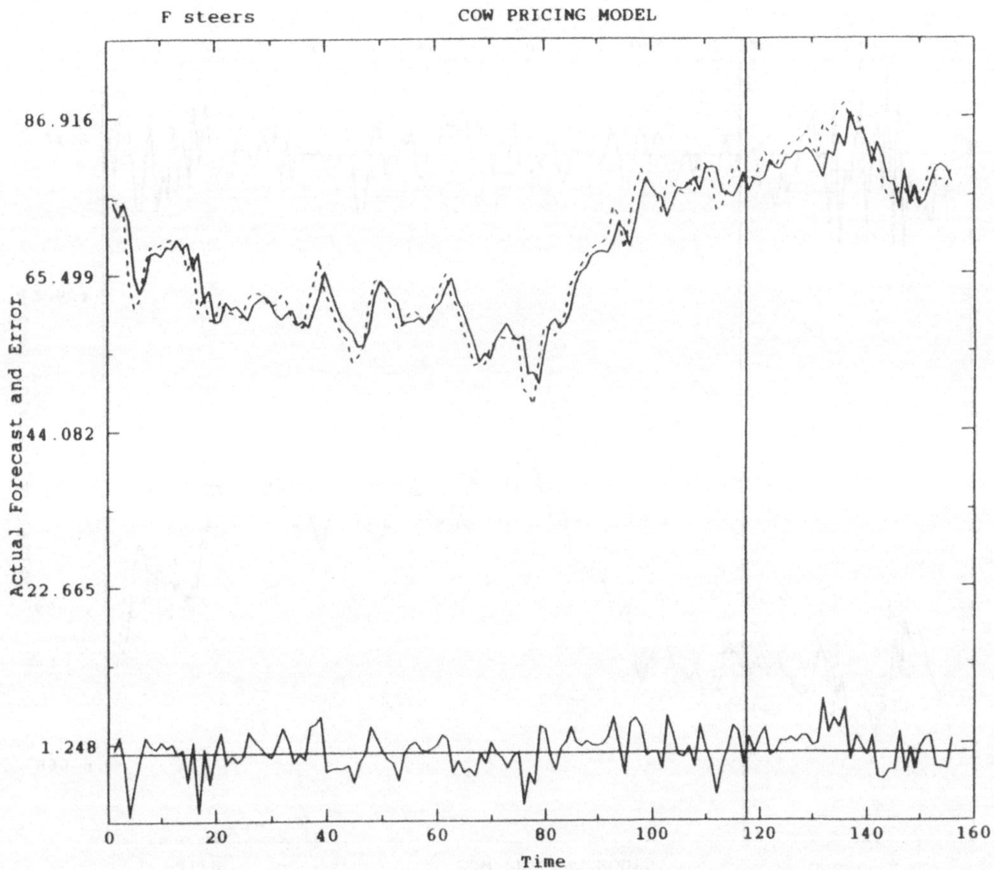
F steers
COW PRICING MODEL
Actual Forecast and Error
86.916
65.499
44.082
22.665
1.248
0
20
40
60
80
100
120
140
160
Time

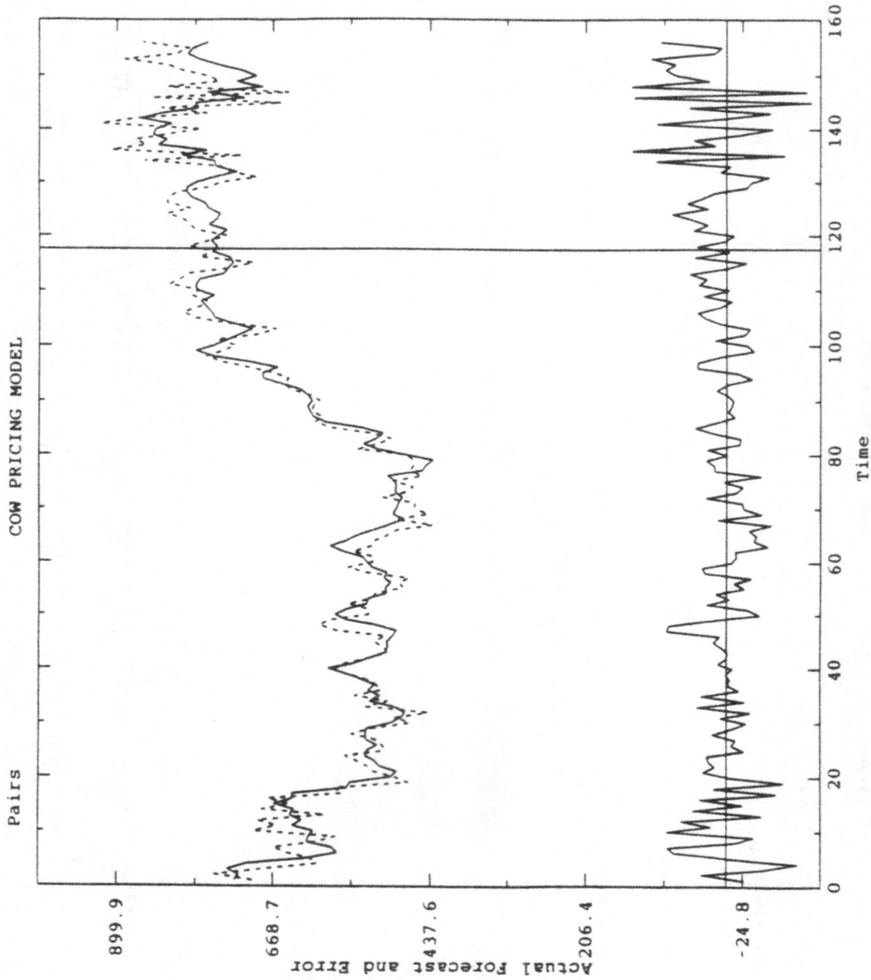
COW PRICING MODEL
Pairs
Actual Forecast and Error
899.9
668.7
437.6
206.4
-24.8
Time
0
20
40
60
80
100
120
140
160

References

Aoki, Masanao. 1983. *Notes on Time Series Modeling*. Heidelberg: Springer-Verlag.

Aoki, Masanao. 1987, 1990 2nd ed. *State Space Modeling of Time Series*. Berlin: Springer-Verlag.

Aoki, Masanao and Arthur M. Havenner. 1991. "State Space Modeling of Multiple Time Series." *Econometric Reviews*. (10)1: 1-59.

Feuz, Dillon, John J. Wagner and Larry J. Held. 1992. "Retained ownership Revisited: The Economic Significance of Genetic Variability."

Foster, Kenneth, and K. Hendrix. 1995. "Economic analysis of calf retention strategies," presented at the third Annual Beef Integrated Resource Management Intensive Workshop, January 1995, Bedford, IN.

Gilliam, Henry C. 1984. "The U.S. Beef Cow-Calf Industry." U.S. Department of Agriculture, Economic Research Service, Economic Report No 51. Washington D.C.

Hamilton, James D. 1994. *Time Series Analysis*. Princeton University Press. Princeton, New Jersey.

Havenner, Arthur M. and Zhiqiang Leng. 1996. "Improved Estimates of the Parameters of State Space Time Series Models" *Journal of Economic Dynamics and Control*. (20)5: 767-790.

Henriksson, Roy D. and Robert C. Merton. 1981. "On Market Timing and Investment Performance. II. Statistical Procedures for Evaluating Forecasting Skills" *Journal of Business*. (54): 513-533.

Just, Richard E., and Gordon Rausser. 1991. "Commodity Price Forecasting with Large-scale Econometric Models and the Futures Market." *American Journal of Agricultural Economics*. (63): 197-208.

Sims, Danny D. 1993. "Kansas Steer Futurities: An Analysis of the Retained Ownership Program." Cooperative Extension Service. Kansas State University, Manhattan Kansas.

Labor Market and Cyclical Fluctuations[1]

Riccardo Fiorito
Department of Economics
Piazza S. Francesco 7
53100 Siena, Italy

1 Introduction

The purpose of this paper is to investigate how labor and good market comovements account for aggregate fluctuations in the US economy after the first oil shock. Labor supply and labor demand are approximated by labor force and employment respectively since we do not assume market clearing at cyclical frequencies. Real wages link labor and good markets by affecting labor quantities and aggregate supply of goods. Aggregate demand for goods is expressed in terms of income determination, which is largely dominated by labor income. Finally, a money reaction function is introduced to assess the role of nominal shocks.

Handling both sides of labor and goods markets aims at disentangling technology and labor supply shocks, two sources that are inevitably mixed when extracted from unemployment and GNP variables only (Blanchard-Quah, 1989; Aoki-Fiorito, 1993). This problem was perceived though not settled by Shapiro-Watson (1988) who proxied labor supply by worked hours, i.e. by a variable that at business cycle frequencies should reflect labor demand instead.

Here business cycles are made observable and then estimated by using Aoki's (1987, 1990) two-step state space algorithm: residuals of the first step are stationary cyclical variables to be modeled in the second stage. Identification is achieved by contemporaneous restrictions that extend to the economy the intertemporal decision rules found for an optimizing representative agent. The model dynamics has no restriction in order to fully use time series evidence. Thus, no long-run constraint such as those pioneered by Blanchard and Quah (1989) is introduced, our emphasis being on testing rather than assuming the dynamics of the structural shocks.

The plan of the paper is as follows: In Section 2 we illustrate the theoretical model. Section 3 describes the state space modeling and its econometric identification. In Section 4 the properties of the business cycles data and the second-step results are presented. Section 5 evaluates the impulse responses and the variance decomposition of the structural model. Section 6 concludes.

2 The Theoretical Model

Preferences

Let a representative agent maximize over time the expected discounted utility function

[1]Support by 40% MURST funds is gratefully acknowledged. Flavio Padrini and Roger Farmer provided useful comments. Remaining errors are mine.

$$E_0 \sum_{t=0}^{\infty} \beta^t U(c_t, \ell_t) \tag{1}$$

where $\beta \in [0,1]$ is the constant time preference rate and where E_0 is the expectation operator conditional on the information available at time 0. The instantaneous utility function is:

$$U(c_t, \ell_t) = \begin{cases} \dfrac{[c_t^\alpha \ell_t^{1-\alpha}]^\gamma}{\gamma}, & 0 < \alpha < 1,\ \gamma < 1 \\ \alpha \log(c_t) + (1-\alpha)\log(\ell_t), & \gamma = 0 \end{cases} \tag{2}$$

where γ is the constant relative risk aversion parameter and where $\sigma \equiv 1/(1-\gamma)$ is the intertemporal substitution parameter; c and ℓ denote real consumption and leisure, respectively. The representative agent will maximize equation (1) subject to the following constraints, expressed as physical units and for non-negative values:

$$\begin{aligned} T &= h_t + \ell_t && (\textit{time endowment}) && (3)\\ y_t &= w_t h_t + r m_{t-1} && (\textit{sources of disposable income}) && (4)\\ y_t &= c_t + s_t && (\textit{uses of disposable income}) && (5)\\ m_t &= m_{t-1} + s_t && (\textit{money holdings transition}) && (6) \end{aligned}$$

The time endowment (T) refers to the total number of hours in a given period. With no taxes and government transfers, the disposable income (y_t) is allocated between consumption and savings and is generated by supplying labor services (h_t) at the real wage (w_t) and by receiving interest payments (rm_{t-1}) on the previous money stock. As the discount rate, the real interest rate is assumed to be constant, while real wages fluctuate around a constant value w:

$$w_t = w + \epsilon_t, \quad \epsilon_t \sim (0, \frac{\sigma_u^2}{1-\rho_w^2}), 0 < \rho_w < 1. \tag{7}$$

Positive first-order serial correlation in wage shocks is intended to reflect positive serial correlation in productivity. Money holdings evolve according to (6). The representative agent will maximize over time the utility function with respect to the decision variables h_t, c_t and m_t. This implies solving the following Lagrangean:

$$\mathcal{L} = \sum_{t=0}^{\infty} \beta^t \{ \frac{[c_t^\alpha (T-h_t)^{1-\alpha}]^\gamma}{\gamma} + \mu_t [w_t h_t - m_t + (1+r) m_{t-1} - c_t] \} \tag{8}$$

where $\mu_t \equiv \lambda_t \beta^{-t}$.

The relevant first-order conditions are:

$$\frac{\partial \mathcal{L}}{\partial h_t} = -\frac{[c_t^\alpha (T-h_t)^{1-\alpha}]^{\gamma-1} (1-\alpha) c_t^\alpha}{(T-h_t)^\alpha} + \mu_t w_t = 0 \tag{9}$$

$$\frac{\partial \mathcal{L}}{\partial c_t} = [c_t^\alpha (T-h_t)^{1-\alpha}]^{\gamma-1} \alpha c_t^{\alpha-1} (T-h_t)^{1-\alpha} - \mu_t = 0 \tag{10}$$

$$\frac{\partial \mathcal{L}}{\partial m_t} = -\mu_t + \beta \mu_{t+1}(1+r) = 0 \tag{11}$$

$$\frac{\partial \mathcal{L}}{\partial \mu_t} = w_t h_t - m_t + (1+r) m_{t-1} - c_t = 0, \tag{12}$$

from which we find the decision rule for supplying hours:

$$h_t = \alpha T + \frac{(1-\alpha)}{w_t}[m_t - (1+r)m_{t-1}] \tag{13}$$

and for individual consumption:

$$c_t = \alpha T w_t - \alpha[m_t - (1+r)m_{t-1}], \tag{14}$$

since no solution for the money variable can be found.[2] The solution value for μ_t is:

$$\mu_t = [\beta(1+r)]^{-t}\mu_0. \tag{15}$$

First, note that the term in brackets in equations (13)-(14) roughly corresponds to the pertinent nominal stock variables (M_t,M_{t-1}) deflated by nominal wages. Moreover we can see from (13) that

$$\frac{\partial h}{\partial w} \geq (\leq)\, 0 \text{ if } s_t \leq (\geq)\, r m_{t-1}, \tag{16}$$

where labor supply will have a positive slope if interest payments exceed current savings. Whenever this condition is not met, the slope will be negative, i.e. the income effect will prevail over the substitution effect. Finally, when money holdings grow at the $(1+r)$ rate, the supply of hours will be proportional to the consumption share in the utility function.

Consumption is positively affected in equation (14) both by the real wage and by the real interest rate as implied by the sources of disposable income and by intertemporal substitution motives. The corresponding non-stochastic steady-state expressions are:

$$h^* = \alpha T + (\alpha - 1)\frac{r}{w}m^* \tag{17}$$

$$c^* = \alpha T w + \alpha r m^* = y^*, \tag{18}$$

where the steady-state value of the disposable income (y^*) is obtained by combining equations (4) and (14).

Technology

Let aggregate output be produced by a collection of identical units according to the following technology:

$$Y_t = \theta_t K_t^{\beta} N_t^{1-\beta}, \qquad 0 < \beta < 1, \tag{19}$$

$$\log(\theta_t) = \rho \log(\theta_{t-1}) + v_t, \qquad v_t \sim iid(0, \sigma_v^2) \tag{20}$$

where Y_t, K_t and N_t denote aggregate output, capital and labor respectively, and where θ_t is the serially correlated technology level. The AR(1) parameter in (20) must be smaller than one in absolute value but is intended to be positive and close to one since we expect that technology shocks last for a long time. If we approximate labor costs by real wages (w_t), the employment level consistent with profit maximization will be $N_t = [(1-\beta)Y_t/w_t]$. Then, by allowing output to change and by assuming for simpler analysis that capital stock is fixed,[3] we find:

[2]This could, perhaps, justify the choice of considering the money stock econometrically exogenous in the aggregate model.

[3]This is also consistent with the assumption that the real interest rate is constant.

$$N_t = \mu_N \theta_t^{\frac{1}{\beta}} w_t^{-\frac{1}{\beta}}, \tag{21}$$

where $\mu_N \equiv K(1-\beta)^{1/\beta}$. In the market for goods also, equilibrium is not assumed at cyclical frequencies. Demand for labor depends on changes in Solow's residual and in real wages. From (19) and (21) we obtain the aggregate supply of output as:

$$Y_t = \mu_Y \theta_t^{\frac{1}{\beta}} w_t^{\frac{\beta-1}{\beta}}, \tag{22}$$

where $\mu_Y \equiv [K(1-\beta)^{(1-\beta)/\beta}]$. The implied aggregate supply curve will be affected negatively by real wages and positively by changes in total factor productivity. Thus, the corresponding demand for labor and supply of goods can be expressed in logs as:

$$\log(N_t) = log(\mu_N) - \frac{1}{\beta}\log(w_t) + \frac{1}{\beta}(\frac{v_t}{1-\rho L}), \tag{23}$$

$$\log(Y_t) = log(\mu_Y) + (\frac{\beta-1}{\beta})\log(w_t) + \frac{1}{\beta}(\frac{v_t}{1-\rho L}), \tag{24}$$

where $L^j z_t \equiv z_{t-j}$ is the lag operator for any integer j. An equivalent dynamic representation of labor demand and output supply is obtained by multiplying both sides of equations (23)-(24) by the common factor $(1-\rho L)$.

3 Dynamic Specification and Identification

Fast and slow modes

VAR models and especially structural VAR models (Sims, 1986; Bernanke, 1986) are most commonly used in evaluating business cycles comovements. Yet, state space models are more general than VAR models (Aoki, 1987) and can also be identified in the econometric sense to produce a vector of structural shocks.

In this section we shall briefly refer to Aoki's two-step modeling of the state space that we used to detect first cycles from non-stationary data and then to model cycles. Details on the algorithm can be found in the original literature (Aoki, 1987, 1989a,b, 1990).

Let the aggregate state space model be

$$\begin{cases} X_{t+1} &= AX_t + Be_t \\ Y_t^T &= CX_t + e_t, \end{cases} \tag{1}$$

where X_t is a $n \times 1$ unobservable state vector and where Y_t^T is a $p \times 1$ observation vector. A, B and C are conformable matrices of parameters that can be estimated by selecting the minimal dimension of the state vector (n) and the size (k) of the estimated Hankel matrix that arranges the autocovariances of the observation vector (Aoki, 1987).[4]

In most economic applications Y_t^T is not stationary, therefore high-frequency comovements cannot be analyzed without prior detrending of the individual data. Aoki's two-step modeling divides the eigenvalues of the $n \times n$ transition matrix A into two mutually exclusive classes, $C_1 = \{\lambda_i : \lambda_i > \lambda^*, i = 1,..,k\}$ and $C_2 = \{\lambda_i : \lambda_i \leq \lambda^*, i =$

[4]Theoretically, the rank of the Hankel matrix is equal to the dimension of the state vector. Aoki's algorithm uses the singular value decomposition of the Hankel matrix to retain only those eigenvalues that are significantly different from zero. To evaluate this, a Bartlett's type of test is developed by Aoki (1989a).

$1, .., n-k\}$, where C_1 corresponds to the slow dynamic modes and C_2 to the fast decaying or cyclical dynamics; λ^* is the threshold value associated with the stochastic trend, i.e. a value close to but smaller than one unless a random walk dynamics is assumed. Let P be a $n \times k$ matrix spanning the subspace of the row eigenvectors associated to the eigenvalues in C_1 and let R be a $n \times (n-k)$ matrix spanning the left subspace of A associated to the eigenvalues in C_2:

$$\begin{aligned} AP &= P\Lambda, \quad P'P = I_k \\ R'A &= NR', \quad R'R = I_{n-k} \\ R'P &= 0, \end{aligned} \tag{2}$$

where the last equality is because Λ and N have by construction no eigenvalues in common. In the new coordinate system the aggregate state vector x_t becomes:

$$X_t = P\tau_t + Rz_t. \tag{3}$$

Given equations (1)-(2) and noting that:

$$[P, R]^{-1} = \begin{bmatrix} P' \\ R' \end{bmatrix},$$

the state vector dynamics can be expressed in the triangular form:

$$\begin{bmatrix} \tau_{t+1} \\ z_{t+1} \end{bmatrix} = \begin{bmatrix} \Lambda & P'AR \\ 0 & N \end{bmatrix} \begin{bmatrix} \tau_t \\ z_t \end{bmatrix} + \begin{bmatrix} P' \\ R' \end{bmatrix} Be_t, \tag{4}$$

where slow modes are affected by fast modes while the opposite does not hold. The companion observation vector

$$Y_t^T = CP\tau_t + CRz_t + e_t, \tag{5}$$

consists of three components, the first one corresponding to the stochastic trend, the second to business cycle comovements and the third to white-noise residuals. When τ is a scalar denoting a common trend, λ is the only eigenvalue in Λ transmitting long-lasting dynamics.

Empirically, the first stage procedure can be set as follows:

$$\begin{cases} \tau_{t+1} = \Lambda\tau_t + Gy_t^c \\ Y_t^T = D\tau_t + y_t^c, \end{cases} \tag{6}$$

where the residual vector $y_t^c = Y_t^T - CP\tau_t = CRz_t + e_t$ denotes stationary deviations around the trend(s) that are typically serially and cross-correlated. In the second stage, the algorithm applies to the weakly stationary vector y_t^c that should be modeled so as to produce a white-noise innovation vector e_t:

$$\begin{cases} z_{t+1} = Fz_t + Je_t \\ y_t^c = Hz_t + e_t. \end{cases} \tag{7}$$

The transfer function for the non-stationary data combines two different expressions:

$$Y_t^T = [I_p + D(L^{-1}I_k - \Lambda)^{-1}G][I_p + H(L^{-1}I_{n-k} - F)^{-1}J]e_t \equiv \Phi(L)e_t, \tag{8}$$

where $\Phi(L) = \sum_{i=0}^{\infty} \Phi_i L^i$, $\Phi_0 = I_p$ and where $E(ee') = \Omega$.

The first term in brackets shows how trended variables are affected by cyclical variables whereas the second term is the Wold representation of the business cycle components:

$$y_t^c = \Pi(L)e_t, \tag{9}$$

where $\Pi(L) = [I_p + H(L^{-1}I_{n-k} - F)^{-1}J]$ is a two-sided matrix polynomial in the lag operator. The corresponding dynamic multipliers can be calculated in terms of the system matrices (7) as:

$$\Pi_{i+1} = HF^iJ, \qquad \Pi_0 = I_p \qquad (i = 0, 1, 2, \ldots) \tag{10}$$

while the total multipliers relating innovations to the non stationary observation vector are:

$$\Phi_{i+1} = [D \;\; H] \begin{bmatrix} \Lambda & GH \\ 0 & F \end{bmatrix}^i \begin{bmatrix} G \\ J \end{bmatrix}. \tag{11}$$

Identification

If the $\Pi(L)$ matrix is invertible, business cycle variables can be expressed as the VAR representation:

$$\Gamma(L)y_t^c = e_t, \qquad \Gamma_0 = I_p, \tag{12}$$

where $e_t = y_t^c - E[y_t^c/y_{t-1}^c, y_{t-2}^c, \ldots]$. Equation (12) is also the reduced form of the structural model:

$$A_0 y_t^c = A_1(L)y_y^c + \eta_t, \qquad E(\eta_t\eta_t') = \Sigma, \tag{13}$$

where deterministic terms are omitted without loss of generality and where $A_1(L) = A_{11}L + A_{12}L^2 + \ldots$; A_0 is a nonsingular sparse matrix whose content is dictated by economic theory. Similarly, Σ does not need to be diagonal since structural shock can also be correlated because of economic theory.

By equating (12) and (13) it appears that the polynomial $\Gamma(L) = I - A_o^{-1}A_1(L)$ and that the structural shocks η_t are a combination of the innovation vector e_t and the identifying matrix A_o:

$$\eta_t = A_0 e_t. \tag{14}$$

The unknown variances and covariances of the structural shocks and the parameters of the A_o matrix can be recovered by solving the following system of $[p \times (p+1)]/2$ algebraic equations:

$$\Sigma = A_0 \Omega A_0'. \tag{15}$$

To match the number of known terms in Ω, we need to impose $p \times (p-1)$ zero restrictions on the off-diagonal elements of A_0 and Σ.[5] By substituting (12) in (9), the transfer function refers to the structural shocks, becoming:

$$y_t^c = [\Pi(L)A_0^{-1}]e_t = [\Pi^*(L)]\eta_t, \tag{16}$$

while the structural multipliers for the business cycle components are: $\Pi_{i+1}^* = HF^iJA_0^{-1}$.

[5] In the special case that Σ is diagonal the number of zero restrictions will be $[p \times (p-1)]/2$ (Aoki-Fiorito, 1993).

Similarly, it is possible to evaluate a structural counterpart of the total multiplier (11) as $\Phi^*_{i+1} = \Phi_{i+1} A_0^{-1}$, though it is not clear how to give an economic interpretation to these impulses unless an explicit growth model is formulated.

Complete specification

The contemporaneous specification reflects the optimal decision rules found in Section 2: labor supply depends on real wages and on a money stock acting as a wealth variable. Labor demand also depends on real wages and on the serially correlated Solow residual. The same specification applies to the aggregate supply of goods (2.24) normalized on the real wage. The structural shocks (η_{2t}, η_{3t}) reflect productivity shifts and are therefore serially and cross-correlated (σ_{23}).

To avoid the ambiguity by which observed hours per worker can denote either supply or demand for labor, we approximated in the aggregate model labor supply and labor demand by two observationally different variables: total civilian labor force (L) and total employment (N), respectively.[6] The simplifying assumption is that the variables explaining the supply of individual hours and the size of the labor force are the same. This should also rule out the possibility that changes in hours reflecting at cyclical frequencies changes in labor utilization be wrongly ascribed to labor supply changes.

Real wages (w) link labor and good markets by affecting labor quantities and the aggregate supply of goods. Aggregate demand for goods is expressed in terms of expenditure components, stemming as in equation (2.4) from labor income and interest payments only.

In the aggregate model money enters as a nominal variable and is econometrically exogenous for the representative agent. Money stock does not enter contemporaneous matrix A_0 because of informational and decision lags in the reaction function.

By specifying matrices A_0 and Σ, letting the observation vector for cyclical data be: $y^c_t = [L^c_t, N^c_t, w^c_t, Y^c_t, M^c_t]'$ and expanding the lag polynomial matrix $A_1(L)$, equation (13) becomes:

$$\begin{bmatrix} 1 & 0 & a_1 & 0 & a_2 \\ 0 & 1 & a_3 & 0 & 0 \\ 0 & 0 & 1 & a_4 & 0 \\ 0 & a_5 & a_5 & 1 & a_6 \\ 0 & 0 & 0 & 0 & 1 \end{bmatrix} \begin{bmatrix} L_t \\ N_t \\ w_t \\ Y_t \\ M_t \end{bmatrix} = \begin{bmatrix} A_{11}(L) & A_{12}(L) & A_{13}(L) & A_{14}(L) & a_{15}(L) \\ A_{21}(L) & A_{22}(L) & A_{23}(L) & A_{24}(L) & a_{25}(L) \\ A_{31}(L) & A_{32}(L) & A_{33}(L) & A_{34}(L) & a_{35}(L) \\ A_{41}(L) & A_{42}(L) & A_{43}(L) & A_{44}(L) & a_{45}(L) \\ A_{51}(L) & A_{52}(L) & A_{53}(L) & A_{54}(L) & a_{55}(L) \end{bmatrix} \begin{bmatrix} L_t \\ N_t \\ w_t \\ Y_t \\ M_t \end{bmatrix} + \begin{bmatrix} \eta_{1t} \\ \eta_{2t} \\ \eta_{3t} \\ \eta_{4t} \\ \eta_{5t} \end{bmatrix} \tag{17}$$

where

$$E(\eta_t \eta_t') = \Sigma = \begin{bmatrix} \sigma_{11} & \sigma_{12} & 0 & \sigma_{14} & 0 \\ - & \sigma_{22} & \sigma_{23} & 0 & 0 \\ - & - & \sigma_{33} & 0 & \sigma_{35} \\ - & - & - & \sigma_{44} & 0 \\ - & - & - & - & \sigma_{55} \end{bmatrix}.$$

We complete the specification by inserting into Σ matrix three further nonzero covariance terms $(\sigma_{12}, \sigma_{14}, \sigma_{35})$: σ_{12} says that labor force and employment must be positively correlated since both variables reflect population changes, while σ_{14} handles the procyclicality of labor participation, i.e. the so-called *discouragement effect* (Bowen-Finegan, 1969). Finally, σ_{35} accounts for the possibility that money injection and productivity are negatively correlated as long as they have an opposite effect on real wages.[7]

Unlike covariance terms, contemporaneous a_i's may have any sign in a dynamic setting. Meanwhile, their long-run signs are too obvious to be reported. Given this

---

[6]This amounts to assuming, as in Hansen(1985), that hours are fixed or, equivalently, that labor is indivisible.

[7]This assumption combines the procyclicality of the real wages set forth by Real Business Cycle theory on one side and Keynes' idea that wages are more rigid than prices on the other.

system of 15 unknown $(a_1, a_2, a_3, a_4, a_5, a_6, \sigma_{11}, \sigma_{22}, \sigma_{33}, \sigma_{44}, \sigma_{55}, \sigma_{12}, \sigma_{14}, \sigma_{23}, \sigma_{35})$ in 15 nonlinear equations, structural parameters can be identified. However, there is no guarantee that equation (14) can be solved, that solutions are unique and all real rather than complex.

4 The empirical model

Detecting cycles

The trended data (Y_t^T) analyzed in the first stage model are seasonally adjusted and range from 76Q1 to 90Q3. We did not use a longer series to avoid the risk of estimating data belonging to different structures. Its being widely recognized that the first oil shock was the critical breaking point, we initiated our sample after this date. Yet, since 1975 seems too atypical to provide the initial conditions, we began our estimates in 1976.

As stated before, supply and demand for labor are approximated by civilian labor force and employment, respectively. In order to match the coverage of the real GNP, real wages are constructed by dividing the ratio between the "National Income Wages and Salaries" and total employment (N) by the GNP deflator. Moreover, we used M2 as the pertinent money stock variable since the theoretical model requires an interest-bearing money definition.[8] To assess the role of the nominal shocks we used in the empirical model M2 rather than its deflated stock (2.6) as outlined in the theoretical model. To eliminate the effects of nonzero initial conditions all data have been normalized on their initial value and then logged.

The first stage model identifies both the stochastic trend(s) and the cyclical variables. In selecting the trend some requirements should be met:

- Trends should not overfit actual data, i.e. trends should be notably smoother or much less jagged than actual data. This implies that the k parameter (order of the Hankel matrix) should be small (one or two).
- Eigenvalue(s) should be close to one and real rather than complex.
- Residuals from the first stage should be serially and cross-correlated.
- Residuals from the first stage should be modeled in the second stage by notably smaller eigenvalues.

On the basis of these criteria we selected for the first stage a one-dimensional state vector ($n = 1$) and also the Hankel matrix dimension was chosen to be one ($k = 1$).

This common trend model has the following parameters:

$$\lambda = Eig(\Lambda) = 0.979, \qquad D = [0.215, 0.226, 0.040, 0.283, 0.906]'$$

where symbols are defined as in (3.6). λ is the only nearly-integrated eigenvalue of the transition matrix Λ representing a common stochastic trend while D is the vector that disaggregates the common trend to the individual series: actually, the nominal value M2 increases more than real series do, while real wages increase very little over the relevant sample.

---

[8] The relevant Citicorp's Citibase data bank code are: LHC(L) = Civilian Labor Force, total; LHEM(N) = Civilian Labor Force: employed, total; ((GWY/LHEM)/GD) = w, where GWY = National Income: Wages and Salaries and GD = Implicit Price deflator: Gross National Product(GD); GNP82(Y) = Gross National Product at 1982$; FM2(M) = Money Stock M2.

The 1st stage residuals are plausible in terms of commonly accepted characteristics of business cycle comovements as we can see from Tables 1 and 2. All variables are serially correlated, though monetary cycles seem less persistent and more asymmetric than labor market cycles.

TABLE 1. Cross Correlations of 1st Stage Residuals

L	N	w	Y	M2
1.00	.59	-.47	.01	-.19
	1.00	.28	.78	-.66
		1.00	.76	-.43
			1.00	-.65
				1.00

TABLE 2. Selected Statistics of 1st Stage Residuals

	L	N	w	Y	M2
$\rho(1)$	.94	.97	.95	.94	.80
$\rho(2)$	.90	.91	.90	.86	.65
$\rho(4)$	.80	.74	.80	.65	.34
$\rho(8)$	.58	.32	.51	.22	-.04
cv	2.73	5.02	3.38	16.26	7.69
sk	.24	-.02	.43	-.57	.65
ku-3	-1.29	-.79	-.81	-.33	.35
LB(19)	369.6	318.8	341.1	255.0	185.0

***Legenda*: ρ's are aucorrelations at the given lag; cv is the coefficient of variation; sk= skewness; ku= kurtosis; LB is the Ljung-Box statistics.**

Synchronous comovements are notable except for labor force and GNP variables that are, however, much more correlated with real GNP at several leads and lags. Also by comparing relative variabilities of these residuals we obtain a ranking that conforms to recognized stylized facts of the business cycles: real wages are less volatile than employment that in turn is less volatile than output (Fiorito-Kollintzas, 1994). M2 also is less volatile than output, a fact that seems to confirm the evidence that prices are countercyclical (Kydland-Prescott, 1990). Finally, labor force cycles are — as expected — less volatile than any other component.

Modeling cycles

Residuals from the first stage have been modeled in the second in the light of the requirements that second-stage eigenvalues should be lower than first-stage eigenvalues and that second-stage residuals should be white. Finally, fit should be adequate. The estimated model ($k = 11$, $n = 5$) satisfying by and large these requirements is the following:

$$F = \begin{bmatrix} 0.920 & -0.170 & -0.200 & 0.002 & 0.137 \\ 0.065 & 0.951 & -0.031 & 0.028 & -0.003 \\ 0.017 & 0.012 & 0.786 & -0.090 & 0.006 \\ 0.053 & 0.109 & 0.464 & 0.701 & -0.181 \\ -0.013 & -0.012 & -0.068 & -0.050 & 0.536 \end{bmatrix},$$

$$Eig(F) = \begin{bmatrix} 0.936 \pm 0.107i \\ 0.762 \pm 0.208i \\ 0.497 \end{bmatrix},$$

$$J = \begin{bmatrix} 26.732 & -33.548 & -8.753 & -2.688 & 1.232 \\ -16.698 & -10.720 & 17.456 & -2.202 & 4.071 \\ -30.759 & 11.442 & -35.935 & 27.584 & 15.657 \\ -62.052 & 15.726 & -22.092 & -12.209 & -44.181 \\ 58.817 & -71.142 & -20.471 & 46.520 & -11.132 \end{bmatrix},$$

$$H = \begin{bmatrix} -0.003 & -0.020 & -0.007 & -0.004 & 0.002 \\ -0.022 & -0.018 & 0.002 & -0.002 & -0.003 \\ -0.025 & 0.019 & -0.008 & -0.002 & -0.001 \\ -0.039 & 0.003 & 0.007 & -0.002 & 0.002 \\ 0.018 & 0.008 & 0.003 & -0.010 & -0.001 \end{bmatrix},$$

$$\Omega = \begin{bmatrix} 0.089 & 0.073 & -0.100 & -0.012 & 0.078 \\ -- & 1.155 & -0.037 & 0.185 & 0.065 \\ -- & -- & 0.383 & 0.252 & -0.071 \\ -- & -- & -- & 0.720 & -0.020 \\ -- & -- & -- & -- & 0.744 \end{bmatrix} \times 10^{-4},$$

where matrices have the same meaning as in equation (3.7) and where Ω is the innovation covariance matrix.

From the eigenvalues of F it is possible to calculate the implied length of cycles: while the only real root reveals cycles decaying in about one year ($.497^4 = 0.06$), the two complex roots produce cycles that - at least in one case - last too long, their duration being about 11 and 6 years respectively. Our results seem - however - robust under a variety of experimentations that never produced notably lower eigenvalues and residuals that remain serially uncorrelated. Thus, in the light of the Ljung-Box statistics and of a satisfactory fit, the selected model seems adequate to provide structural impulse responses (Fig. 2.1–2.5) and variance decompositions s-steps ahead.

$$LB(19) = \begin{bmatrix} 22.9 \\ 15.0 \\ 23.2 \\ 37.1 \\ 20.3 \end{bmatrix}, R^2 = \begin{bmatrix} 0.923 \\ 0.951 \\ 0.967 \\ 0.931 \\ 0.749 \end{bmatrix},$$

5 Results

Given the estimated covariance matrix Ω we solved numerically the system (3.15) by obtaining all real and unique values for the unknown elements of A_0 and Σ:

$$a_1 = 0.216; a_2 = -0.084; a_3 = 0.934; a_4 = 0.263; a_5 = -0.953; a_6 = 0.028;$$

$$\sigma_{11} = 0.058; \sigma_{22} = 0.420; \sigma_{33} = 0.565; \sigma_{44} = 0.309; \sigma_{55} = 0.744,$$

being monetary shocks by construction equal to the estimated money innovations.

The covariance terms are: $\sigma_{12} = 0.049; \sigma_{14} = -0.001; \sigma_{23} = 0.431; \sigma_{35} = -0.078$, implying the following correlations among structural shocks:

$$\rho_{12} = 0.31; \rho_{14} = -0.01; \rho_{23} = 0.88; \rho_{35} = -0.12.$$

From equation (3.14) we recover the structural shocks for which we present in Table 3 a few descriptive statistics. As expected, the employment and real wages shocks are serially correlated, in the theoretical model reflecting the productivity/aggregate supply nexus. The fact that these two shocks are not too persistent (the estimated AR1 parameters are .31 and .36) should be related to the stationary nature of the business cycle data.

Blanchard and Watson (1986) suggested looking at the fourth moments of the structural shock distribution to assess whether fluctuations are symmetric and whether they arise from large infrequent shocks or from a sequence of small shocks. From Table 3 we can see that money shocks are the most asymmetric and also those characterized by large infrequent shocks or episodes. To a much smaller extent this applies also to aggregate demand shocks and labor supply while wage and employment have a flatter distribution: again, this confirms the idea that productivity shocks are effective not because of their size but because they are sequentially transmitted.

TABLE 3. Selected Statistics of Structural Shocks

	η^L	η^N	η^w	η^Y	η^M
$\rho(1)$	.03	.30	.33	.00	.05
$\rho(2)$	-.15	.13	.01	.06	.15
sk	-.48	.20	.25	.12	1.16
ku-3	.67	-.36	-.38	.87	1.46
LB(19)	26.5	35.7	41.6	13.2	20.3

Legenda: see Table 2.

Impulse response functions

The responses to a labor supply shock are relatively smooth as we can see from Figure 2.1. As expected, real wages are decreased by a higher labor force. Yet, we cannot disentangle in this case which part of the shock is due to changes in participation rate and which part is due to demography. Actually, employment and output also fall presumably because in the short run decreasing labor costs reduce income and expenditure, especially if they do not originate from a productivity boost. As a result, unemployment is notably increased by the rising labor force and — to a lesser extent — by the falling employment level. Thus, there is a monetary reaction function aiming at offsetting the real GNP decline.

The second impulse (see Fig. 2.2) pertains to a labor demand shock that we interpret in the light of equation (2.23) as related to productivity. The output response is higher than employment response as it is typical of labor hoarding models where labor is less volatile than production (Fiorito-Kollintzas, 1994). Real wages increase in the short-run and then decrease as long as output and labor decrease. Hence, real wages are procyclical as implied by Real Business Cycle models or RBC models allowing for some labor hoarding behavior (Burnside-Eichenbaum-Rebelo, 1992). Labor force response is mildly procyclical as suggested by the "discouraged worker" view despite the negligible value of the contemporaneous correlation σ_{14}. Money response is again countercylical and almost reverse of output fluctuations.

Reactions to real wage surprises (Fig. 2.3) are more difficult to interpret because this shock, though highly correlated with productivity, stems also from other sources[9] that we cannot investigate in a small aggregative model. Unlike other impulses, the real wage shock is very sustained though not monotonically. Its seesaw pattern suggests that the numerator and the denominator interact over the business cycle because of productivity and price/earnings catch ups. Increasing labor costs reduce aggregate supply while increasing productivity does the opposite. As a result, cycles in output and labor input originate while there is again a procyclical labor force and a countercyclical money stock response.

As one would expect of aggregate demand shocks (Fig. 2.4), real GNP impulses are not sustained. Labor quantities are procyclical whereas real wages fluctuate in this case countercyclically. The economy responses are moderate. In particular, employment response is again smaller than real GNP and M2 is again countercyclical. By comparing labor force and employment multipliers, it seems that aggregate demand shocks can reduce unemployment in the short-run only and that this effect is small because of the increasing labor force.

Finally, we evaluate in Fig. 2.5 the model response to an M2 surprise. First of all, the shock is not sustained enough to last more than one year so that all impulses decay very quickly and are limited in size. As usual, output responds more than employment does while real wages and labor force are weakly procyclical.

Variance decomposition

A second group of results deals with the variance decomposition s-steps-ahead of the structural shocks that we calculated as in equation (3.16). Our findings are that the most influential sources of business cycless are the shocks associated to wages and employment. In our case, however, a cautionary word is in order because these shocks are by construction cross-correlated and actually they are highly correlated (ρ_{23}=0.88) in the empirical model.

Yet, the fact that these two shocks cannot be disentangled in our decomposition should not prevent us from asserting that labor income — i.e. the product of wages and employment — is a major source of business cycle fluctuations. Since labor income components reflect productivity being also a major source of expenditure (2.4), it seems inappropriate to overemphasize 'supply' or 'demand' sources of business cycles rather than to evaluate more direct and recognizable sources.

Labor force seems to be in the short-run econometrically exogenous while it later reflects more wages and employment determinants. Money and real GNP innovations play a minor role in this decomposition.

Employment is strongly affected by its own shocks but — unlike what is found for labor force — this exogeneity feature seems to last for a long time. Only wages seem to influence employment, while real GNP and M2 contribution is again negligible. Real wages reflect both labor supply and labor demand determinants as in textbook explanations. Moreover, they are also strongly affected — as the other labor market variables — by their own shocks. In the short-run, real wages reflect also some real GNP influence, showing that expenditure impulses decay rather quickly. In turn, real GNP seems to be dominated by 'supply' shocks and very little of the variance decomposition is left to its own component. However, by taking once more into account that labor share combines employment and real wages, the importance of expenditure cannot be ruled out, though changes in productivity seem to be the originating source.

M2 does not operate as an econometrically exogenous variable reflecting instead employment and wage shocks, i.e. sources of expenditure that typically character-

---

[9]Slack in labor market and changing composition of the labor force are obvious candidates.

TABLE 4. Variance decomposition s-steps ahead

Labor Force

s	L	N	w	Y	M2
1	0.58	0.04	0.37	0.01	0.00
2	0.41	0.12	0.46	0.01	0.00
4	0.16	0.29	0.53	0.01	0.01
8	0.05	0.44	0.49	0.02	0.00
12	0.03	0.49	0.44	0.03	0.01
20	0.04	0.51	0.40	0.03	0.02
30	0.04	0.51	0.40	0.03	0.02

Employment

s	L	N	w	Y	M2
1	0.00	0.57	0.41	0.02	0.00
2	0.01	0.57	0.40	0.02	0.00
4	0.02	0.58	0.37	0.02	0.01
8	0.03	0.59	0.33	0.03	0.02
12	0.03	0.60	0.31	0.03	0.03
20	0.04	0.58	0.32	0.03	0.03
30	0.04	0.57	0.33	0.03	0.03

Real Wages

s	L	N	w	Y	M2
1	0.12	0.11	0.59	0.17	0.01
2	0.22	0.23	0.39	0.15	0.01
4	0.33	0.36	0.21	0.10	0.00
8	0.41	0.29	0.21	0.06	0.03
12	0.28	0.25	0.40	0.05	0.02
20	0.11	0.38	0.45	0.05	0.01
30	0.09	0.43	0.41	0.05	0.02

Real GNP

s	L	N	w	Y	M2
1	0.04	0.75	0.15	0.06	0.00
2	0.05	0.73	0.17	0.04	0.01
4	0.08	0.70	0.17	0.03	0.02
8	0.11	0.68	0.13	0.03	0.05
12	0.12	0.64	0.15	0.03	0.06
20	0.09	0.57	0.27	0.03	0.04
30	0.08	0.56	0.29	0.03	0.04

Money Stock

s	L	N	w	Y	M2
1	0.05	0.35	0.28	0.00	0.32
2	0.06	0.44	0.33	0.01	0.16
4	0.07	0.50	0.34	0.02	0.07
8	0.07	0.53	0.31	0.03	0.06
12	0.08	0.52	0.30	0.03	0.07
20	0.07	0.51	0.32	0.03	0.07
30	0.07	0.51	0.32	0.03	0.07

ize money demand behavior, though this can also be compatible with some reaction function.

6 Conclusions

We analyze business cycle comovements in the US ranging from 1976 to 1990 by using Aoki's state space algorithm that does not require prior detrending of the individual series. The model is identified by contemporaneous restrictions that extend to the economy the intertemporal optimal decisions found for the representative agent. Markets for goods and labor do not clear at cyclical frequencies. We separate the overall supply shock into technology and labor supply components. Labor force and employment approximate labor supply and labor demand, respectively. Real wages connect labor and goods markets by affecting labor quantities and the aggregate supply of goods. Aggregate demand for goods reflects income determination and is normalized on real GNP. Labor income and interest payments are the disposable income sources in a simplified environment where no taxation and government are considered. We let M2 be the interest-bearing money stock definition entering the empirical model.

Our results show that productivity shocks are important while the shocks associated to the chosen money variable are not important in shaping macroeconomic fluctuations. Unlike the findings of Shapiro and Watson (1988), labor supply seems to be more a product than an instigator of the business cycle. Real GNP surprises have virtually no power in accounting for business cycle fluctuations.

Since real wages and employment are strongly correlated, their specific roles cannot be evaluated with precision. All results indicate, however, that technology and productivity are the most important sources of comovements and especially of the real GNP cycles. Yet, labor income being nothing but the combination of these two variable and expenditure being strongly dependent on labor income, no clear-cut distinction between 'supply' and 'demand' sources of business cycles should be asserted.

References

Aoki M. (1987), *State Space Modeling of Time Series*, Heidelberg-New York, Springer-Verlag

Aoki M. (1989a): *Singular Values, Canonical Correlation and the Dimension of State Space Models of Time Series*, presented at the 1989 NSF/NBER Seminar, Madrid

Aoki M. (1989b): *A Two-step State Space Time Series Modeling Method*, "Computer Mathematics Applications", 17(8/9), 1165–76

Aoki M. (1990), *State Space Modeling of Time Series*, 2nd revised and enlarged edition, Heidelberg-New York, Springer-Verlag

Aoki M. and R.Fiorito (1993),*Analysis of US Real GNP and Unemployment Interactions: State Space Approach*, "Journal of Economic Dynamics and Control", 17, 385–99

Bernanke B.S. (1986): *Alternative Explanations of the Money-Income Correlation*, "Carnegie-Rochester Series on Public Policy", 25, 49–100

Blanchard O.J. and M.W.Watson (1986), *Are Business Cycles All Alike?*, in: R.J.Gordon (ed.), *The American Business Cycle. Continuity and Change*, NBER, Studies in Business Cycle, vol. 25, The University of Chicago Press, 1986, 123–82

Blanchard O.J. and D.Quah (1989), *The Dynamic Effects of Aggregate Demand and Supply Disturbances*, "American Economic Review", 79(4), 655–73

Bowen W.G. and T.A.Finegan (1969), *The Economics of Labor Force Participation*, Princeton N.J., Princeton University Press

Burnside C., M.Eichenbaum and S.Rebelo (1993), *Labor Hoarding and the Business Cycle*, "Journal of Political Economy", 101(2), 245–73

Fiorito R. and T.Kollintzas (1994), *Stylized Facts of Business Cycles in the G7 from a Real Business Cycles Perspective*, "European Economic Review", 38, 235–69

Hansen, G.D. (1985): *Indivisible Labor and the Business Cycle*, "Journal of Monetary Economy", 16, 309–28

Kydland F. and E.C.Prescott (1990): *Business Cycles: Real Facts and a Monetary Myth*, Federal Reserve Bank of Minneapolis "Quarterly Review", Spring, 3–18

M.Shapiro and M.W.Watson (1988), *Sources of Business Cycle Fluctuations*, in: S.Fischer (ed.), *NBER Macroeconomic Annual 1988*, NBER and MIT Press, 111–48

C.Sims (1986), *Are Forecasting Models Usable for Policy Analysis?*, "Federal Reserve Bank of Minneapolis Quarterly Review", Winter, 2–16

Figure 1.1: Labor forces cycles and prediction.

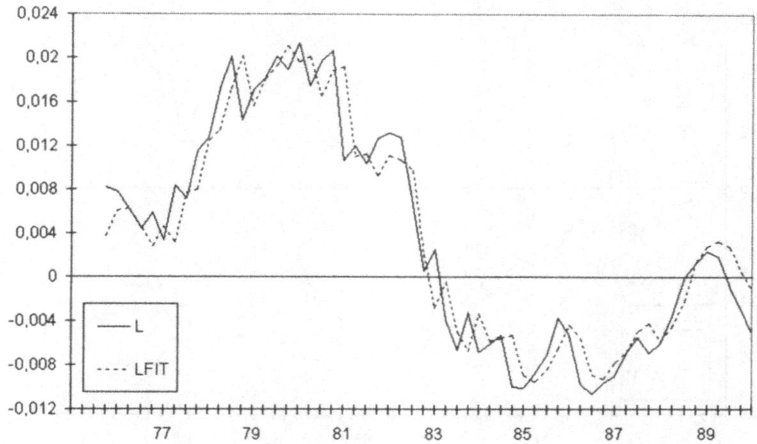

Figure 1.2: Employment cycles and prediction.

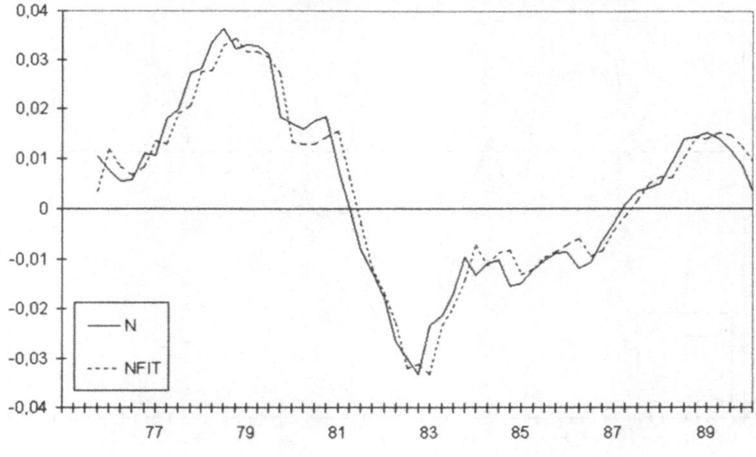

Figure 1.3: Real wage cycles and prediction.

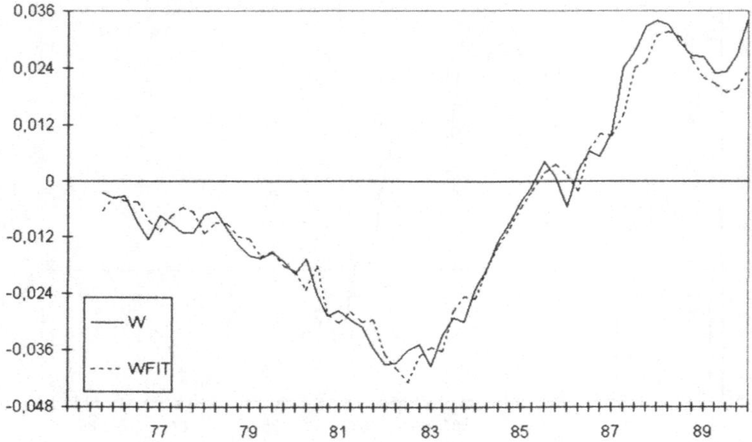

Figure 1.4: Real GNP cycles and prediction.

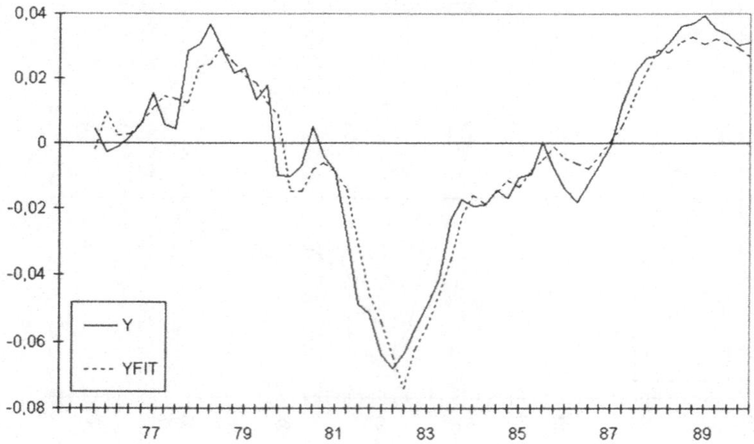

Figure 1.5: M2 cycles and prediction.

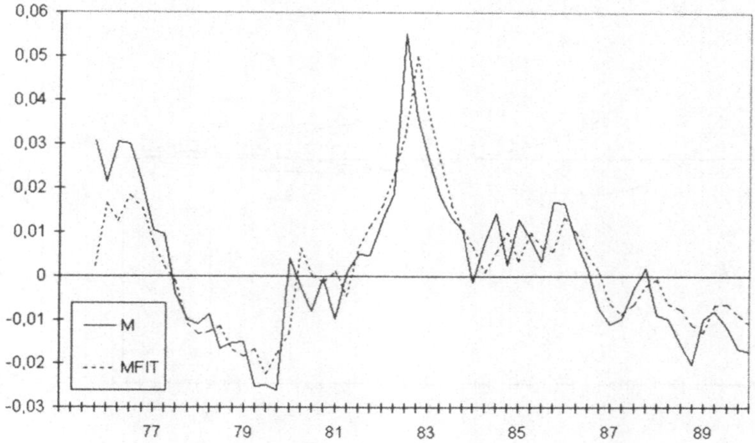

Figure 2.1: Responses to a labor supply shock.

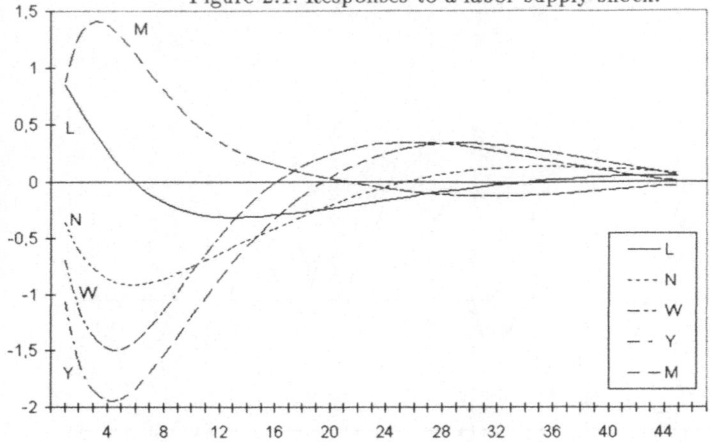

Figure 2.2: Responses to a labor demand shock.

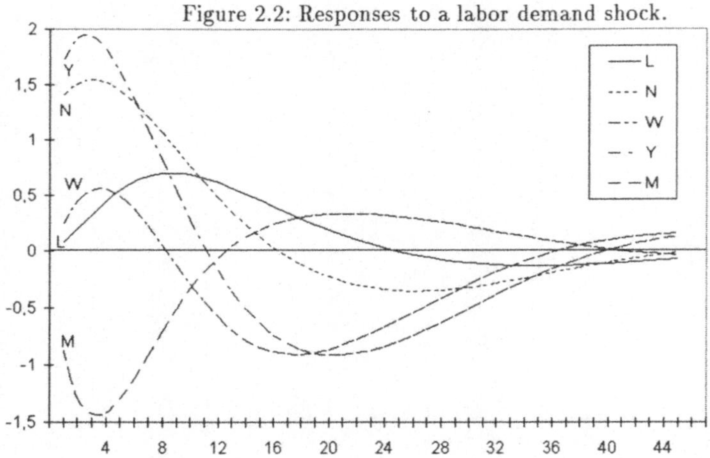

Figure 2.3: Responses to a real wage shock.

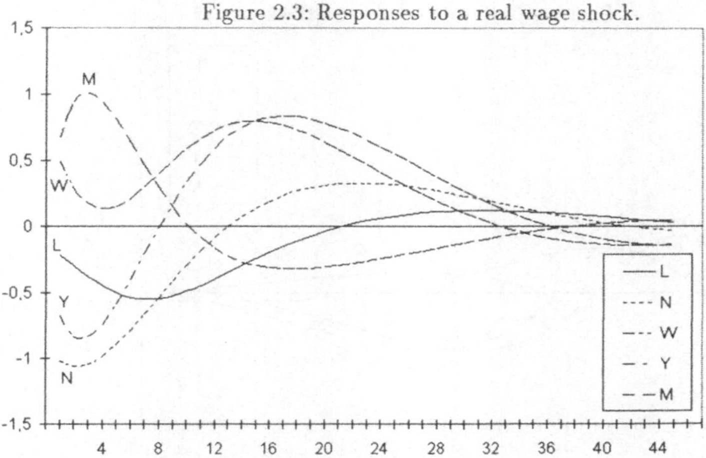

Figure 2.4: Responses to a real GNP shock.

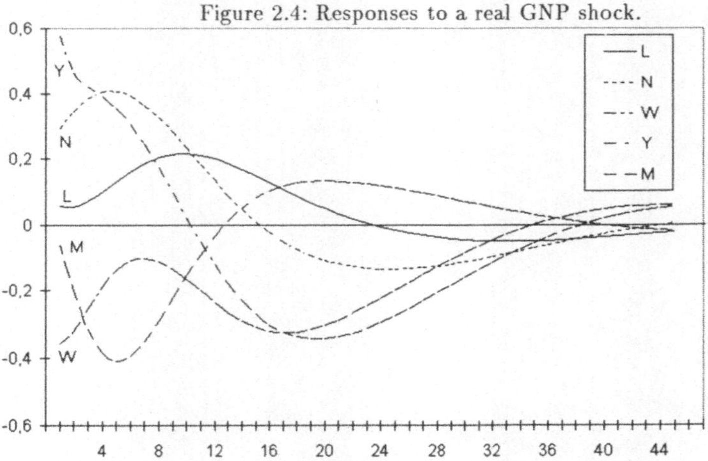

Figure 2.5: Responses to an M2 shock

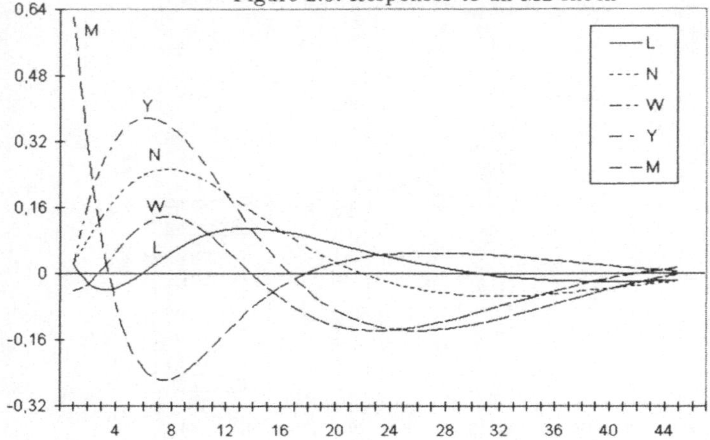

Modeling Cointegrated Processes by a Vector-Valued State Space Algorithm - Evidence on The Impact of Japanese Stock Prices on The Finnish Derivatives Market[1]

Ralf Östermark
Department of Business Administration
Åbo Akademi University
Henriksgatan 7, FIN-20500 Åbo, Finland

Abstract

In the present study we provide new evidence on the impact of the Japanese stock market on the Finnish derivatives. Initially, we test for the presence of unit roots among the observed input-output processes. Next, cointegration in the system is explicitly tested, to justify the estimated vector-valued state space model. The trend and cyclical components of the endogenous vector are extracted and analyzed. The content of the cyclical component is analyzed by spectral analysis. Finally, the error processes are subjected to some statistical tests and used as signal variables in nonlinear model building.

1 Introduction

The objective of the present paper is to provide new evidence on the dynamic relationships between the Japanese and Finnish financial economies. The endogenous vector consists of the Finnish daily FOX and Futures market price indexes, and the exogenous vector of the daily price and volume indexes of Tokyo stock exchange.

Methodologically, the present study is organized as follows:

(1) Testing for unit roots, cointegration and causality. These aspects are critical in modeling vector time series. Previous evidence (cf., *e.g.*, Östermark and Aoki (1992), Malkamäki (1993)) suggests the presence of unit roots in the Finnish financial economy.

---

[1]Acknowledgement: I wish to thank Dr. Esko Torsti (Economist at The Research Institute of the Finnish Economy) for fruitful discussions on cointegration. I am grateful to Research Assistant, Lic.Sc. (Econ.) Jaana Aaltonen for discussions and her invaluable help in monitoring the computer systems. I am also indebted to my friend and colleague, M.A. Rune Höglund for good advice in statistical testing. I thank The Academy of Finland for its financial support.

The Schur, eigenvalue, and singular value decompositions were carried out by LAPACK- and EISPACK-code. The Unit Root and Cointegration tests were computed by the SHAZAM-package.

(2) Estimating a state space model with exogenous variables and separating the trend/cycle dynamics of the endogenous vector.

(3) Analyzing the spectral dynamics of the state space model by frequency domain methodology. The idea of utilizing spectral information in state space modeling was presented by Ng and Young (1990).

(4) Tests for linearity of state space residuals.

Step (1) constitutes a statistical justification for the estimation effort in step (2). Step (3), again, provides a basis for ex post evaluation of the trend and cyclical components isolated by rotating the state vector. Thus, we attempt to relate the system theoretic state space model of Aoki (1987) empirically to some central econometric model specification techniques. (cf. Diebold (1989), where such an effort is explicitly called for). In step (4) the estimated state space residuals are subjected to tests for linearity within the framework recently suggested by Granger and Teräsvirta (1993).

Recent advances in state space theory provide powerful tools for modeling vector-valued time series processes, both in cases where exogenous inputs are present and in cases where only the endogenous vector is available. Several recent studies reveal an increasing interest in state space modeling of economic time series (cf., *e.g.* Aoki (1976ab, 1987, 1988ab, 1989ab, 1990, 1991), Mittnik (1990), Otter (1985), Otter and van Dal (1987, 1989), Cerchi and Havenner (1988), Bordignon and Trivellato (1989), and Östermark and Aoki (1992)). The approach followed in these papers and also in the present study yields an internally balanced state space representation, in which any lower-dimensional approximation can be obtained without re-estimation by eliminating the corresponding elements of the system matrices. If the original model is stable, then so will be any lower-dimensional model.

When forecasting economic time series, some exogenous variables are often included to improve forecasting precision. Furthermore, there is an interest in evaluating the dynamic effects of changes in exogenous variables on observed data. Aoki (1989b,1991) proposed various alternative techniques for incorporating exogenous variables in vector-valued state space models. A particularly interesting approach is to use the state vector as an instrumental variable for the exogenous process (Aoki (1991)).

Decomposition of economic and business time series into separate components such as trend, cyclic and seasonal factors is necessary for determining seasonally adjusted series (cf. Sastri (1989)). The ARIMA model-based approaches seem to be by far the most popular seasonal adjustment procedures. For example, an ARIMA-based procedure proposed by Hillmer and Tiao (1982) is based on decomposing a given time series into three mutually independent seasonal, trend-cycle and noise components. These components are assumed to follow some ARIMA models, under certain restrictions which assume that the autoregressive polynomials of the three components may be uniquely determined. The final results are linear filters for the seasonal and trend-cycle components. Alternative signal extraction methods based on the Kalman filtering technique have been proposed by, *e.g.*, Athans (1974),

Mehra (1978), and Sastri (1985). Since the Kalman filter is recursive, it can be used to revise model parameters in real time, given the availability of an appropriate state-space model for the dynamic system. Sastri (1989) demonstrated that the Kalman filter may be employed as a unified time series decomposition, forecasting and seasonal-adjustment algorithm. Ng and Young (1990) presented a recursive state space approach to modeling non-stationary time series. The basic time series model was represented in structural form, to separate the trend and seasonal components.

2 Unit Roots and Cointegration

Stationarity and cointegration are critical issues in modeling financial time series. Malkamäki (1993) applied an augmented Dickey-Fuller (ADF)-test to examine the stationarity of Finnish, Swedish, British, German and U.S. stock prices and returns in local currencies (cf. Bos *et al.* (1995)). Among other things, he demonstrated that, in the time interval 1974-1989, the Finnish monthly stock market return is stationary after first-order differencing. In the present study we will apply the ADF-test to both the input and output processes of the state space model.

The augmented Dickey-Fuller test is based on the following regression model (cf. Dickey and Fuller (1979), SHAZAM (1993)):

$$\nabla y_t = \alpha_0 + \phi\, y_{t-1} + \sum_{i=1}^{n} \beta_i \nabla y_{t-i} + \lambda t + e_t, \tag{2.1}$$

where ∇ is the difference operator
t is the trend component.

Under the null hypothesis, y_t is first-order integrated $I(1)$, allowing for standard distributional tests of the parameter estimates. If H_0: $\lambda = 0$ is rejected then y_t is $I(0)$. This conclusion follows from the fact that an $I(1)$-process cannot explain an $I(0)$. The critical values for the t-statistic for ϕ are tabulated in Fuller (1976). Phillips and Perron (1988) suggest some non-parametric corrections to the conventional test statistics which eliminate the nuisance parameter depencies asymptotically. The Dickey-Fuller critical values can still be directly used.

Given the presence of $I(1)$-processes in both the signal and response variables, the next step is to formulate and test hypotheses concerning possible cointegration in the data (cf. Johansen (1988)). Consider the VAR-model for the data:

$$y_t = \phi_1 y_{t-1} + \phi_2 y_{t-2} + \ldots + \phi_p y_{t-p} + \varepsilon_t,\ t = 1,\ldots,T \tag{2.2}$$

where $y_t \in \mathbb{R}^{k\times l}$.

We can transform y_t to error correction form by adding the zero-valued vector expression

$$(I, I, \ldots, I)\begin{bmatrix} I-I & I-I & I-I & \cdots \\ 0 & \phi_1-\phi_1 & \phi_1-\phi_1 & \cdots \\ 0 & 0 & \phi_2-\phi_2 & \cdots \\ 0 & 0 & 0 & \cdots \\ \vdots & \vdots & \vdots & \ddots \end{bmatrix}\begin{pmatrix} y_{t-1} \\ y_{t-2} \\ y_{t-3} \\ \vdots \end{pmatrix}$$

to the right-hand side of (2.2), giving

$$\nabla y_t = \Gamma_1 \nabla y_{t-1} + \ldots + \Gamma_{p-1} \nabla y_{t-(p-1)} + \Pi y_{t-p} + \varepsilon_t,\ t = 1,\ldots,T \tag{2.3}$$

where $\Gamma_i = -(I - \phi_1 - \ldots - \phi_i),\ i = 1,\ldots, p-1$
$\Pi = -(I - \phi_1 - \ldots - \phi_p)$.

Under the null hypothesis of no cointegration, the rank of Π, *i.e.*, $r(\Pi)$ is zero. If $r(\Pi) = c,\ 0 < c < p$, then c linear cointegrating vectors exist. A rough estimate of $r(\Pi)$ can be obtained by singular value decomposition of Π, *i.e.*,

$$\Pi = USV^T, \tag{2.4}$$

where $U, V \in \mathbb{R}^{k\times c},\ S \in \mathbb{R}^{c\times c}$
$S \geq 0,\ U^TU = I,\ V^TV = I,\ U^TV = 0$
and S is diagonal with the singular values of Π, s_{ii}, arranged in descending order of magnitude.

The rank of Π equals the number of nonzero singular values. If the system is cointegrated, then the State Space-model is expected to produce stationary white noise residuals and standard techniques of statistical inference will apply. A statistical test of $r(\Pi)$ is described in Johansen and Juselius (1990). Their procedure is applied in the empirical part below. The Johansen-Juselius procedure decomposes matrix Π of (2.3) into $\Pi = \alpha\beta^T$, where α and β are $p\times r$ matrices ($r \leq p$). The vector $\beta^T y_t$ represents the long-run equilibrium for the system.

Even if cointegration tests are usually based on the assumption that all data series are $I(1)$, stationary ($I(0)$) processes can be included in the data vector as well (Johansen and Juselius (1990)). Therefore, the implication that the Tokyo market price is stationary during the study period (cf. Table 5.1 below) does not preclude the possibility of cointegration between the Finnish and Japanese economies.

Given that two $I(1)$-processes y_t and x_t are cointegrated, they can be written in the following error correction form (cf. Engle and Granger (1987), Malliaris and Urrutia (1992)).

$$\nabla y_t = \alpha_0 + \sum_{i=1}^{p} \beta_i \nabla y_{t-i} + \sum_{j=0}^{m} \gamma_j \nabla x_{t-j} + \delta \hat{e}_{t-1} + \varepsilon_t \tag{2.5a}$$

where the $I(0)$-error correction term $\hat{e}_t$ is obtained from the cointegrating regression

$$y_t = \alpha_0 + \beta x_t + e_t. \tag{2.5b}$$

Equation (2.5a) provides information both on the short run and long run dynamic linkages between x_t and y_t through the coefficients on ∇x_{t-j} and $\hat{e}_{t-1}$ respectively. The joint significance of m lagged values of k exogenous variables can be tested by means of the following F-statistic (cf. Granger (1969), Malkamäki *et al.* (1993), Martikainen and Puttonen (1992)):

$$F = \frac{(ESS_R - ESS_U)/(p+1)}{ESS_U/(N-(p+km+k+1))} \sim F[p+1, N-(p+km+k+1)], \tag{2.6}$$

where ESS_R, ESS_U are the error sum of squares for the restricted and unrestricted models respectively.
N = number of observations
$km + k$ = number of restrictions.

Simulation results reported by Nelson and Schwert (1980) suggest a strong preference for the Granger (1969) and Sims (1972) tests over a test proposed by Haugh (1976) and Pierce (1977). On the other hand, Geweke *et al.* (1983) indicate a preference for variants of either the Granger test or the Sims test.

Nelson and Schwert (1982) observed a serious loss of power in the tests when they are based on unnecessarily heavily parameterized models. Due to this indeterminacy, we decided also to conduct a completely different causality test based on structural relations as follows:

$$\eta = \Gamma\xi + \zeta,$$

where (2.7)

$\Gamma \in \mathbb{R}^{m\times n}$, $\eta \in \mathbb{R}^{m\times 1}$, $\xi \in \mathbb{R}^{n\times 1}$

$\zeta = (\zeta_1, \ldots, \zeta_m)$ is a random vector of residuals

such that $\text{cov}(\zeta, \xi) = 0$.

The unobserved vectors η and ξ are connected to observed vectors y and x respectively by the measurement (factor analysis) models

$$\mathbb{R}^{p\times 1} \ni y = \Lambda_y \eta + \varepsilon \quad \text{and} \tag{2.8}$$

$$\mathbb{R}^{q\times 1} \ni x = \Lambda_x \xi + \delta \tag{2.9}$$

such that $\text{cov}(\varepsilon,\eta) = 0$, $\text{cov}(\delta,\xi) = 0$ and ζ, ε and δ are mutually uncorrelated. Hence,

$$y = \Lambda_y(\Gamma\xi + \zeta) + \varepsilon. \tag{2.10}$$

Assuming that $\Lambda_x^T\Lambda_x$ is invertible, we may premultiply eq. (2.9) by Λ_x^T to extract ξ:

$$\begin{aligned} \Lambda_x^T x &= \Lambda_x^T\Lambda_x\xi + \Lambda_x^T\delta \\ \Rightarrow \xi &= (\Lambda_x^T\Lambda_x)^{-1}\Lambda_x^T(x - \delta) \end{aligned} \tag{2.11}$$

Substituting in (2.10) gives

$$y = \Lambda_y(\Gamma(\Lambda_x^T\Lambda_x)^{-1}\Lambda_x^T(x - \delta) + \zeta) + \varepsilon. \tag{2.12}$$

The parameters of (2.12) can be estimated by the LInear Structural RELation (LISREL) system designed by Jöreskog (1967).

In the present study the endogenous vector is composed of the FOX and Futures indexes, whereas the exogenous vector consists of the Japanese price and volume data. If the Japanese and Finnish markets are cointegrated, we attempt to capture the relationship through state space modeling.

The stationarity of the residuals ($\hat{\varepsilon}_t$) of the state space-model may be tested in the Dickey-Fuller regression

$$\nabla\hat{\varepsilon}_t = \alpha + \phi\hat{\varepsilon}_{t-1} + v_t. \tag{2.13}$$

Under the null hypothesis $\hat{\varepsilon}_t$ is $I(1)$ and, hence, $\nabla\hat{\varepsilon}_t \sim I(0)$. H_0 is rejected if $\phi \neq 0$, indicating that $\hat{\varepsilon}_t$ is $I(0)$.

3 A State Space Model with Exogenous Variables

In the present section we first present the key notation and the basic steps of the algorithm. We then discuss state rotation and a decomposition of the observed vector process into slow and fast dynamic modes (cf. Aoki (1988b), Östermark and Aoki (1992)).

3.1 State Vector as Instrumental Variables

The exogenous variables augmented state space model (SPMX) in innovation representation is

$$\begin{cases} y_t = Cz_t + \theta x_t + e_t \\ z_{t+1} = Az_t + Dx_t + Be_t, \end{cases} \tag{3.1}$$

where $z_t \in \mathbb{R}^{n\times 1}$, $x \in \mathbb{R}^{m\times 1}$, $\{y,e\} \in \mathbb{R}^{k\times 1}$ and all matrices have conformable dimensions.

In (3.1), y_t is the observed endogenous vector process, x_t the exogenous vector process and z_t the state vector. We now define the finite-dimensional stacked vectors y_t^+, y_t^-, y_{t-1}^-, x_{t-1}^- truncated at some finite points in time, *i.e.*,

$$y_t^+ = \begin{bmatrix} y_{1t} & y_{2t} & \cdots & y_{kt} & | & y_{1,t+1} & y_{2,t+1} & \cdots & y_{k,t+1} & | & \cdots & | & y_{1,t+J} & y_{2,t+J} & \cdots & y_{k,t+J} \end{bmatrix}^T$$

$$y_t^- = \begin{bmatrix} y_{1t} & y_{2t} & \cdots & y_{kt} & | & y_{1,t-1} & y_{2,t-1} & \cdots & y_{k,t-1} & | & \cdots & | & y_{1,t-J} & y_{2,t-J} & \cdots & y_{k,t-J} \end{bmatrix}^T$$

$$y_{t-1}^- = \begin{bmatrix} y_{1,t-1} & y_{2,t-1} & \cdots & y_{k,t-1} & | & y_{1,t-2} & y_{2,t-2} & \cdots & y_{k,t-2} & | & \cdots & | & y_{1,t-K} & y_{2,t-K} & \cdots & y_{k,t-K} \end{bmatrix}^T$$

$$x_{t-1}^- = \begin{bmatrix} x_{1,t-1} & x_{2,t-1} & \cdots & x_{m,t-1} & | & x_{1,t-2} & x_{2,t-2} & \cdots & x_{m,t-2} & | & \cdots & | & x_{1,t-L} & x_{2,t-L} & \cdots & x_{m,t-L} \end{bmatrix}^T.$$

where $J = K - 1$ for convenience. The estimation procedure for the matrices C, θ, A, D and B consists of six basic steps. We note that Di Ruscio and Ljungquist (1992) presented several alternative formulas for determining the system matrices. An important future empirical research issue is to compare the spectral decomposition due to these alternative formulas.

Step 1: Define

$$\begin{aligned}
\Omega_1 &= E[x_t y_{t-1}^{-T}] \in \mathbb{R}^{m\times Kk} \\
H &= E[y_t^+ y_{t-1}^{-T}] \in \mathbb{R}^{Kk\times Kk} \\
H_x &= E[y_t^+ x_{t-1}^{-T}] \in \mathbb{R}^{Kk\times Lm} \quad (\Gamma_l = E[y_{t+l} x_t^T]) \\
R_- &= E[y_t^- y_t^{-T}] \in \mathbb{R}^{Kk\times Kk} \\
Y &= E[y_{t-1}^- x_{t-1}^{-T}] \in \mathbb{R}^{Kk\times Lm} \\
X &= E[x_t x_{t-1}^{-T}] \in \mathbb{R}^{m\times Lm} \\
X^+ &= X^T (XX^T)^{-1} \in \mathbb{R}^{Lm\times m}.
\end{aligned}$$

Step 2:

$$\mathbb{R}^{Kk\times Kk} \ni Q_2\Omega_2 = (H - H_x X^+ \Omega_1)(I - R_-^{-1} Y X^+ \Omega_1)^{-1}$$

Step 3: Compute the singular value decomposition of $Q_2\Omega_2 = U\Sigma V^T$ and assign

$$\mathbb{R}^{Kk\times n} \ni Q_2 = U\Sigma^{1/2}, \quad \mathbb{R}^{n\times Kk} \ni \Omega_2 = \Sigma^{1/2} V^T.$$

Step 4: Compute $\mathbb{R}^{Kk\times m} \ni Q_1 = (H_x - Q_2\Omega_2 R_-^{-1} Y) X^T (XX^T)^{-1}$

Step 5: Compute C and θ,

$$\mathbb{R}^{k\times m} \ni \theta = [I \; 0 \cdots 0] Q_1$$
$$\mathbb{R}^{k\times n} \ni C = [I \; 0 \cdots 0] Q_2.$$

Step 6: Compute A, D and B. For this, we

define the instrument $\hat{z}_t = \Omega_2 R_-^{-1} y_{t-1}^- \in \mathbb{R}^r$ and (3.2)
estimate the residual process $\hat{e}_t = y_t - C\hat{z}_t - \theta x_t$.

In step 6, the matrices of the process equation are obtained by solving a set of linear equations, derived as follows. Insert the instrument $\hat{z}_t$ defined in (3.2) in the process equation:

$$\hat{z}_{t+1} = A\hat{z}_t + Dx_t + Be_t. \tag{3.3}$$

Postmultiply by $\hat{z}_t^T$ and take expectations:

$$E[\hat{z}_{t+1}\hat{z}_t^T] = = AE[\hat{z}_t\hat{z}_t^T] + DE[x_t\hat{z}_t^T]. \tag{3.4}$$

Next, postmultiply (3.3) by x_t and take expectations:

$$\begin{aligned} E[\hat{z}_{t+1}x_t^T] &= AE[\hat{z}_t x_t^T] + DE[x_t x_t^T] + BE[e_t x_t^T] \\ &= AE[\hat{z}_t x_t^T] + DE[x_t x_t^T] \end{aligned}$$

$$\Rightarrow [A, D] = \left[E[\hat{z}_{t+1}\hat{z}_t^T], E[\hat{z}_{t+1}x_t^T]\right] \begin{bmatrix} E[\hat{z}_t\hat{z}_t^T] & E[\hat{z}_t x_t^T] \\ E[x_t\hat{z}_t^T] & E[x_t x_t^T] \end{bmatrix}^{-1}. \tag{3.5}$$

Finally, matrix B is solved from the process equations as shown in Östermark and Aoki (1992):

$$B = E[(\hat{z}_{t+1} - A\hat{z}_t - Dx_t)\hat{e}_t^T]E[\hat{e}_t\hat{e}_t^T]^{-1}. \tag{3.6}$$

3.2 State Rotation and Residual Filtering

Aoki (1987) proposed three analytical decompositions of the time series into trends and cycles, one based on spectral decomposition and two based on the Schur decomposition of the transition matrix A in the process equation. In Östermark and Aoki (1992), we extend the results in Aoki (1988b) to the exogenous variables augmented state space model, approximating state rotation by the numerically robust singular value decomposition.

The singular value decomposition of the state transition matrix is written as follows:

$$A = USV^T = [P_1 \;\; P_2]\begin{bmatrix} \Lambda & \\ & \Gamma \end{bmatrix}\begin{bmatrix} Q_1^T \\ Q_2^T \end{bmatrix}, \tag{3.7}$$

where U, S, and V are decomposed in correspondence to c large and $n\text{-}c$ small singular values of A. In the singular value decomposition, $U^TU = V^TV = I$. If A is symmetric and positive definite, then

$$U^TV = I, \text{ so}$$

$$P_1^TAP_1 = \Lambda \text{ and}$$

$$Q_2^TAQ_2 = \Gamma.$$

The more A deviates from symmetry, the more Q_2^T and P_1 will deviate from orthogonality.

We first consider rotating the standard model given by (3.1) with no exogenous variables, *i.e.*, $\theta = 0$. Change coordinates by defining

$$z_t = P_1\tau_t + Q_2\eta_t \tag{3.8}$$

$$\Rightarrow z_{t+1} = Az_t + Be_t = A(P_1\tau_t + Q_2\eta_t) + Be_t. \tag{3.9}$$

Multiply (3.8) and (3.9) from the left by P_1^T:

$$P_1^Tz_t = P_1^TP_1\tau_t + P_1^TQ_2\eta_t = \tau_t$$

$$\Rightarrow \tau_{t+1} = P_1^TAP_1\tau_t + P_1^TBe_t$$

$$= \Lambda Q_1^TP_1\tau_t + P_1^TBe_t \qquad \text{(trend component)} \tag{3.10}$$

Multiply (3.8) and (3.9) from the left by Q_2^T:

$$Q_2^Tz_t = \eta_t$$

$$\Rightarrow \eta_{t+1} = Q_2^TAQ_2\eta_t + Q_2^TBe_t \text{ (cyclical component).}$$

If A is (close to) symmetric, then the matrices P and Q generated by singular value decomposition produce a decomposition of matrix A analogous to that obtained by eigenvalue analysis (cf. Aoki and Havenner (1991)), with the trend- and cyclical components interconnected only through the error process e_t.

To carry out state rotation of the complete model (3.1) we, once again, use the singular value decomposition of A in (3.7) and change coordinates exactly as in (3.8). Substituting (3.9) into (3.1) then yields

$$z_{t+1} = A(P_1\tau_t + Q_2\eta_t) + Dx_t + Be_t. \tag{3.11}$$

Given (3.7b) we obtain

$$\boxed{\tau_{t+1} = P_1^T A P_1 \tau_t + P_1^T D x_t + P_1^T B e_t.} \quad \text{(trend component).} \tag{3.12}$$

On premultiplying (3.11) by Q_2^T, we get

$$\boxed{\eta_{t+1} = Q_2^T A Q_2 \eta_t + Q_2^T D x_t + Q_2^T B e_t} \quad \text{(cyclical component).} \tag{3.13}$$

Substituting (3.8) in the observation equation $y_t = Cz_t + \theta x_t + e_t$ yields

$$\boxed{y_t = CP_1 \tau_t + CQ_2 \eta_t + \theta x_t + e_t.} \tag{3.14}$$

From (3.12) and (3.13) we now see that the trend and cyclical components are interconnected through the error process e_t and the exogenous process x_t. Substituting the error process of (3.14) in the trend- and cyclical components finally yields

$$\tau_{t+1} = P_1^T(A - BC)P_1 \tau_t + P_1^T(D - B\theta)x_t + P_1^T B y_t - P_1^T BCQ_2 \eta_t \tag{3.15a}$$

and

$$\eta_{t+1} = Q_2^T(A - BC)Q_2 \eta_t + Q_2^T(D - B)x_t + Q_2^T B y_t - Q_2^T BCP_1 \tau_t. \tag{3.15b}$$

Having obtained the rotated representation we may - on occasion - filter the residual process in order to improve the predictions. The complete rotated model with error correction then becomes:

$$\begin{cases} \tau_{t+1} = P_1^T A P_1 \tau_t + P_1^T D x_t + P_1^T B e_t \\ \eta_{t+1} = Q_2^T A Q_2 \eta_t + Q_2^T D x_t + Q_2^T B e_t \\ e_{t+1} = \rho e_t + a_t \\ y_t = CP_1 \tau_t + CQ_2 \eta_t + \theta x_t + e_t . \end{cases} \tag{3.16}$$

4 Spectral Analysis

Spectral Analysis is a stochastic version of the discrete Fourier transform. The purpose is to estimate the spectral density function (spectrum) of one or several time series (cf., *e.g.*, Chatfield (1987), p. 127). The spectrum of a series z_t is defined as the discrete Fourier transform of its autocovariance function, *i.e.*,

$$f_z(\omega) = \sum_{k=-\infty}^{\infty} \gamma_k e^{-ik\omega}, \text{ where} \tag{4.1}$$

$\gamma_k = \operatorname{cov}(z_t, z_{t-k})$, and $\omega \in [0,\pi]$ is the given frequency.

The corresponding empirical quantity is called the periodogram. Analysis in the frequency domain may provide important insights into the dynamic components of the time series. Since the state rotation procedure followed in the present study extracts only the common trend- and cyclical components for the endogenous vector process, complementary spectral information is particularly valuable with series containing multiple cyclical variations. In the present study, we analyze some central auto- and cross-spectral characteristics of the observed processes, as described below (cf., *e.g.*, Höglund and Östermark (1993)):

(i) The autoperiodogram for each input and output series respectively, as well as for the trend- and cyclical components obtained by rotating the state space model. The periodogram is defined as:

$$I_z(\omega_j) = c_0 + 2\sum_{k=1}^{n-1} c_k \cos(k\omega_j), \text{ where}$$
$$c_k = \hat{\gamma}_z(k),\ \omega_j = \frac{j\pi}{F},\ j = 0,1,\ldots,F, \tag{4.2}$$

and F is the spectral computation number.

(ii) The coherence and phase between the dependent variables and each input variable. The coherence is defined as

$$b_{xy}(\omega) = |f_{xy}(\omega)|^2 \Big/ \left[f_x(\omega) f_y(\omega)\right],$$
$$\text{where } |f_{xy}(\omega)|^2 = c_{xy}^2(\omega) + q_{xy}^2(\omega) \text{ is the squared cross-} \tag{4.3}$$

amplitude (gain) between the input- and output series.

The phase is defined as

$$\phi_{xy}(\omega) = \arctan\left[-q_{xy}(\omega)/c_{xy}(\omega)\right]. \tag{4.4}$$

The periodograms of the individual series indicate at which frequencies the contribution to the variance of the series is pronounced. (High loadings at low frequencies are typical of trend components, whereas high frequency loadings indicate the presence of cyclical effects.) The coherence can be interpreted as the squared correlation between the corresponding frequency components (trend or cyclical) of the input-output series. The phase spectrum, again, may reveal information about the lead-lag relationship between the series. If the phase spectrum diagram shows linear trends, the angle of the trend corresponds to the length in time points of the lead (lag). Several different trend angles in the same phase spectrum may indicate multiple lead (lag) relationships between the series.

Various statistical tests and confidence intervals for the above spectral characteristics have been derived in, *e.g.*, Shimoshni (1971), Coates and Diggle (1986). In the present study we are especially interested in the Granger causality between the input and output processes. Therefore, we will test for significance of the coherence at a given frequency by the Koopmans (1974) F-statistic:

$$F = \frac{m\hat{b}_{xy}(\omega)}{1-\hat{b}_{xy}(\omega)} \sim F_{2,2m;\alpha}. \tag{4.5}$$

5 Empirical Findings

Our database consists of daily readings between the time period 900712-910830 of:

(i) The regressor variables: the market price index and trading volume of the Tokyo stock exchange.

(ii) The endogenous variables: The Finnish FOX and Futures price indexes (cf. Östermark (1991ab,1992), Östermark and Hernesniemi (1994)).

5.1 Testing for Unit Roots and Cointegration

The results of the augmented Dickey-Fuller test (2.1) are presented in Table 5.1. The table reveals the presence of $I(1)$-processes in both the signal and response variables. In the study period, the Finnish FOX and Futures as well as the Japanese market volume index are $I(1)$, whereas the Japanese market price index is stationary. However, since both the signal and response variables contain $I(1)$-processes, there may exist a linear combination of these variables that produces a stationary residual.

In order to check this conjecture, we conducted the reduced rank test of Johansen and Juselius (1990), concerning r cointegration vectors in the data. The maximal eigenvalue and trace statistics are presented below (the critical values under the assumption of a possible trend are taken from table A2 in Johansen and Juselius (1990)):

Table 5.1.: Augmented Dickey-Fuller tests of the response variables FOX, Futures and the signal variables Tokyo Price Index and Tokyo Trading Volume.

(1) Differenced variables		
Entering variable	Test (cf. eq. (2.1) with discussion)	
	H_0: $\alpha_0 = \phi = \lambda = 0$ (critical value at α = 10% = 4.03)	H_0: $\phi = \lambda = 0$ (critical value at α = 10% = 5.34)
FOX Index	3.758 ⇒ $I(1)$	5.35 ⇒ $I(0)$
Futures Index	3.254 ⇒ $I(1)$	4.653 ⇒ $I(1)$
Tokyo Price Index	4.574 ⇒ $I(0)$	6.34 ⇒ $I(0)$
Tokyo Trading Volume	2.86 ⇒ $I(1)$	4.28 ⇒ $I(1)$
(2) Twice differenced variables		
Entering variable	Test (cf. eq. (2.1) with discussion)	
	H_0: $\alpha_0 = \phi = \lambda = 0$ (critical value at α = 10% = 4.03)	H_0: $\phi = \lambda = 0$ (critical value at α = 10% = 5.34)
FOX Index	4.063 ⇒ $I(0)$	6.09 ⇒ $I(0)$
Futures Index	5.053 ⇒ $I(0)$	7.578 ⇒ $I(0)$
Tokyo Price Index	5.768 ⇒ $I(0)$	8.598 ⇒ $I(0)$
Tokyo Trading Volume	11.04 ⇒ $I(0)$	16.858 ⇒ $I(0)$

Table 5.2: Reduced rank test of cointegrating vectors in the data

	Trace statistic	Critical value			Maximal eigenvalue	Critical value		
		$\alpha = 10\%$	$\alpha = 5\%$	$\alpha = 1\%$		$\alpha = 10\%$	$\alpha = 5\%$	$\alpha = 1\%$
$r = 0$	82.882	65.956	69.977	77.911	43.141	30.818	33.262	38.858
$r \leq 1$	39.740	45.248	48.419	55.551	20.599	24.917	27.341	32.616
$r \leq 2$	19.141	28.436	31.256	37.291	14.135	18.959	21.279	26.154
$r \leq 3$	5.007	15.583	17.844	21.962	5.007	12.783	14.595	18.782

The Johansen test, when conducted on all four series, implies rejection of $r = 0$ (*i.e.*, an ordinary VAR-system), but acceptance of the hypothesis that $r \leq 1$. The same result emerges irrespective of the assumption of a possible linear trend in the data (cf. Johansen and Juselius (1990)). We therefore conclude that one cointegrating factor is present in the data. The normalized vectors α and β of the reduced rank test are presented in Table 5.3 along with some short run dynamics and residual analysis.

Table 5.3. Short run dynamics and residual analysis of error correction model with $r = 1$.

a) α- and β-vectors

Variable	Eigenvalue: 0.188	
	α	β
FOX	-0.033	1
Futures	-0.025	-0.847
TokyoPrice	-0.041	0.124
TokyoVolume	1.262	-0.164

b) Matrix Π (standard deviations in parenthesis)

-0.033 (0.010)	0.028 (0.009)	-0.004 (0.001)	0.005 (0.002)
-0.025 (0.015)	0.021 (0.013)	-0.003 (0.002)	0.004 (0.003)
-0.041 (0.019)	0.034 (0.016)	-0.005 (0.002)	0.007 (0.003)
1.262 (0.284)	-1.069 (0.241)	0.156 (0.035)	-0.207 (0.047)

c) Variance/correlation matrix for the residuals (diagonal elements are the variance, off-diagonal elements are correlations); d) Univariate residual analysis

c) Variance/correlation matrix					d) Box-Pierce test for residual autocorrelation (χ^2_1)	ARCH(2)-test for homoscedasticity (χ^2_2)	Skewness	Kurtosis	Jarque-Bera test of normality (χ^2_2)
FOX	0.0001				65.526	3.89	0.259	0.995	13.107
Futures	0.7370	0.0002			51.474	1.284	0.298	1.859	31.903
TokyoPrice	0.1437	0.2007	0.0004		80.241	4.787	1.126	7.052	570.76
Tokyo Volume	0.1183	0.1528	0.3375	0.8096	78.623	3.948	-0.166	1.109	13.963
Critical values ($\alpha = 5\%$)					3.8	5.99	-	-	5.99

e) Matrix Γ_1 (t-values in parentheses)				
FOX	0.031 (0.4)	0.275 (4.3)	-0.040 (1.0)	-0.003 (1.3)
Futures	0.146 (1.1)	0.147 (1.5)	-0.068 (1.2)	-0.003 (0.8)
TokyoPrice	-0.052 (0.3)	0.088 (0.8)	-0.014 (0.2)	0 (0.1)
TokyoVolume	-1.614 (0.7)	3.16 (1.9)	0.119 (0.1)	-0.185 (2.9)

f) Constant (t-values in parentheses)
-0.006 (3.3)
-0.005 (1.7)
-0.008 (2.4)
0.214 (4.1)

Table 5.3 reveals that nearly all entries in matrix Π are significant, with those for the Futures series being slightly less pronounced. The univariate residual analysis indicates that the residuals are homoscedastic, yet heavily autocorrelated and nonnormal. The poor fit of the dynamic system (2.3) makes the model inadequate for prediction purposes. (Similar results have also been presented by Johansen and Juselius). Since our purpose here is not to derive an optimal error correction model, we accept the results generated by the Johansen-Juselius (1990) algorithm. Interestingly enough, only Tokyo trading volume seems to have a short term dynamic influence on the Finnish stock market (cf. matrix Γ_1), whereas in the long run equilibrium relationship (matrix Π) both Japanese series affect the Finnish economy.

For the purposes of the present study, the significance of the Japanese information in the cointegrating relationship is a crucial issue. We therefore formulate the following hypotheses:

H_1: The Japanese information is weakly exogenous for the long run relationship. If this holds, then the last two rows of matrix α - corresponding to Tokyo price and volume data - are zero.

H_2: The Japanese information is both weakly exogenous and irrelevant in the cointegrating relationship. If this is true, then the last two rows of matrix β are zero.

Even if tests H_1 and H_2 are nested (H_1 is false $\Rightarrow H_2$ is obsolete a fortiori), we conducted both independently. The results are presented in Table 5.4.

H_1 (and H_2) is clearly rejected, indicating that - during the study period - Japanese information is relevant for the long-run cointegrating relationship on the Finnish market. (A detailed analysis of the specific impact of the price and volume information is left for future research.)

The ADF-test indicated that $\hat{e}_t$ obtained from the cointegrating regression (2.5a) is $I(0)$ for both FOX and Futures. The estimated unrestricted error correction equations are presented in vector form in (5.1), with variables significant at $\alpha = 0.1$ shown as nonzero.

Table 5.4: Tests of weak exogeneity (H_1) and irrelevance (H_2) of the Japanese series in the cointegrating relationship

Test:		H_1	Matrix Π (standard deviations in parentheses)			
χ^2_2-value:		21.36				
p-value:		0.00				
Eigenvalue:		0.083				
	α	β				
FOX	-0.102	1	-0.102 (0.023)	0.077 (0.017)	0.026 (0.006)	0.003 (0.001)
Futures	-0.064	-0.761	-0.064 (0.034)	0.049 (0.026)	0.016 (0.009)	0.002 (0.001)
TokyoPrice	0	-0.252	0	0	0	0
TokyoVolume	0	-0.028	0	0	0	0
Test:		H_2	Matrix Π (standard deviations in parentheses)			
χ^2_2-value:		23.73				
p-value		0.00				
Eigenvalue		0.075				
	α	β				
FOX	-0.014	1	-0.114 (0.034)	0.099 (0.030)	0	0
Futures	-0.044	-0.869	-0.049 (0.051)	0.042 (0.044)	0	0
TokyoPrice	-0.13	0	-0.130 (0.062)	0.113 (0.054)	0	0
TokyoVolume	-0.173	0	-0.173 (0.967)	0.150 (0.840)	0	0

$$\begin{pmatrix}\nabla FOX\\ \nabla Futures\end{pmatrix}_t = \begin{bmatrix}0 & -0.002 & 0 & 0.51\\ 0.08 & 0 & 1.15 & 0\end{bmatrix}\begin{pmatrix}\text{Tokyo PI}\\ \nabla\text{Tokyo VI}\\ \nabla FOX\\ \nabla Futures\end{pmatrix}_t + \begin{bmatrix}0 & 0 & 0 & 0.20\\ -0.13 & 0 & 0 & -0.17\end{bmatrix}\begin{pmatrix}\text{Tokyo PI}\\ \nabla\text{Tokyo VI}\\ \nabla FOX\\ \nabla Futures\end{pmatrix}_{t-1}$$

$$+\begin{bmatrix}0 & 0 & -0.19 & 0.14\\ 0 & 0 & 0.18 & 0\end{bmatrix}\begin{pmatrix}\text{Tokyo PI}\\ \nabla\text{Tokyo VI}\\ \nabla FOX\\ \nabla Futures\end{pmatrix}_{t-2} + \begin{bmatrix}0.10 & -0.002 & 0 & 0\\ -0.14 & 0 & 0.12 & 0\end{bmatrix}\begin{pmatrix}\text{Tokyo PI}\\ \nabla\text{Tokyo VI}\\ \nabla FOX\\ \nabla Futures\end{pmatrix}_{t-3} \tag{5.1}$$

$$+\begin{bmatrix}-0.07 & -0.001 & 0 & 0.06\\ 0.12 & 0 & 0 & -0.09\end{bmatrix}\begin{pmatrix}\text{Tokyo PI}\\ \nabla\text{Tokyo VI}\\ \nabla FOX\\ \nabla Futures\end{pmatrix}_{t-4} + \begin{pmatrix}-0.08\\ -0.09\end{pmatrix}\hat{e}_{t-1} + \varepsilon_t.$$

The *F*-statistics for the error correction models (2.5) are shown in Table 5.5. (the critical value at $\alpha = 0.01$ is $F(5,271) = 3.02$).

Table 5.5: *F*-statistic for Granger causality

FOX_t	$Futures_t$
106.49	90.47

The *F*-statistics indicate - consistently with the Johansen-Juselius test - a strong causality between the Japanese and Finnish financial markets. The FOX index is affected by the Japanese stock market at lags 0, 3 and 4, whereas the Finnish Futures react to Japanese price and volume impulses at lags 0, 1, 3, and 4. We note that the Japanese impulse at lag 2 does not seem to affect the Finnish financial market at all.

The observed Granger causality is confirmed by the LISREL-test reported in Table 5.6. below. Only significant parameters are presented. (The significance levels for the coefficients are indicated in the tables as follows: $* \rightarrow \alpha = 10\%$; $** \rightarrow \alpha = 5\%$; $*** \rightarrow \alpha = 1\%$).

Table 5.6: Coefficients for LISREL-model: Tokyo Price Index, Tokyo Trading Volume ⇒ FOX and Futures indexes (*t*-values are shown in parentheses).

Structural equation model: $\eta = \Gamma\xi + \zeta$				
η	γ_1	γ_2	σ_ζ^2	R^2
FinDeriv	0.83*** (15.68)	-0.20*** (-4.53)	0.37	0.63

Measurement model for *y*: $\Lambda_y\eta + \varepsilon$			
y	λ_{y1}	σ_ε^2	R^2
FOX	1.00*** (24.98)	0.0030 (0.036)	1.00
Futures	0.99*** (23.82)	0.025 (0.30)	0.97

Measurement model for *x*: $\Lambda_x\xi + \delta$				
x	λ_{x1}	λ_{x2}	σ_δ^2	R^2
TokyoPrice	0.99*** (190.55)		0.011 (0.13)	1.00
TokyoVolume		0.77*** (18.70)	0.40*** (3.83)	0.076
$TokyoPrice_{t-1}$	0.99*** (162.92)		0.025 (0.29)	0.97
$TokyoVolume_{t-1}$		0.95*** (22.41)	0.10 (0.90)	0.12

Table 5.7: Covariance matrix of latent variables

	FinDeriv	TokyoPrice	TokyoVolume
FinDeriv	1.00		
TokyoPrice	0.77	1.00	
TokyoVolume	0.07	0.33	1.00

Table 5.8: Error covariances for the pairs of variables, for which errors were allowed to covariate.

Variables	Error Covariance
TokyoPrice ⇔ TokyoVolume	0.043*** (5.42)

To summarize, the results of this section indicate the presence of a significant unit root in both the signal and response variables. There exists Granger causality from the Japanese to the Finnish financial economy. The Johansen and Juselius (1990) test indicate the presence of one cointegrated vector in the data.

5.2 Estimated State Space Models

A fundamental assumption of the state space model is that all entering variables are stationary. Therefore, any $I(1)$ series should be differenced before estimation. However, the above tests demonstrate that the Japanese and Finnish markets are cointegrated and the the former Granger causes the latter. Thus, a stationary linear combination of the Japanese and Finnish variables should exist. We therefore wish to test, whether this cointegrating relationship can be effectively captured by state space modeling. Hence, all variables will be entered in undifferenced form.

The estimated system matrices obtained under alternative state rotations are presented in Table 5.9. The fairly strong symmetry of the transition matrix A (cf. Table 5.9a) implies that state rotation by singular value decomposition should yield an acceptable approximation of the Schur rotation. This conjecture is confirmed by the results in Table 5.9c. The alternative rotation methods yield practically the same decomposition of the series into the trend and cyclical dynamics (cf. Figures 4-5 below). The sensitivity of the SVD-rotation to deviations from symmetry in the transition matrix is left for future research. Tables 5.9e and 5.9f present the error covariance matrix of the state vector, $\Pi = E[z_t z_t^T]$, obtained by two different techniques, an iterative one and a noniterative one as described in Aoki (1990). Preliminary tests - using 1000, 10000 and 100000 iterations - indicated that the iterative solution approaches the symplectic solution in the limit. These results provide encouraging evidence on the feasibility of the state space algorithm for modeling the time series processes under study.

Table 5.9: Coefficient matrices of state space models for the Finnish FOX and Futures indexes regressed on Tokyo price index and trading volume

a) Unrotated System Matrices

$y_t = Cz_t + \theta x_t + e_t$

Dim(A)	C	θ
3	-0.8404 0.2318 -0.1010 -0.8259 0.2257 0.1287	0.0459 0.0254 0.0567 0.0274

$z_{t+1} = Az_t + Dx_t + Be_t$

Dim(A)	A	D	B
3	1.0090 -0.3179 0.0081 0.0527 0.2024 0.0843 0.0105 0.0678 0.9310	-0.0126 0.0123 -0.0048 0.0620 0.0039 0.0008	1.2749 -1.4674 2.3874 -2.1723 -3.5505 3.0412

b) Eigenvalue Analysis of Matrix A

Dim(A)	Eigenvalues	Left Eigenvectors	Right Eigenvectors
3	0.9769 0.9504 0.2151	0.7675 -0.3788 -0.0643 -0.3614 0.2420 0.9912 -0.5295 0.8933 -0.1160	0.9255 0.6041 0.3709 0.1027 0.1312 0.9240 0.3646 0.7860 -0.0930

c) Rotated System Matrices

$y_t = CP_1\tau_t + CQ_2\eta_t + \theta x_t + e_t$

Rotation		CP_1	CQ_2	θ
Schur		-0.7908 0.2399 -0.6942 0.4462	0.2955 0.2619	0.0459 0.0254 0.0567 0.0274
SVD		-0.8411 0.0908 -0.8306 -0.1371	-0.0237 -0.0473	

$\tau_{t+1} = P_1'AP_1\tau_t + P_1'Dx_t + P_1'Be_t$

Rotation	Dim(τ_t)	$P_1'AP_1$	$P_1'D$	$P_1'B$
Schur	2	0.9769 -0.0444 0.0000 0.9505	-0.0108 0.0181 0.0079 0.0014	0.1305 -0.4722 -3.5560 3.1753
SVD	2	1.0116 0.0259 -0.0178 0.9387	-0.0126 0.0116 -0.0031 -0.0079	1.3116 -1.4972 3.2393 -2.7533

$\eta_{t+1} = Q_2'AQ_2\eta_t + Q_2'Dx_t + Q_2'Be_t$

Rotation	Dim(η_t)	$Q_2'AQ_2$	$Q_2'D$	$Q_2'B$
Schur	1	0.2151	-0.0044 0.0605	2.6961 -2.4115
SVD	1	0.1922	-0.0087 0.0624	2.9914 -2.7963

d) Residual Filters

Dim(A)	Residual Component	Unrotated Residuals	Schur Rotated Residuals	Singular Value Rotated Residuals
3	e_1 e_2	0.8272 0.9020 -0.0787 0.0018	1.0423 1.0913 -0.1917 -0.1025	1.5867 1.6299 -0.7163 -0.6156

e) Error covariance matrix (Π) of the state vector obtained as an iterative solution of a Riccati matrix equation in 100000 iterations (cf. Aoki (1990))		
Π		
1.1225	0.0664	0.0053
0.0664	0.7175	-0.1546
0.0053	-0.1546	0.0689

f) Error covariance matrix (Π) of the state vector obtained as a noniterative solution of an algebraic Riccati equation exploiting properties of symplectic matrices (cf. Aoki (1990), p. 79)		
Π		
1.1225	0.0665	0.0053
0.0665	0.6735	-0.1314
0.0053	-0.1314	0.0566

Figures 1 and 2 present the endogenous and exogenous data series along with the estimated state space models and residual processes. The state space model seems to effectively capture the dynamics of the series.

Figure 1a: Filtered unrotated model of the daily FOX index between 900712-910830.

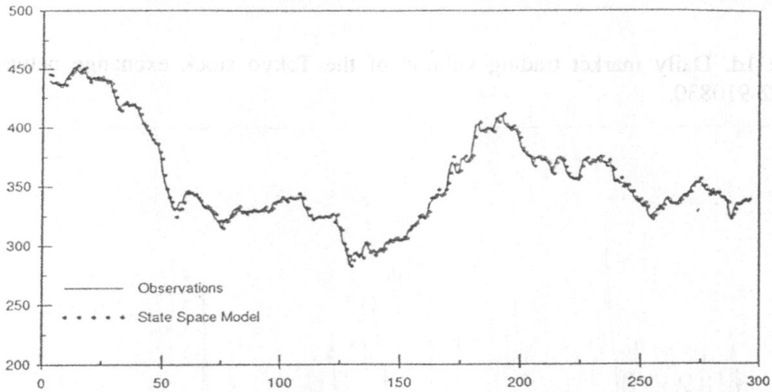

Figure 1b: Filtered unrotated model of the daily Futures index between 900712-910830.

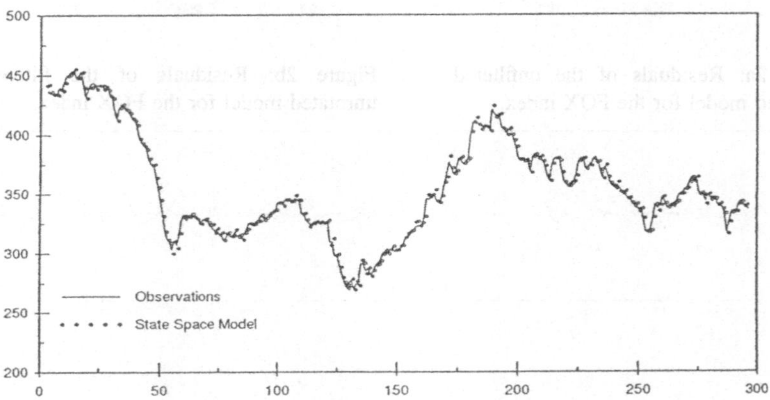

Figure 1c: Daily market price index of the Tokyo stock exchange between 900712-910830.

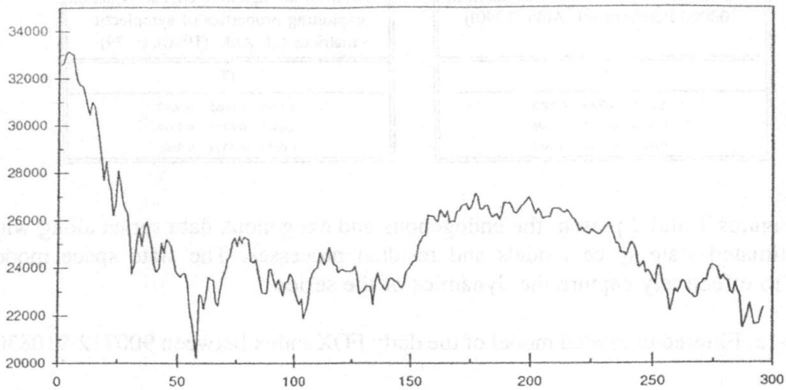

Figure 1d: Daily market trading volume of the Tokyo stock exchange between 900712-910830.

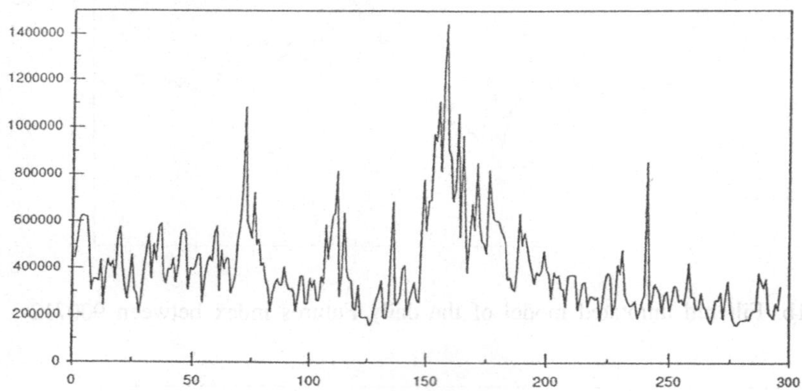

Figure 2a: Residuals of the unfiltered unrotated model for the FOX index.

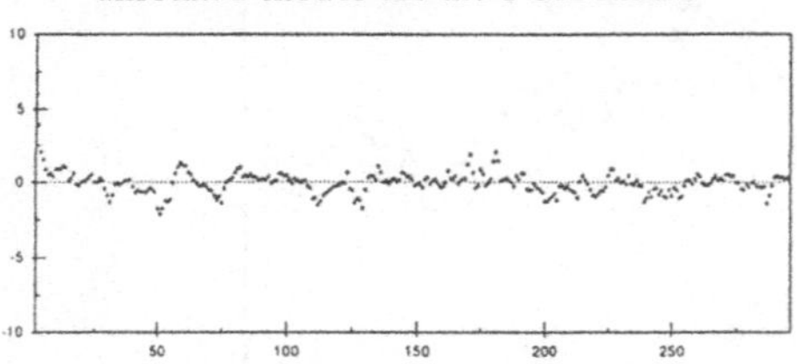

Figure 2b: Residuals of the filtered unrotated model for the FOX index.

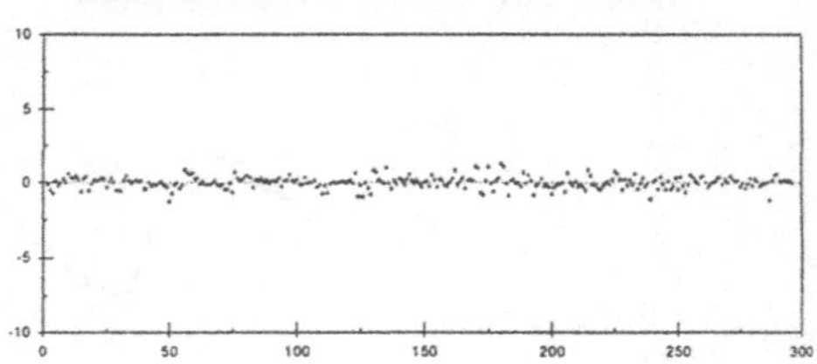

Figure 2c: Residuals of the unfiltered SVD-rotated model for the FOX index.

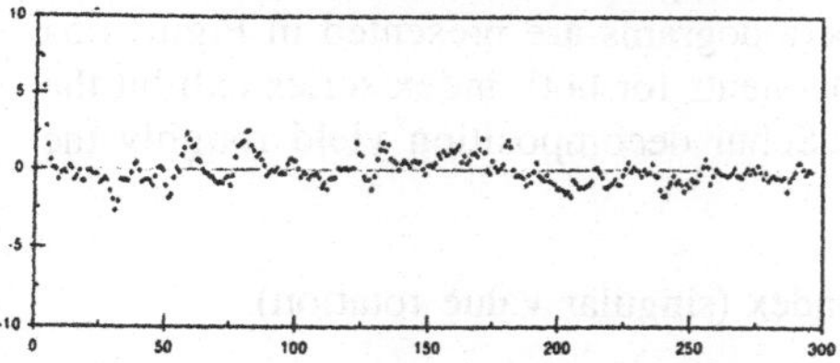

Figure 2d: Residuals of the filtered SVD-rotated model for the FOX index.

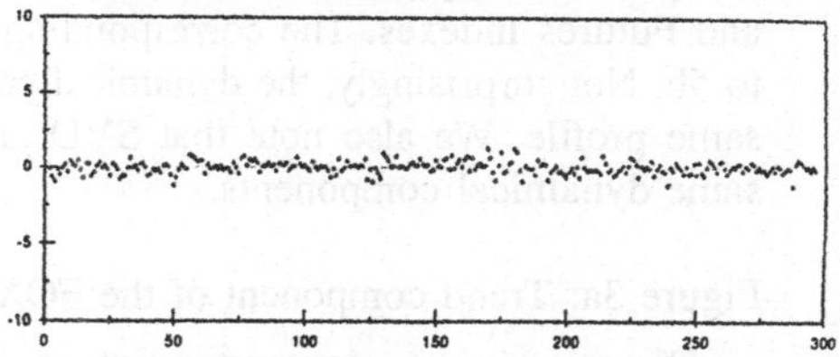

Figure 2e: Residuals of the unfiltered Schur-rotated model for the FOX index.

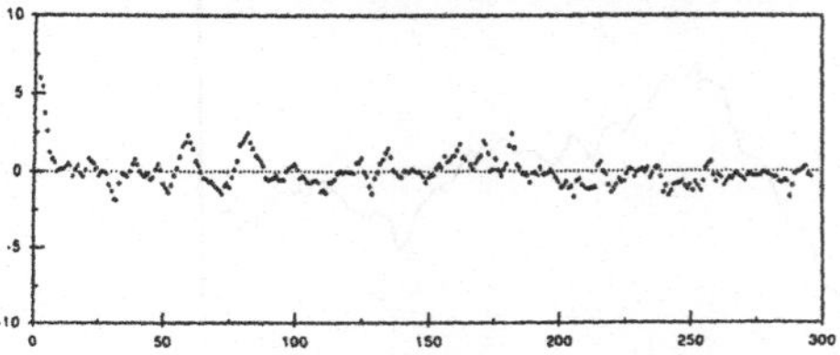

Figure 2f: Residuals of the filtered Schur-rotated model for the FOX index.

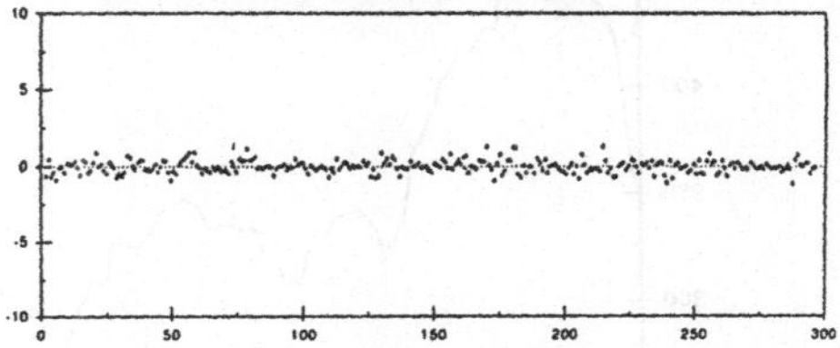

Figure 2g: Residuals of the unfiltered unrotated model for the Futures index.

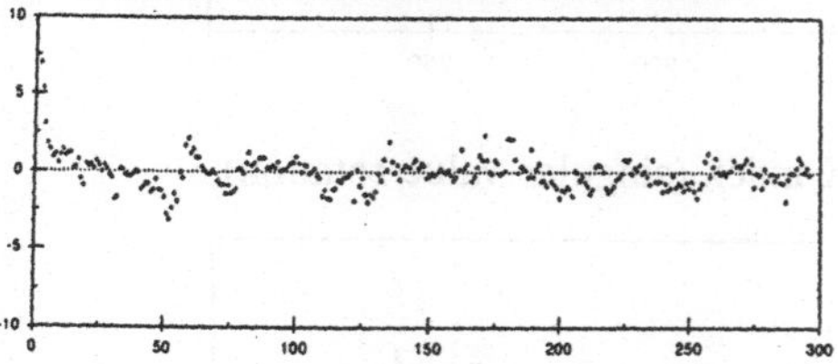

Figure 2h: Residuals of the filtered unrotated model for the Futures index.

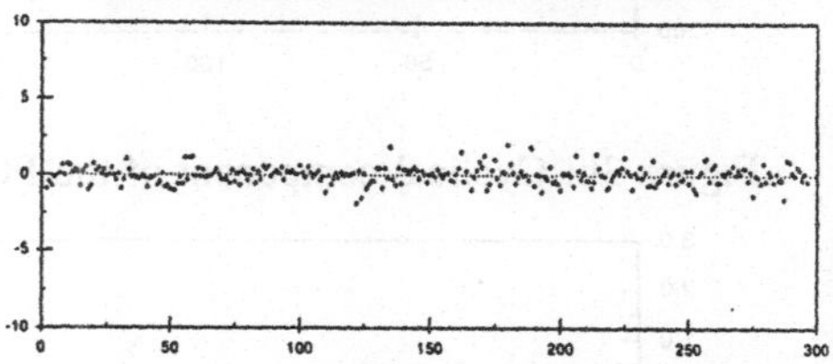

Figure 2i: Residuals of the unfiltered SVD-rotated model for the Futures index.

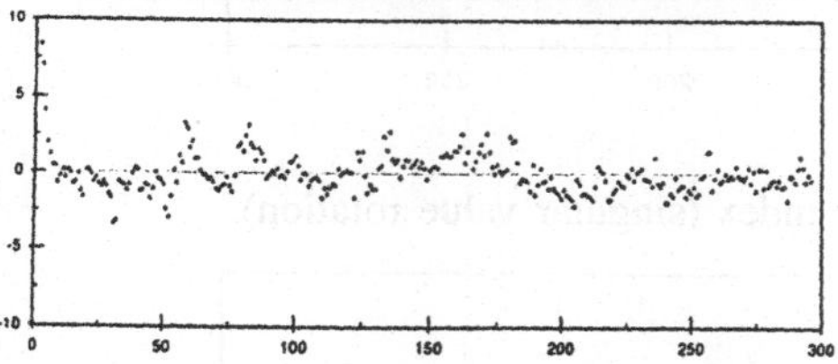

Figure 2j: Residuals of the filtered SVD-rotated model for the Futures index.

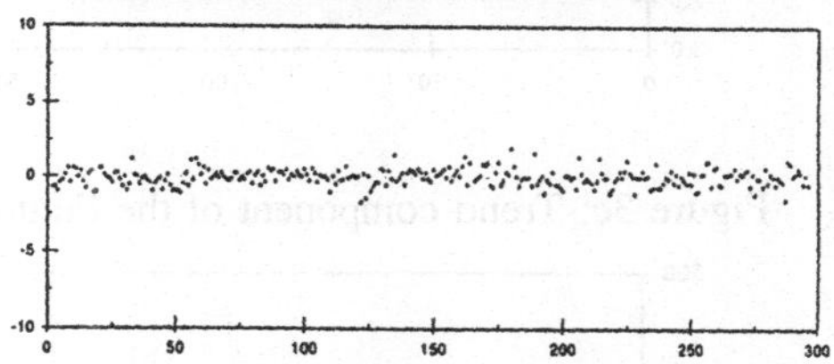

Figure 2k: Residuals of the unfiltered Schur-rotated model for the Futures index.

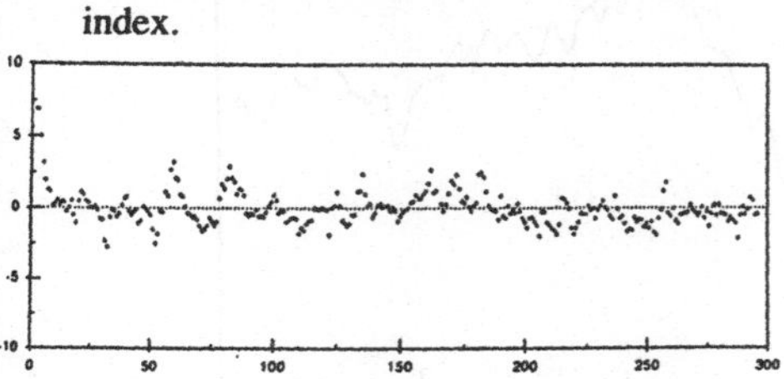

Figure 2l: Residuals of the filtered Schur-rotated model for the Futures index.

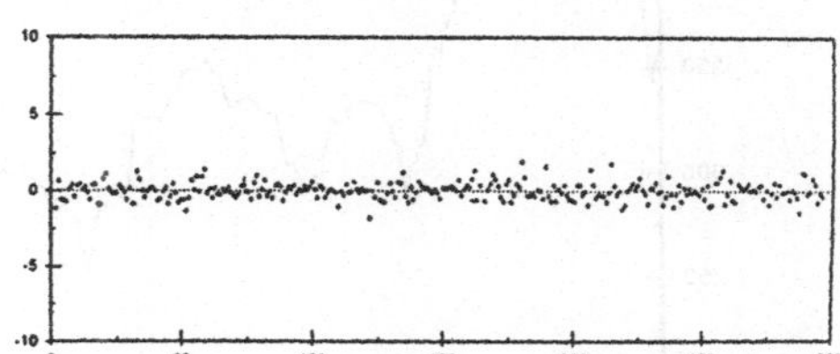

In Figure 3 we present the rotated trend components $CP_1\tau_t$ and the cyclical components $CQ_2\eta_t$ of the three-dimensional state space model for the Finnish FOX and Futures Indexes. The corresponding periodograms are presented in Figures 5a to 5b. Not surprisingly, the dynamical components for both index series exhibit the same profile. We also note that SVD and Schur decomposition yield roughly the same dynamical components.

Figure 3a: Trend component of the FOX index (singular value rotation).

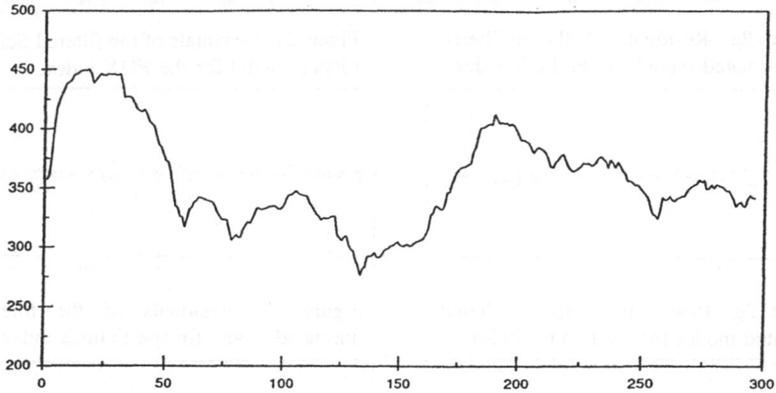

Figure 3b: Cyclical component of the FOX index (singular value rotation).

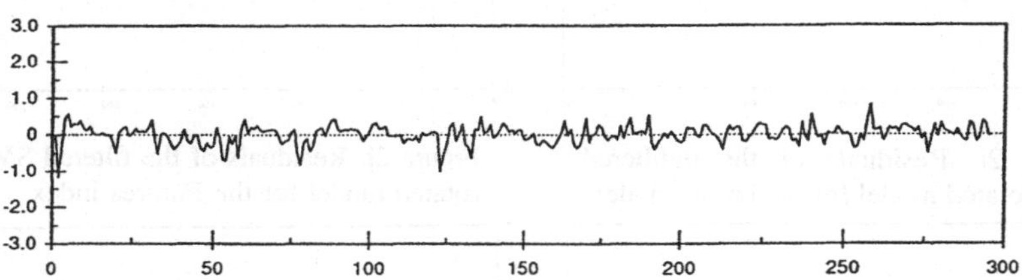

Figure 3c: Trend component of the Futures index (singular value rotation).

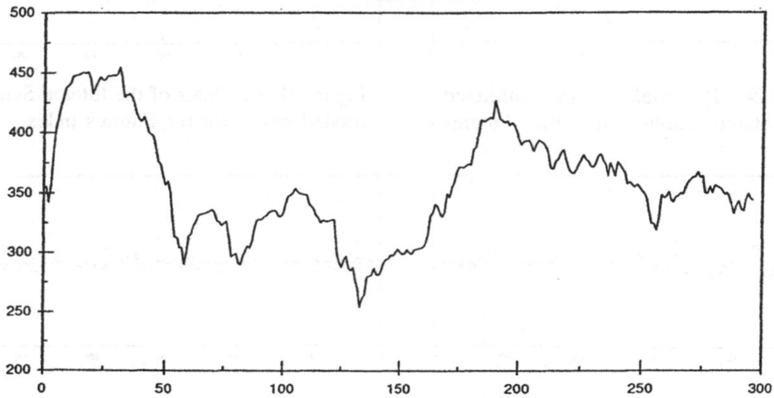

Figure 3d: Cyclical component of the Futures index (singular value rotation).

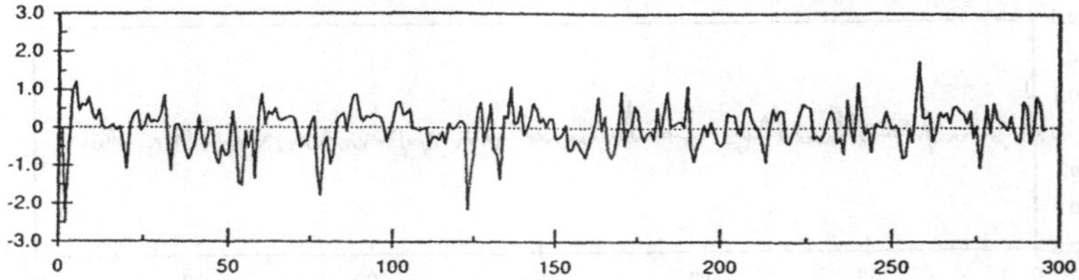

Figure 3e: Trend component of the FOX index (Schur rotation).

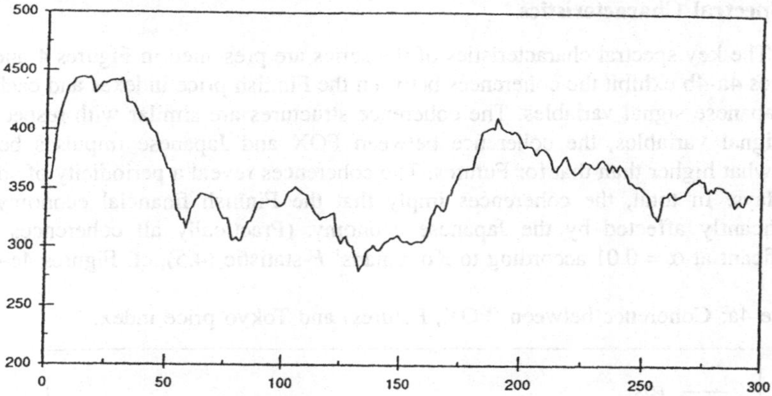

Figure 3f: Cyclical component of the FOX index (Schur rotation).

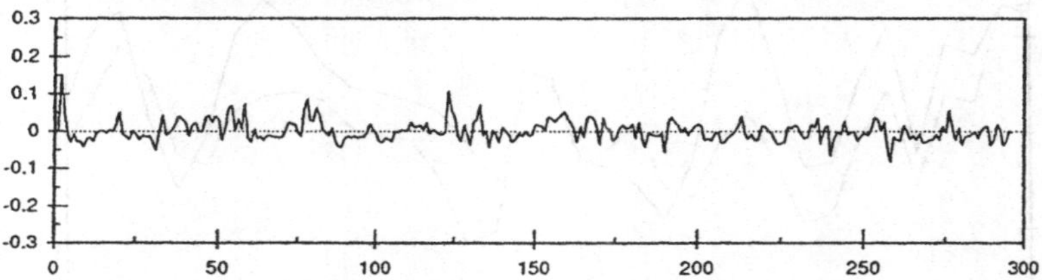

Figure 3g: Trend component of the Futures index (Schur rotation).

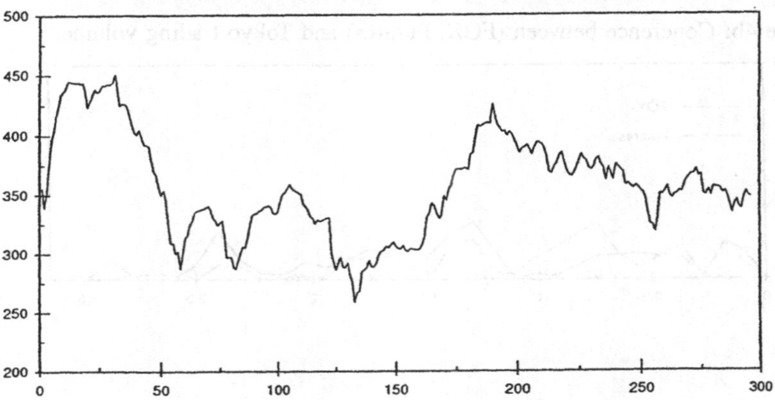

Figure 3h: Cyclical component of the Futures index (Schur rotation).

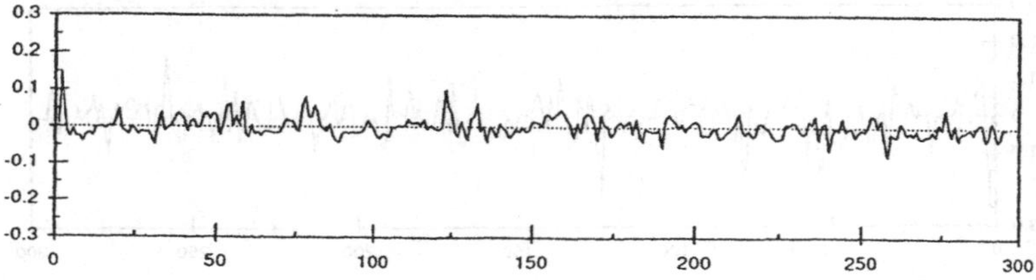

5.3 Spectral Characteristics

The key spectral characteristics of the series are presented in Figures 4 and 5. Figures 4a-4b exhibit the coherences between the Finnish price indexes and each of the Japanese signal variables. The coherence structures are similar with respect to the signal variables, the coherence between FOX and Japanese impulses being somewhat higher than that for Futures. The coherences reveal a periodicity of some 2-3 days. In total, the coherences imply that the Finnish financial economy is significantly affected by the Japanese economy. (Practically all coherences are significant at $\alpha = 0.01$ according to Koopmans' F-statistic (4.5), cf. Figures 4e-4f).

Figure 4a: Coherence between (FOX, Futures) and Tokyo price index.

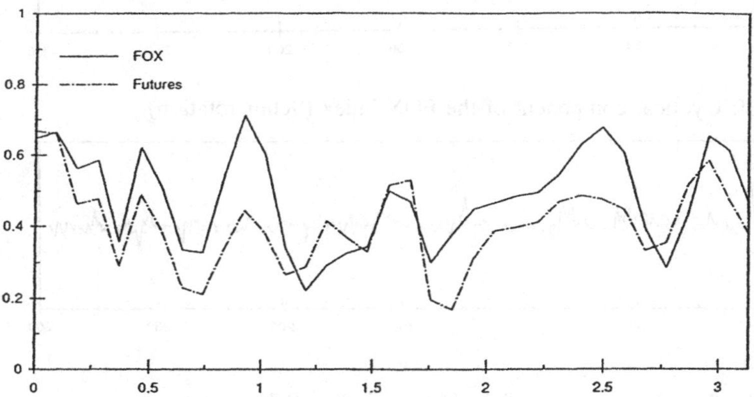

Figure 4b: Coherence between (FOX, Futures) and Tokyo trading volume.

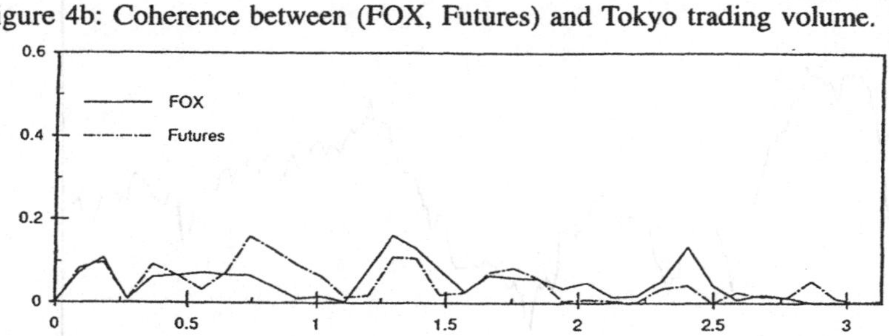

Figure 4c: Phase between (FOX, Futures) and Tokyo price index.

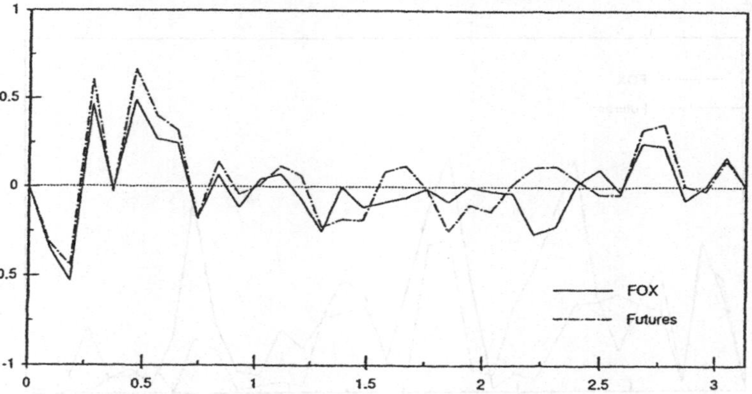

Figure 4d: Phase between (FOX, Futures) and Tokyo trading volume.

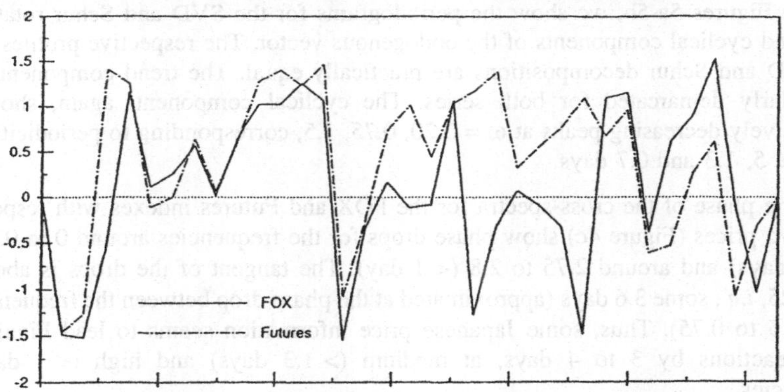

Figure 4e: F-statistics for coherence between (FOX, Futures) and Tokyo price index.

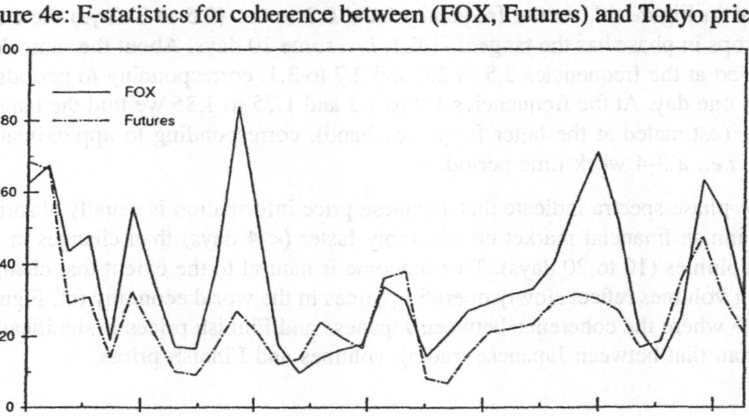

Figure 4f: F-statistics for coherence between (FOX, Futures) and Tokyo trading volume.

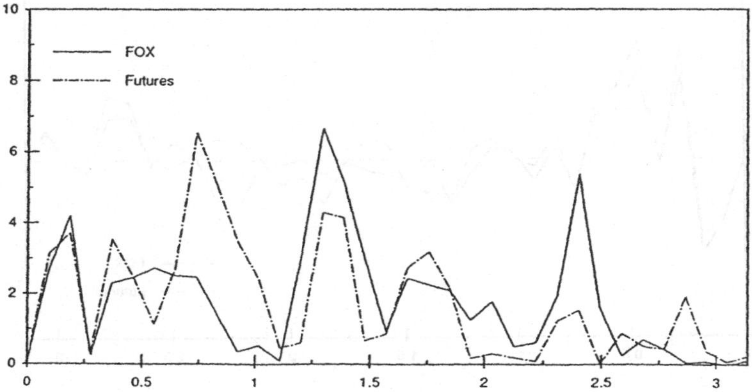

In Figures 5a-5h, we show the periodograms for the SVD and Schur rotated trend and cyclical components of the endogenous vector. The respective profiles of the SVD and Schur decompositions are practically equal. The trend component is particularly demarcated for both series. The cyclical component, again, shows successively decreasing peaks at $\omega = 0.20, 0.75, 1.5$, corresponding to periodicities of some 5, 1.3 and 0.7 days.

The phase of the cross-spectra for the FOX and Futures indexes with respect to Tokyo prices (Figure 4c) show phase drops for the frequencies around 0 to 0.75 (> 1.3 days) and around 2.75 to 2.8 (< 1 day). The tangent of the drops is about -0.9/0.25, *i.e.*, some 3.6 days (approximated at the phase drop between the frequency band 0.5 to 0.75). Thus, some Japanese price information seems to lead Finnish price reactions by 3 to 4 days, at medium (> 1.3 days) and high (< 1 day) frequencies.

The phase spectra for FOX and Futures relative to Tokyo trading volumes are presented in Figure 4d. At the frequency band 0.3 to 0.4 (2.5 to 3 days) the angle of the drops in phase has the tangent -1/0.1, *i.e.*, some 10 days. About the same drop is observed at the frequencies 2.5 to 2.6 and 2.7 to 3.1, corresponding to periods of less than one day. At the frequencies 1.2 to 1.3 and 1.75 to 1.85 we find the tangent -2.25/0.1 (estimated at the latter frequency band), corresponding to approximately 23 days, *i.e.*, a 3-4 week time period.

The phase spectra indicate that Japanese price information is usually absorbed by the Finnish financial market considerably faster (< 4 days) than changes in the trading volumes (10 to 20 days). This outcome is natural to the extent that changes in trading volumes reflect slowly operating forces in the world economy (cf. Figures 4a and 4b where the coherence between Japanese and Finnish prices is significantly higher than that between Japanese trading volumes and Finnish prices.

Figure 5a: Periodogram of SVD-rotated FOX trend component.

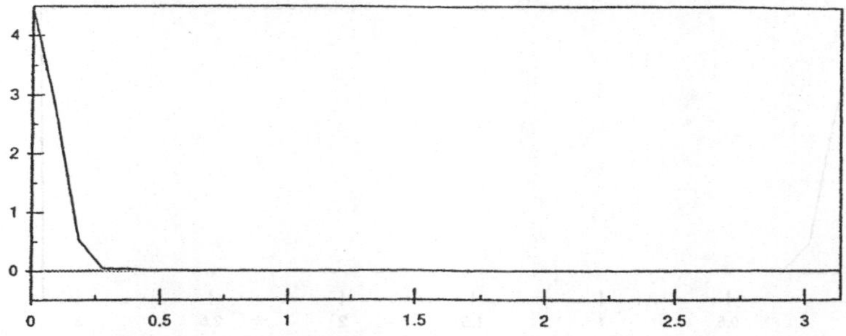

Figure 5b: Periodogram of SVD-rotated FOX cyclical component.

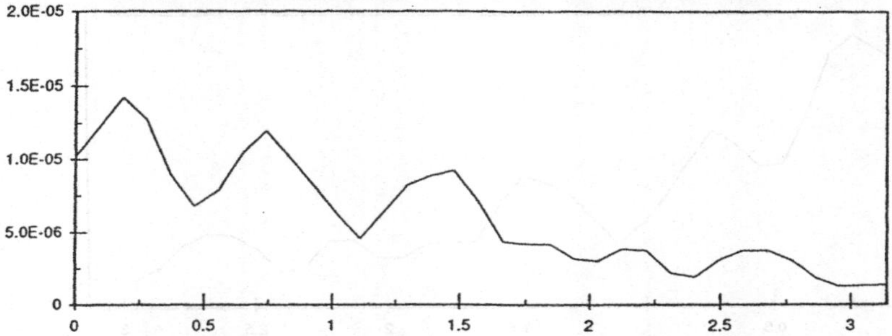

Figure 5c: Periodogram of SVD-rotated Futures trend component.

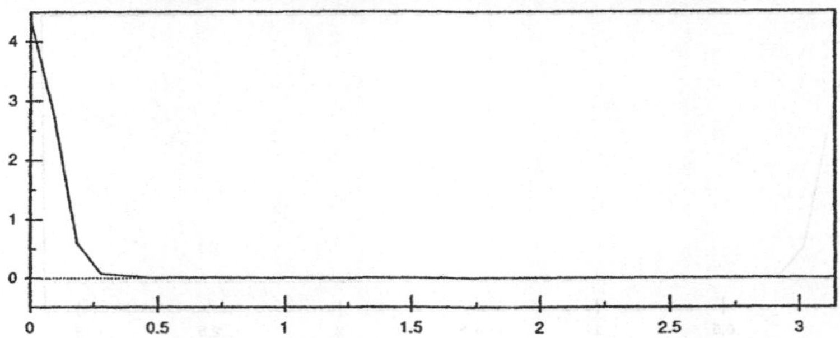

Figure 5d: Periodogram of SVD-rotated Futures cyclical component.

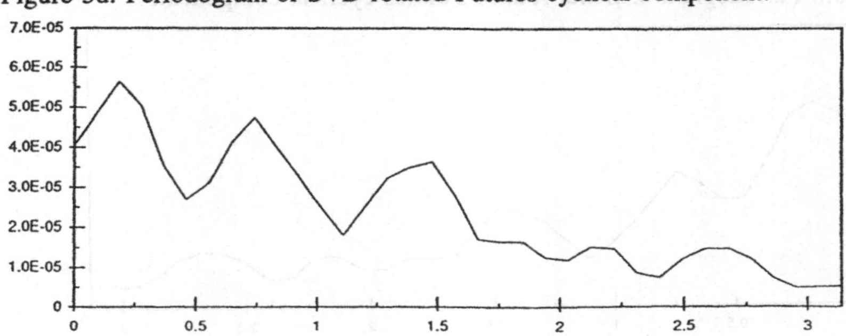

Figure 5e: Periodogram of Sçhur-rotated FOX trend component.

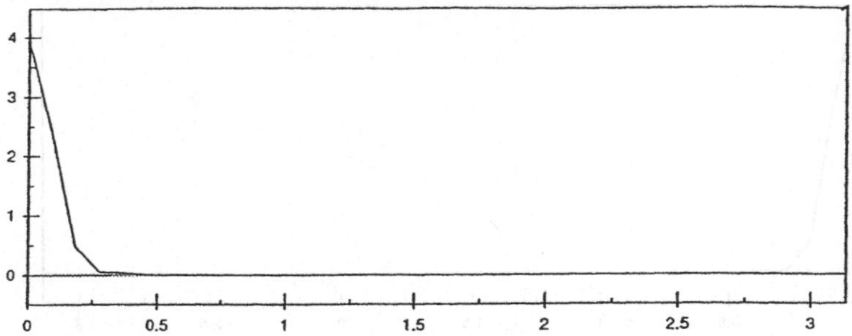

Figure 5f: Periodogram of Schur-rotated FOX cyclical component.

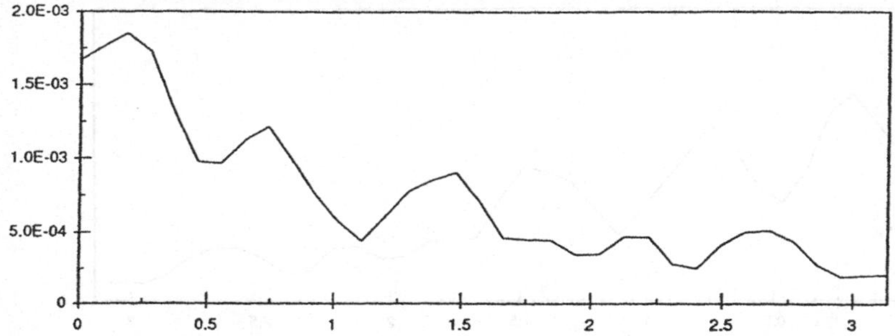

Figure 5g: Periodogram of Schur-rotated Futures trend component.

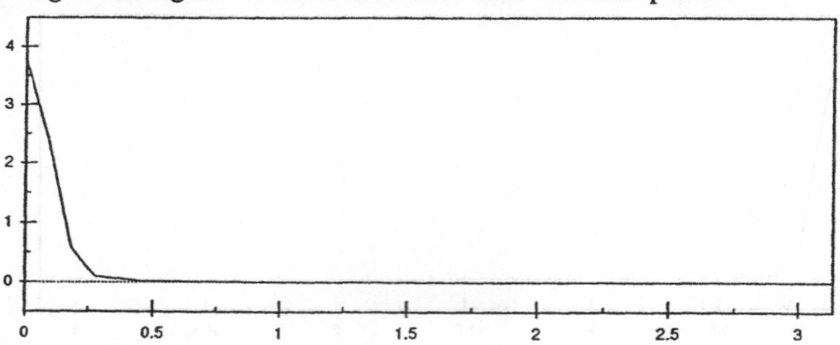

Figure 5h: Periodogram of Schur-rotated Futures cyclical component.

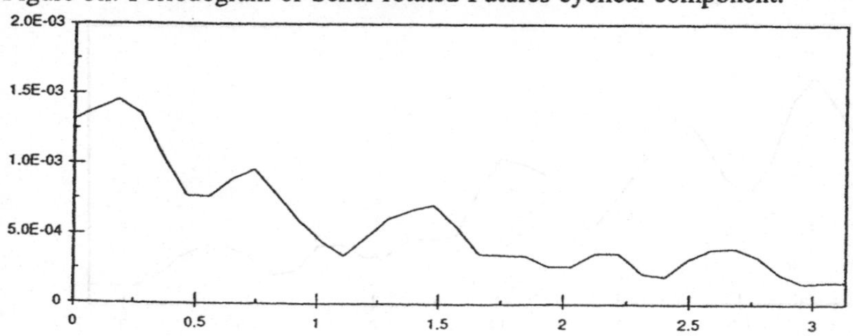

5.4 Analysis of the Error Processes

Table 5.10 presents the mean absolute percentage errors (MAPEs) and the runs test on the residuals of the estimated state space models.

Table 5.10: Summary statistics of state space models for the Finnish FOX and Futures indexes regressed on Tokyo price index and Tokyo trading volume.

Rotation Method	Unfiltered					Filtered				
	FOX Index		Futures Index		Overall MAPE	FOX Index		Futures Index		Overall MAPE
	MAPE	Runs	MAPE	Runs		MAPE	Runs	MAPE	Runs	
Unrotated model	0.0115	-4.6540	0.0160	-3.2310	0.0138	0.0073	-0.9107	0.0130	0.5890	0.0102
Schur rotation	0.0152	-2.8874	0.0207	0.3635	0.0179	0.0081	0.8295	0.0138	0.7652	0.0109
Singular Value rotation	0.0128	-0.2921	0.0183	-0.2921	0.0156	0.0074	0.1899	0.0134	0.3345	0.0106
Dim(*A*) = 3										

Interestingly enough, singular value rotation performs even better than the more complicated Schur decomposition in terms of the above statistics.

The Dickey-Fuller test (2.13) computed on the filtered unrotated residuals indicates stationarity (cf. Table 5.11).

Table 5.11: Dickey-Fuller test of state space residuals

Entering variable	Test (cf. eq. (2.7))	
	H_0: $\alpha_0 = \phi = 0$ (critical value at α = 10% = 3.78)	H_0: $\phi = 0$ (critical value at α = 10% = 2.57)
FOX residual	21.59	-6.56
Futures residual	28.80	-7.59

As indicated by the residuals in Figures 2a-2l, however, especially the unfiltered residuals are noisy. This impression is confirmed by the Lagrange Multiplier (LM) test for ARCH(1), conducted on the residuals for the alternative models (cf. Table 5.12). A test for ARCH(*q*) can be constructed by regressing the squared residuals on a constant and *q* lags. The LM-test NR^2 - where N is the number of observations and R^2 is the coefficient of determination - is χ^2_q-distributed. The critical χ^2_1-value at $\alpha = 5\%$ is 5.024.

Table 5.12: LM-test results

Rotation method	Unfiltered		Filtered	
	FOX Index	Futures Index	FOX Index	Futures Index
Unrotated models	218.7866	212.9390	5.9062	0.1334
Schur rotation	243.9764	221.8165	3.5565	2.9773
Singular Value rotation	229.8528	208.6341	4.7419	0.5516

In Table 5.13 we present the correlations between the residual processes of the alternative state space models. The SVD-rotated residuals are seen to correspond to the unrotated residuals in all cases better than the Schur-rotated residuals.

Table 5.13: Correlations between the residual processes of the alternative state space models

FOX Index - Unfiltered Models			
Unrotated	1.0000		
SVD	0.8117	1.0000	
Schur	0.7840	0.8789	1.0000
	Unrotated	SVD	Schur

FOX Index - Filtered Models			
Unrotated	1.0000		
SVD	0.8898	1.0000	
Schur	0.7192	0.8203	1.0000
	Unrotated	SVD	Schur

Futures Index - Unfiltered Models			
Unrotated	1.0000		
SVD	0.8305	1.0000	
Schur	0.7751	0.8446	1.0000
	Unrotated	SVD	Schur

Futures Index - Filtered Models			
Unrotated	1.0000		
SVD	0.9289	1.0000	
Schur	0.8402	0.8882	1.0000
	Unrotated	SVD	Schur

We note that the correlation coefficients for SVD- and Schur-residuals are closest to each other in the unrotated/unfiltered case. Hence, if the difference between these coefficients is significant, so it will be in the other cases. A simple test for the difference of two correlation coefficients under the assumption of independent estimates is provided by Neave (1981), p. 35 using Fisher's Z-transform.

The test statistic is $\Delta Z \sim \left(0, \frac{2}{n-3}\right)$, where

$$\Delta Z = \frac{1}{2}\left[\ln\left(\frac{1+r_1}{1-r_1}\right) - \ln\left(\frac{1+r_2}{1-r_2}\right)\right] = \frac{1}{2}\ln\left(\frac{(1+r_1)(1-r_2)}{(1-r_1)(1+r_2)}\right),$$

and r_1 and r_2 are the empirical correlation coefficients under study. Under the null hypothesis,

$$\frac{\Delta Z}{\sqrt{\frac{2}{n-3}}} \sim N(0,1).$$

In the unfiltered and unrotated case of table (5.13) the correlation coefficients give

$$\frac{\Delta Z}{\sqrt{\frac{2}{n-3}}} = \frac{\frac{1}{2}\ln\left(\frac{(1+0.8117)}{(1-0.8117)}\frac{(1-0.784)}{(1+0.784)}\right)}{\sqrt{\frac{2}{268}}} = 8.8352.$$

This result indicates that the SVD-rotated residuals are significantly closer to the unrotated ones than are the Schur-rotated residuals. Thus, in specific practical situations, Schur-rotation may distort the residual processes more than does SVD-rotation. It should be emphasized, however, that the observed correlation estimates were assumed to be independent in the above test. More research on this topic is therefore needed in the future.

5.5 Tests for Linearity

Since the late eighties, the Finnish financial economy has undergone significant changes, in particular through a marked increase in turnover on the stock market. In the early nineties restrictions on foreign stock ownership were abolished. At the same time, the Finnish economy got into a severe recession and, because of that, the Finnish markka was floated. As a consequence, the Finnish financial economy is particularly suitable for regime-switching studies, say, from the year 1989 onward. This conjecture is supported by the high values on the LM-statistic obtained in our tests for the unfiltered state space residuals (cf. table 5.12).

An interesting class of nonlinear models is provided by smooth transition regression (STR) models (cf., *e.g.*, Bacon and Watts (1971)). In this study we will use the unrotated and unfiltered residuals of the state space model (cf. table 5.9a) as input signals to nonlinear scalar valued STR-models. Thus, the residuals serve as deviational transition variables for the equations, denoted by STR_D.

Granger and Teräsvirta (1993), p. 115, suggest the following specification procedure for STR_D-models:

(i) Given the residuals of a linear model (here: the state space model), test linearity against STR_D using the linear model as H_0.

(ii) If H_0 is rejected, choose between, *e.g.*, a logistic (LSTR_D) and an exponential (ESTR_D) transition model specified as follows:

$$y_t = \pi^T z_t + F(\hat{u}_{t-1})\theta^T z_t + \varepsilon_t, \text{ where} \tag{5.2}$$

$z_t = (1, \tilde{z}_t^T)^T$

$\tilde{z}_t$ is a vector of properly lagged values of the endogenous and exogenous variables (cf. table 5.15)

$\hat{u}_t = (\hat{u}_t, \hat{u}_{t-1}, \ldots, \hat{u}_{t-h+1})^T$ is the vector of lagged state space residuals for FOX or Futures, corresponding to y_t (In subsequent tests, $h = 2$ has been used)

$$F(\hat{u}_{t-1}) = \begin{cases} (1 + \exp(-\gamma(a^T\tilde{u}_{t-1})))^{-1} - \frac{1}{2}, \ \gamma > 0 \quad \text{(LSTR\_D)} \\ \text{or} \\ 1 - \exp(-\gamma(a^T\tilde{u}_{t-1})^2), \ \gamma > 0 \quad \text{(ESTR\_D)}\,. \end{cases}$$

π, θ, and a are parameter vectors of proper length to be estimated together with the scalars γ and c, by some suitable nonlinear optimization algorithm. (In this study the estimates were obtained by the Davidon-Fletcher-Powell algorithm using the NL-procedure of the SHAZAM (1993) package (cf. Judge *et al.* (1985), appendix B)).

Following Granger and Teräsvirta (1993), p. 75, we test linearity against a logistic STR-model separately for FOX and Futures in the auxiliary regression

$$\hat{u}_t = \beta^T z_t + \sum_{i=1}^{m} \sum_{j=1}^{h} \varphi_{ij} \tilde{z}_{ti} \hat{u}_{t-j} + v_t, \tag{5.3}$$

where $\hat{u}_t$ is the state space residual process for FOX or Futures respectively. The estimated residuals $\hat{v}_t$ are compared to the state space residuals $\hat{u}_t$ in the Lagrange multiplier test

$$\begin{aligned} \text{LM} &= T(SSR_0 - SSR)/SSR_0, \\ \text{where} \quad SSR_0 &= \sum \hat{u}_t^2 \\ SSR &= \sum \hat{v}_t^2 . \end{aligned} \tag{5.4}$$

The test statistic has an asymptotically χ^2_{mh} distribution. When replacing $\hat{u}_t$ by $\hat{u}_t^2$ in (5.3), the alternative hypothesis becomes an ESTR_D model (cf. Granger and Teräsvirta (1993), p. 76). The unfiltered and unrotated state space residuals for FOX and Futures (cf. table 5.9 and figures 2a, 2g) produced the following LM-statistics.

Table 5.14: Test for linearity of unrotated and unfiltered state space residuals

Series:	LM-test (5.4):	
	LSTR_D	ESTR_D
FOX-residuals	214.09	152.17
Futures-residuals	185.19	140.44

While the LM-test clearly indicates nonlinearity in the unfiltered residuals, it is inconclusive as to the type of model. (Indeterminacy is not surprising with LM-type tests, cf. Granger and Teräsvirta (1993), p. 101). Since the null is rejected with approximately equal strength by the two alternatives, we initially estimated both logistic and exponential models for the residual processes. The exponential models outperformed the logistic for both series, however, hence we discarded the latter. The obtained parameter estimates and some tests on the residuals are presented in Table 5.15. The residuals and *F*-functions are plotted in Figure 6. The lag structures are the result of extensive and time-consuming test runs.

Table 5.15: Parameter estimates and tests on residuals of ESTR_D-models for FOX and Futures (cf. (5.2)).

Coefficient	FOX			Futures		
	Variable	Estimate	t-value	Variable	Estimate	t-value
π_0		0.0005	0.2054		-0.0062	-1.5569
π_1	FOX_{t-1}	0.8815	9.8182	$Futures_{t-1}$	0.8746	12.954
π_2	FOX_{t-2}	-0.1617	-2.1564	$Futures_{t-2}$	-0.2370	-3.9704
π_3	$Futures_t$	0.2560	10.501	FOX_t	0.3561	6.3259
π_4	Tokyo $Price_t$	0.0462	0.8100	Tokyo $Price_t$	0.0295	1.1707
π_5	Tokyo $Volume_t$	0.0002	0.0908	Tokyo $Volume_t$	0.0027	1.3001
π_6	Tokyo $Price_{t-1}$	-0.1030	-1.3343			
π_7	Tokyo $Volume_{t-1}$	0.0022	1.2491			
π_8	Tokyo $Price_{t-2}$	0.0863	1.6156			
π_9	Tokyo $Volume_{t-2}$	-0.0008	-0.4830			
θ_0		51.695	4.0690		-83.073	-1.3360
θ_1	FOX_{t-1}	48.327	0.2543	$Futures_{t-1}$	464.13	1.1304
θ_2	FOX_{t-2}	-114.99	-0.8880	$Futures_{t-2}$	341.03	0.7003
θ_3	$Futures_t$	52.846	0.6961	FOX_t	-520.74	-1.1184
θ_4	Tokyo $Price_t$	-12.093	-0.0485	Tokyo $Price_t$	-15.18	-0.0708
θ_5	Tokyo $Volume_t$	-3.6646	-0.6224	Tokyo $Volume_t$	-58.414	-1.0865
θ_6	Tokyo $Price_{t-1}$	89.242	0.3127			
θ_7	Tokyo $Volume_{t-1}$	-11.498	-1.7883			
θ_8	Tokyo $Price_{t-2}$	-239.85	-1.4043			
θ_9	Tokyo $Volume_{t-2}$	6.5430	0.8878			
a_1	u_{t-1}	13.433	8.0986	u_{t-1}	-0.3956	-0.6200
a_2	u_{t-2}	-7.5662	-7.1334	u_{t-2}	-1.2706	-1.6308
γ		0.0480	3.3714		0.2265	3.8383
ARCH(1)-test (5.4)		5.3670			0.6814	
Jarque-Bera χ^2_2-test		3.997			22.951	

According to the Jarque-Bera test the FOX-residuals are normal, whereas the ARCH(1)-test indicates that there may be some heteroscedasticity left. The residuals of the Futures model are homoscedastic, but nonnormal. Since most coefficients in the nonlinear part are not significant, the interpretation of this model is problematic. A thorough search for an acceptable nonlinear representation of Futures prices is called for in future research.

Figure 6a: Residuals of ESTR_D model for FOX.

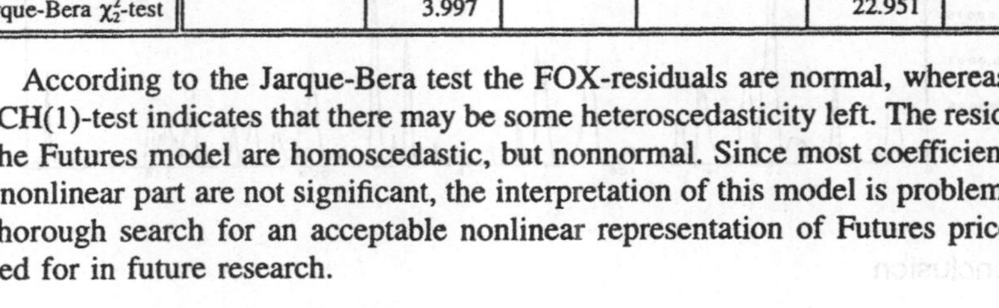

Figure 6b: Residuals of ESTR_D model for Futures.

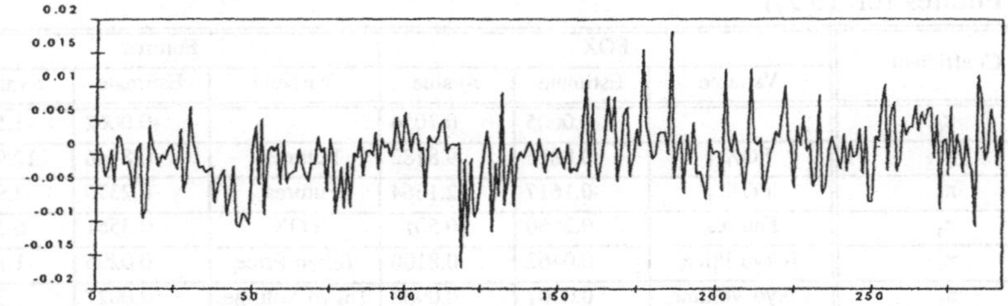

Figure 6c: Exponential transition function for FOX.

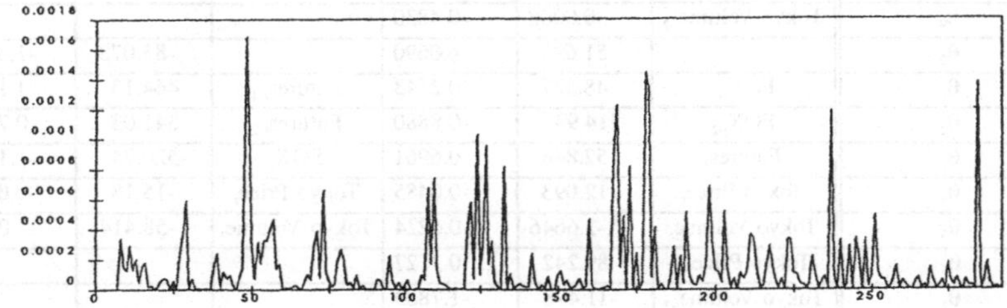

Figure 6d: Exponential transition function for Futures.

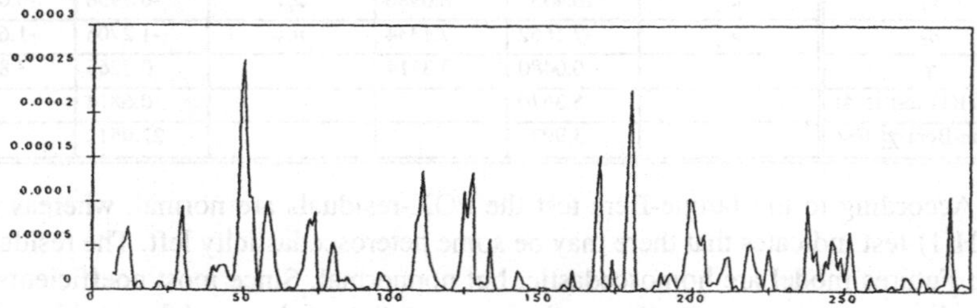

6 Conclusion

In the paper, we have studied the dynamic relationship between the Japanese and Finnish economies. In particular, we tested the signalling effect of the Tokyo stock market price index and trading volume upon the Finnish FOX and Futures indexes. Initially, we studied the stationarity conditions of both the endogenous and exogenous vectors by the augmented Dickey-Fuller test. The ADF-test indicated the presence of $I(1)$-processes, in both the signal and response variables. The financial markets have one cointegration vector, according to Johansen's multivariate test. Next, Granger causality between the Japanese and Finnish economies was tested, as well as the cointegration of the data. The results imply a significant Granger-causality between Japan and Finland.

Our results indicate that singular value decomposition of the state transition matrix provides a numerically robust approximation of the common trend and cyclic components of the endogenous vector. However, in cases where the characteristic roots of the transition matrix are complex, the real part of the eigenvectors may easily distort the residual process such that even an ex post filtering cannot restore white noise residuals. The analysis would be complicated by the presence of mixed $I(0)/I(1)$-processes in the response variables. Experiments carried out on pure $I(0)$- and $I(1)$-processes indicate that the SVD- and Schur-rotation methods usually conform to each other fairly well. This result is also confirmed by the results of the present study.

Spectral analysis reveals that the endogenous vector is related to the exogenous vector at several frequencies. A particularly strong common trend component was found for the endogenous vector, whereas the cyclical component is less pronounced, exhibiting peaks at several frequencies. Future research issues include, *e.g.*, the question of transforming the data for better model performance. The possibility of connecting the state space algorithm to a powerful neural network is also interesting. Furthermore, the joint effect of both the U.S. and Japanese economies on the Finnish financial market is interesting. We note that the present study was carried out on a fairly short time period, which probably explains the $I(0)$-outcome for the Japanese stock price index. It is therefore important to repeat the test on a longer time period in the future.

The unrotated/unfiltered state space residuals turned out to be nonlinear. We therefore used the processes as input signals to nonlinear scalar-valued transition models in the spirit of Granger and Teräsvirta (1993). The elimination of nonlinearity is a tricky business, however. Even if the nonlinear algorithm that we used converged for both logistic and exponential transition models (the gradient vector approximately zero in all tests), most of them failed in the ARCH(1)-test on residuals. A possible reason for this is that a persistent local optimum is found, another that the models are misspecified. Since the empirical testing and specification of nonlinear models is both time consuming and awkward under the present state of the art, there is a need for automated equipment for fast nonlinear modeling. The issue of specifying the optimal nonlinear form to a given data set is still unresolved in the doctrine. More research on this topic is therefore needed in the future. In the presence of nonlinearity the residuals should be handled by proper nonlinear methods.

References

Aoki, M. (1976a), "Control of Large Scale Dynamic Systems by Aggregation", *IEEE Transactions on Automatic Control*, AC-13, pp. 246-253.

Aoki, M. (1976b), *Optimal Control and Systems Theory in Dynamic Economic Analysis*. Amsterdam, North-Holland.

Aoki, M. (1987), *State Space Modeling of Time Series*. Springer-Verlag, Heidelberg, New York.

Aoki, M. (1988a), "State Space Models for Vector-Valued Time Series with Random Walk Components", Research Paper presented at the Eighth International Symposium of Forecasting. Amsterdam, June 1988.

Aoki, M. (1988b), "Cointegration, Error Correction and Aggregation in Dynamic Models", *Oxford Bulletin of Economics and Statistics*, 50, pp. 89-95.

Aoki, M. (1989a), "A State Space Time Series Modeling Method without Individual Detrending", *Journal of Time Series Analysis*, 1, pp. 128-151.

Aoki, M. (1989b), "Incorporating Exogenous Variables in State Space Models of Time Series", University of Los Angeles, USA, March 1989. Presented at the Summer European Econometric Society Meeting, Munich.

Aoki, M. (1990), *State Space Modeling of Time Series*. Springer-Verlag, Heidelberg, New York, 2nd, revised and enlarged edition.

Aoki, M. (1991), "Two Complementary Representations of Multiple Time Series in State Space Innovation Forms", University of Los Angeles. Presented at The Seminar on Recent Advances in Time Series Analysis and Their Impact on Economic Forecasting. Madrid, Spain, December 16-18, 1991.

Aoki, M. and A. Havenner (1991), "State Space Modeling of Multiple Time Series", *Econometric Reviews*, 10(1), pp. 1-59.

Athans, M. (1974), "The Importance of Kalman Filtering Methods for Economic System", *Ann. Economic and Social Measurement*, 3(1), pp. 49-64.

Bacon, D. and D. Watts (1971), "Estimating the Transition between Two Interesting Straight Lines", *Biometrica*, 58, pp. 525-534.

Bordignon, S. and U. Trivellato (1989), "The Optimal Use of Provisional Data in Forecasting with Dynamic Models", *Journal of Business and Economic Statistics*, 7(2), April, pp. 275-286.

Bos, T., T. A. Fetherston, T. Martikainen, and J. Perttunen (1995), "The International Co-movements of Finnish Stocks", *The European Journal of Finance*, 1, pp. 91-111.

Cerchi, M. and A. Havenner (1988), "Co-integration and Stock Prices: The Random Walk on Wall Street Revisited", *Journal of Economic Dynamics and Control*, 12, June/September, pp. 333-346.

Chatfield, C. (1987), *The Analysis of Time Series. An Introduction.* 3rd edition, Chapman and Hall, London.

Coates, D. S. and P. J. Diggle (1986), "Tests for Comparing Two Estimated Spectral Densities", *Journal of Time Series Analysis*, 7, pp. 7-20.

Dickey, D. and W. A. Fuller (1979), "Distribution of the Estimates for Autoregression Time Series with a Unit Root", *Journal of The American Statistical Association*, 74, pp. 427-431.

Diebold, F. (1989), "State Space Modeling of Time Series. A Review Essay", *Journal of Economic Dynamics and Control*, 13, pp. 597-612.

Engle, R. F. and C. W. J. Granger (1987), "Cointegration and Error Correction: Representation, Estimation and Testing", *Econometrica*, 55, pp. 251-276.

Fuller, W. A. (1976), *Introduction to Statistical Time Series*, John Wiley, New York,.

Geweke, J., R. Meese, and W. Dent (1983), "Comparing Alternative Tests of Causality in Temporal Systems", *Journal of Econometrics*, 21, pp. 161-194.

Granger, C. W. J. (1969), "Investigating Causal Relations by Economic Models and Cross-Spectral Methods", *Econometrica*, 37, pp. 424-438.

Granger, C. W. J. and T. Teräsvirta (1993), *Modeling Nonlinear Economic Relationships*, Oxford University Press, New York3.

Haugh, L. D.: "Checking the Independence of Two Covariance-Stationary Time Series: A Univariate Residual Cross Correlation Approach", *Journal of American Statistical Association*, **71**, 1976, pp. 378-385.

Hillmer, S. C. and G. C. Tiao (1982), "An ARIMA-Model Based Approach to Seasonal Adjustment", *Journal of the American Statistical Association*, 77, pp. 63-70.

Höglund R. and R. Östermark (1993), "Multiple Input Transfer Function Noise Modelling in the Time and Frequency Domain: Empirical Evidence from Monte Carlo Simulations", *Journal of Applied Statistics*, 20(1), pp. 69-93.

Johansen, S. (1988), "Statistical Analysis of Cointegration Vectors", *Journal of Economic Dynamics and Control*, 12, pp. 231-254.

Johansen, S. and K. Juselius (1990), "Maximum likelihood Estimation and Inference and Cointegration - with Applications to the Demand for Money", *Oxford Bulletin of Economics and Statistics*, 52(2), pp. 169-210.

Judge, G. G., W. E. Griffiths, R. C. Hill, H. Lütkepohl, and T.-C. Lee (1985), *The Theory and Practice of Econometrics*, 2nd edition, John Wiley & sons, Inc., New York.

Jöreskog, K. G. (1967), "Some Contributions to Maximum Likelihood Factor Analysis", *Psychometrika*, 32, pp. 443-482.

Koopmans, L. H. (1974), *The Spectral Analysis of Time Series.* Academic Press, New York.

Malkamäki, M. (1993), *Essays on Conditional Pricing of Finnish Stocks*, Bank of Finland. (Dissertation)

Malkamäki, M., T. Martikainen, J. Perttunen, and V. Puttonen (1993), "On the Causality and Co-movement of Scandinavian Stock Market Returns", *Scandinavian Journal of Management*, 9(1), pp. 67-76.

Malliaris, A. G. and J. L. Urrutia (1992), "The International Crash of October 1987: Causality Tests", *Journal of Financial and Quantitative Analysis*, 27(3), September, pp. 353-363.

Martikainen, T. and V. Puttonen (1992), "On the Informational Flow between Financial Markets: International Evidence from Thin Stock and Stock Index Futures Markets", *Economic Letters*, 38(2), pp. 213-216.

Mehra, R. (1978), "Practical Aspects of Designing Kalman Filters", in *Appl. of Kalman Filter to Hydrology, Hydraulics and Water Resources*, Chao-Liu Chin (Ed.), University of Pittsburgh, pp. 89-114.

Mittnik, S. (1990), "Forecasting with Balanced State Space Representations of Multivariate Distributed Lag Models", *Journal of Forecasting*, 9, pp. 207-218.

Neave, H. R. (1981), *Elementary Statistics Tables*. Nottingham University, George Allen & UNWIN Ltd., London.

Nelson, C. R. and G. W. Schwert (1982), "Tests for Predictive Relationships between Time Series Variables: a Monte Carlo Investigation", *Journal of American Statistical Association*, 77, pp. 11-18.

Ng, C. and P. Young (1990), "Recursive Estimation and Forecasting of Non-Stationary Time Series", *Journal of Forecasting*, 9, pp. 173-204.

Östermark, R. (1991a), "The Forecasting Performance of Cartesian ARIMA Search and a Vector-Valued State Space Model", Forthcoming in *Kybernetes*.

Östermark, R. (1991b), "Vector Forecasting and Dynamic Portfolio Selection. Empirical Efficiency of Recursive Multiperiod Strategies", *European Journal of Operational Research*, 55, pp. 46-56.

Östermark, R. (1992), "The Forecasting Performance of a VARMAX Search Algorithm and a State Space Algorithm with Exogenous Variables", Åbo Akademi University, Department of Business Administration, Working paper 169.

Östermark, R. and M. Aoki (1995), "Time Series Evidence of Impacts of the U.S. Economy on the Scandinavian Economy (by State Space Modeling)", *International Journal of Systems Science*, 26 (12), pp. 2327-2353.

Östermark R. and H. Hernesniemi (1995), "The Impact of Information Timeliness on the Predictability of Stock Index and Stock Index Futures Returns. An Application of vector Models", *European Journal of Operational Research*, 85(1), pp. 111-131.

Otter, P. W. (1995), *Dynamic Feature Space Modeling, Filtering and Self-tuning Control of Stochastic Systems*. Springer-Verlag, Berlin.

Otter, P. W. and R. Van Dal (1987), "State Space and Distributed Lag Modeling of Dynamic Economic Processes Based on Singular Value Decomposition", *Annales D'Economie et de Statistique*, 6/7, pp. 253-277.

Otter, P. W. and R. Van Dal (1989), "On State Space Approximation of Multi-Input Multi-Output Systems with Stochastic Exogenous Inputs". In: S. Mittnik (ed.) *System-Theoretic Methods in Economic Modeling II*, Oxford: Pergamon Press.

Phillips, P. and P. Perron (1988), "Testing for a Unit Root in Time Series Regression", *Biometrika*, 75(2), pp. 335-346.

Pierce, D. A. (1977), "Relationships - and the Lack Thereof - between Economic Time Series, with Special Reference to Money and Interest Rates", *Journal of American Statistical Association*, 72, pp. 11-22.

Di Ruscio, D. and D. Ljungquist (1992), "State Space Model Realization from Input-Output Time Series", Research Paper, Norwegian Institute of Technology, Division of Engineering and Cybernetics, Trondheim.

Sastri, T. (1985), "A State Space Modeling Approach to Time Series Forecasting", *Management Science*, 31(11), pp. 1451-1470.

Sastri, T. (1989), "Multipass Seasonal Adjustment Filter", *Management Science*, 35(1), January, pp. 100-123.

SHAZAM Econometrics Computer Program (1993), User's Reference Manual, Version 7.0, McGraw-Hill, Canada.

Shimoshni, M. (1971), "On Fisher's Test of Significance in Harmonic Analysis", *Geophyshical Journal of the Royal Astronomical Society*, 23, pp. 373-377.

Sims, C. A. (1972), "Money, Income and Causality", *American Economic Review*, 62, pp. 540-552.

A Method for Identification of Combined Deterministic Stochastic Systems

David Di Ruscio
Department of Process Automation
Telemark Institute of Technology
N-3914 Porsgrunn, Norway

Keywords: System identification; subspace methods; stochastic and deterministic systems; state-space methods; time series analysis.

Abstract

The paper presentes a numerically stable and general algorithm for identification and realization of a complete dynamic linear state space model, including the system order, for combined deterministic and stochastic systems from time series. A special property of this algorithm is that the innovations covariance matrix and the Markov parameters for the stochastic sub-system are determined directly from a projection of known data matrices, without e.g. recursions of non-linear matrix Riccati equations. A realization of the Kalman filter gain matrix is determined from the estimated extended observability matrix and the Markov parameters. Monte Carlo simulations are used to analyze the statistical properties of the algorithm as well as to compare with existing algorithms.

1 Introduction

System identification can be defined as building mathematical models of systems based on observed data. Traditionally a set of model structures with some free parameters are specified and a *prediction error* (PE) criterion measuring the difference between the observed outputs and the model outputs is optimized with respect to the free parameters. In general, this will result in a non linear optimization problem in the free parameters even when a linear time invariant model is specified. A tremendous amount of research has been reported, resulting in the so called *prediction error methods* (PEM).

In our view the field of *subspace identification*, Larimore (1983) and (1990), Verhagen (1994), Van Overschee and De Moor (1994), Di Ruscio (1994), not only resolves the problem of system identification but also deals with the additional problem of *structure identification*. In subspace identification methods a data matrix is constructed from certain projections of the given system data. The observability matrix for the system is extracted as the column space of this matrix and the system order is equal to the dimension of the column space.

Descriptions of the advantages of subspace identification methods over traditional PEM can be found in Viberg (1995) and in Van Overschee (1995).

Aoki (1990) has presented a method for the realization of state space linear discrete time stochastic models on innovations form. See also Aoki (1994) for some further improvements of the method. This method has many interesting numerical properties and many of the numerical tools involved in the method are common with the tools used by the *subspace identification* methods. The method is based on the factorization of the Hankel matrix, constructed from covariance matrices of the output time series, by singular value decomposition. The states are presented relative to an internally balanced coordinate system, Moore (1981), which has some interesting properties when dealing with model reduction.

The *subspace identification* methods which are presented in the above references are using instrumental variables constructed from past input and output data in order to remove the effect of noise from future data. It should be pointed out that the method by Aoki also uses instrumental variables constructed from past data.

The method for system identification and state space model realization which is presented in this work is believed to be a valuable tool for the analysis and modeling of observed input and output data from a wide range of systems, in particular combined deterministic and stochastic dynamical systems. Only linear algebra is applied in order to estimate a complete linear time invariant state space model.

Many successful applications of data based time series analysis and modeling methods are reported. One industrial application is presented in Di Ruscio and Holmberg (1996). One particularly important application is the estimation of econometric models, see Aoki (1990).

The remainder of the paper is organized as follows. Section 2 gives a definition of the system and the problem considered in this work. In Section 3 the data is organized into data matrices which satisfy an extended state space model or matrix equation. Section 4 shows how the system order and the model matrices can be extracted from the known data matrices. A numerically stable and efficient implementation is presented in Section 5. Section 6 gives a comparison of the method presented in this work with other published methods. Real world numerical examples and theoretical Monte Carlo simulations are presented in Section 7 and some concluding remarks follow in Section 8.

2 Preliminary Definitions

2.1 System Definition

Assume that the underlying system can be described by a discrete-time, time invariant, linear state space model (SSM) of the form

$$x_{k+1} = Ax_k + Bu_k + Ce_k \tag{1}$$

$$y_k = Dx_k + Eu_k + e_k \tag{2}$$

where the integer $k \geq 0$ is discrete-time, $x \in \mathbb{R}^n$ is the state vector with initial value x_0, $y \in \mathbb{R}^m$ is the system output, $u \in \mathbb{R}^r$ is the system input, $e \in \mathbb{R}^m$ is an unknown innovations process of white noise, assumed to be covariance stationary, with zero mean and covariance matrix $\mathrm{E}(e_k e_k^T) = \Delta$. The constant matrices in the SSM are of appropriate dimensions. A is the *state transition* matrix, B is the *external input* matrix, C is the *Kalman gain* matrix, D is the *output* matrix and E is the direct *control input to output* (feed-through) matrix. We will assume that (D, A) is an observable pair.

The innovations model, Equations (1) and (2), is discussed in e.g. Faurre (1976) and Aoki (1990).

2.2 Problem Definition

The problem investigated in this paper is to identify a state space model, including the system order (n), for both the deterministic and the stochastic part of the system, i.e. the quadruple matrices (A, B, D, E) and the double matrices (C, Δ) respectively, directly from known system input and output data vectors (or time series) defined as

$$\left.\begin{array}{l} u_k \;\; \forall \;\; k = 0, \ldots, N-1 \\ y_k \;\; \forall \;\; k = 0, \ldots, N-1 \end{array}\right\} \textbf{Known data vectors}$$

In continuous time systems the matrix E in Equation (2) is usually zero. This is not the case in discrete time systems due to sampling. However, E can be forced to be zero by including a structure constraint. This will be commented on later.

2.3 Matrix Definitions

Associated with the SSM, Equations (1) and (2), we make the following definitions:

- The *extended observability* matrix (O_i) for the pair (D, A) is defined as

$$O_i \stackrel{\text{def}}{=} \begin{bmatrix} D \\ DA \\ \vdots \\ DA^{i-1} \end{bmatrix} \in \mathbb{R}^{im \times n} \tag{3}$$

where subscript i denotes the number of block rows. The matrix O_i is denoted the *extended observability* matrix when the number of block rows i is grater than the minimal number of block rows needed in the observability matrix, in order to check if the system is observable.

- The *reversed extended controllability* matrix (C_i^d) for the pair (A, B) is defined as

$$C_i^d \stackrel{\text{def}}{=} \begin{bmatrix} A^{i-1}B & A^{i-2} & \cdots & B \end{bmatrix} \quad \in \mathbb{R}^{n \times ir} \tag{4}$$

  where subscript i denotes the number of block columns.
  A matrix C_i^s for the pair (A, C) is defined similarly to Equation (4), i.e. with C substituted for B in the above definition.

- The *lower block triangular Toeplitz* matrix (H_i^d) for the quadruple matrices (D, A, B, E)

$$H_i^d \stackrel{\text{def}}{=} \begin{bmatrix} E & 0 & 0 & \cdots & 0 \\ DB & E & 0 & \cdots & 0 \\ DAB & DB & E & \cdots & 0 \\ \vdots & \vdots & \vdots & \ddots & \vdots \\ DA^{i-2}B & DA^{i-3}B & DA^{i-4}B & \cdots & E \end{bmatrix} \quad \in \mathbb{R}^{im \times ir} \tag{5}$$

  where subscript i denotes the number of block rows.
  A *lower block triangular Toeplitz* matrix H_i^s for the quadruple matrices (D, A, C, F) is defined as

$$H_i^s \stackrel{\text{def}}{=} \begin{bmatrix} F & 0 & 0 & \cdots & 0 \\ DC & F & 0 & \cdots & 0 \\ DAC & DC & F & \cdots & 0 \\ \vdots & \vdots & \vdots & \ddots & \vdots \\ DA^{i-2}C & DA^{i-3}C & DA^{i-4}C & \cdots & F \end{bmatrix} \quad \in \mathbb{R}^{im \times im} \tag{6}$$

  where $F = I$ for the output model formulation, Equation (2).

2.4 Notation

The projection A/B of two matrices A and B is defined as $AB^T(BB^T)^\dagger B$ where $\dagger$ denotes the Moore-Penrose pseudo-inverse of a matrix.

3 Extended State Space Model

The state space model, Equations (1) and (2), can generally be written as the following extended state space model (ESSM) (Di Ruscio (1994))

$$Y_{k+1|L} = \tilde{A}Y_{k|L} + \tilde{B}U_{k|L+1} + \tilde{C}E_{k|L+1} \tag{7}$$

where the known output and input data matrices $Y_{k|L}$ and $U_{k|L+1}$ are defined as follows

$$Y_{k|L} \stackrel{\text{def}}{=} \begin{bmatrix} y_k & y_{k+1} & y_{k+2} & \cdots & y_{k+K-1} \\ y_{k+1} & y_{k+2} & y_{k+3} & \cdots & y_{k+K} \\ \vdots & \vdots & \vdots & \ddots & \vdots \\ y_{k+L-1} & y_{k+L} & y_{k+L+1} & \cdots & y_{k+L+K-2} \end{bmatrix} \in \mathbb{R}^{Lm \times K} \tag{8}$$

$$U_{k|L+1} \stackrel{\text{def}}{=} \begin{bmatrix} u_k & u_{k+1} & u_{k+2} & \cdots & u_{k+K-1} \\ u_{k+1} & u_{k+2} & u_{k+3} & \cdots & u_{k+K} \\ \vdots & \vdots & \vdots & \ddots & \vdots \\ u_{k+L-1} & u_{k+L} & u_{k+L+1} & \cdots & u_{k+L+K-2} \\ u_{k+L} & u_{k+L+1} & u_{k+L+2} & \cdots & u_{k+L+K-1} \end{bmatrix} \in \mathbb{R}^{(L+1)r \times K} \tag{9}$$

The unknown data matrix $E_{k|L+1}$ of innovations noise vectors is defined as

$$E_{k|L+1} \stackrel{\text{def}}{=} \begin{bmatrix} e_k & e_{k+1} & e_{k+2} & \cdots & e_{k+K-1} \\ e_{k+1} & e_{k+2} & e_{k+3} & \cdots & e_{k+K} \\ \vdots & \vdots & \vdots & \ddots & \vdots \\ e_{k+L-1} & e_{k+L} & e_{k+L+1} & \cdots & e_{k+L+K-2} \\ e_{k+L} & e_{k+L+1} & e_{k+L+2} & \cdots & e_{k+L+K-1} \end{bmatrix} \in \mathbb{R}^{(L+1)m \times K} \tag{10}$$

The scalar integer parameter L defines the number of block rows in the data matrices and the ESSM model matrices. The number of columns in $Y_{k|L}$, $U_{k|L+1}$ and $E_{k|L+1}$ are $K = N - L - k + 1$. Each column in these matrices can be interpreted as extended output, input and noise vectors, respectively. K can be viewed as the number of samples in these extended time series. We also have that $L < K < N$. L is the only necessary parameter which has to be specified by the user. L is equal to the number of block rows in the extended observability matrix ($O_L \in \mathbb{R}^{Lm \times n}$), which will be determined by the algorithm. For a specified L, the maximum possible order of the system to be identified is $n \leq Lm$ (if rank$(D) = m$, i.e. m independent outputs), or $n \leq Ld$ where $1 \leq d = \text{rank}(D) \leq m$, i.e. d independent output variables.

The parameter L can be interpreted as the *identification horizon*. This means that L is the horizon used to recover the present state space vector x_k.

The matrices in the extended state space model, Equation (7), are related to the underlying state space model matrices as follows

$$\tilde{A} = O_L A (O_L^T O_L)^{-1} O_L^T \tag{11}$$

$$\begin{aligned} \tilde{B} &= \left[\, OB - \tilde{A}E_1 \;\; E_1 - \tilde{A}E_2 \;\; E_2 - \tilde{A}E_3 \;\; \cdots \;\; E_{L-1} - \tilde{A}E_L \;\; E_L \,\right] \\ &= \left[\, O_L B \;\; H_L^d \,\right] - \tilde{A}\left[\, H_L^d \;\; 0_{Lm \times r} \,\right] \end{aligned} \tag{12}$$

$$\begin{aligned} \tilde{C} &= \left[\, O_L C - \tilde{A}F_1 \;\; F_1 - \tilde{A}F_2 \;\; F_2 - \tilde{A}F_3 \cdots \;\; F_{L-1} - \tilde{A}F_L \;\; F_L \,\right] \\ &= \left[\, O_L C \;\; H_L^s \,\right] - \tilde{A}\left[\, H_L^s \;\; 0_{Lm \times m} \,\right] \end{aligned} \tag{13}$$

The matrices E_i and F_i, $i = 1, ..., L$, are block columns in the Toeplitz matrices H_L^d and H_L^s defined in Equations (5) and (6), i.e.

$$H_L^d = \begin{bmatrix} E_1 & E_2 & \cdots & E_L \end{bmatrix} \tag{14}$$
$$H_L^s = \begin{bmatrix} F_1 & F_2 & \cdots & F_L \end{bmatrix} \tag{15}$$

The importance of the ESSM, Equation (7), is that the state vector preliminary is eliminated from the problem. Hence, the number of unknowns is reduced. The ESSM also gives us the relationship between the data matrices and the model matrices which at this stage are unknown.

This paper is concerned with the problem of reconstructing the system order and system matrices in the state space model, (1) and (2), from the known data matrices $Y_{k|L}$ and $U_{k|L+1}$ which satisfy Equation (7). We refer to Di Ruscio (1994) and (1995) for a proof of the above results, which are the basis for the method presented in this work.

Note that the matrices H_L^d and H_L^s satisfy the matrix equation

$$Y_{k|L} = O_L X_k + \begin{bmatrix} H_L^d & 0_{Lm\times r} \end{bmatrix} U_{k|L+1} + \begin{bmatrix} H_L^s & 0_{Lm\times m} \end{bmatrix} E_{k|L+1} \tag{16}$$

where

$$X_k = \begin{bmatrix} x_k & x_{k+1} & x_{k+2} & \cdots & x_{k+K-1} \end{bmatrix} \quad \in \mathbb{R}^{n\times K} \tag{17}$$

is a matrix of state vectors. Equation (16) is frequently used in other subspace identification methods, e.g. in Van Overschee and De Moor (1994) and Verhagen (1994).

4 System Identification and Realization

4.1 Identification and Realization of System Dynamics

The basic step in the algorithm is to identify the system order and the extended observability matrix from known data. In order to do so we will in this section derive an autonomous matrix equation from which the system dynamics can be identified. We will show how the system order n and the extended observability matrix O_L are identified. A realization for the system matrices A and D and the ESSM transition matrix $\tilde{A}$ are then computed.

The term $\tilde{B}U_{k|L+1}$ can be removed from Equation (7) by post-multiplying with a projection matrix $U_{k|L+1}^{\perp}$ such that $U_{k|L+1}U_{L|L+1}^{\perp} = 0$. The projection matrix can e.g. be defined as follows

$$U_{k|L+1}^{\perp} = I_{K\times K} - U_{k|L+1}^T(U_{k|L+1}U_{k|L+1}^T)^{-1}U_{k|L+1} \tag{18}$$

Hence, $U_{k|L+1}^{\perp}$ is the orthogonal projection onto the null-space of $U_{k|L+1}$. A numerically well posed way of computing the projection matrix is by use of the singular value decomposition (SVD). The projection matrix is given by the left singular vectors of $U_{k|L+1}$ which are orthogonal to the null-space. However, in order to solve the complete system identification and realization problem, it is more convenient to use the QR decomposition for computing the projection, as will be shown in Section 5. Note that a projection matrix onto the null-space of $U_{k|L+1}$ exists if the number of columns K in the data-matrices satisfies $K > L+1$.

Post-multiplying Equation (7) with the projection matrix $U_{k|L+1}^{\perp}$ gives

$$\begin{aligned} Y_{k+1|L} - Y_{k+1|L}U_{k|L+1}^T(U_{k|L+1}U_{k|L+1}^T)^{-1}U_{k|L+1} = \\ \tilde{A}(Y_{k|L} - Y_{k|L}U_{k|L+1}^T(U_{k|L+1}U_{k|L+1}^T)^{-1}U_{k|L+1}) + \\ \tilde{C}(E_{k|L+1} - E_{k|L+1}U_{k|L+1}^T(U_{k|L+1}U_{k|L+1}^T)^{-1}U_{k|L+1}) \end{aligned} \quad (19)$$

Note that the last noise term in equation (19) is per definition zero as the number of samples approaches infinity, i.e.

$$\lim_{K\to\infty} \tilde{C}\frac{1}{K}E_{k|L+1}U_{k|L+1}^T = 0 \quad (20)$$

Hence, we have the following result

$$Y_{k+1|L}U_{k|L+1}^{\perp} = \tilde{A}Y_{k|L}U_{k|L+1}^{\perp} + \tilde{C}E_{k|L+1} \quad (21)$$

The noise term $\tilde{C}E_{k|L+1}$ can be removed from Equation (21) by post-multiplying with $\frac{1}{K}W_i^T$ where W_i is defined as a matrix of "instrumental" variables which are uncorrelated with $E_{k|L+1}$, i.e., we are seeking for a matrix with the following property

$$\lim_{K\to\infty} \frac{1}{K}E_{k|L+1}W_i^T = 0 \quad (22)$$

An additional property is that W_i should be sufficiently correlated with the informative part in the ESSM in order not to destroy information about e.g. the system order.

An intuitive good choice is to use *past* data as instruments to remove *future* noise. This choice ensures that the instruments are sufficiently correlated with the informative part of the signals and sufficiently uncorrelated with future noise.

Define J as the number of time instants in the *past horizon* which is used for defining the instrumental variable matrix. Define L as the number of time instants in the horizon necessary for identifying the state at time instant k, that is x_k, as well as the extended observability matrix of the system. Define M as a prediction horizon. However, we will restrict ourself to the case $M = 1$ in this work. These horizons are illustrated schematically in Figure 1.

Some alternative instruments, W_i, that will remove the noise and which satisfy Equation (22) are as follows. Define

$$W_i \in \mathbb{R}^{ni\times K} \quad \forall \; i = 1,2,3 \quad (23)$$

where the row dimension ni is the number of instrumental variables and

$$W_1 = \begin{bmatrix} Y_{0|J} \\ U_{l|J} \end{bmatrix} \quad W_2 = Y_{0|J} \quad W_3 = U_{l|J} \quad \forall \; l \geq 0 \quad (24)$$

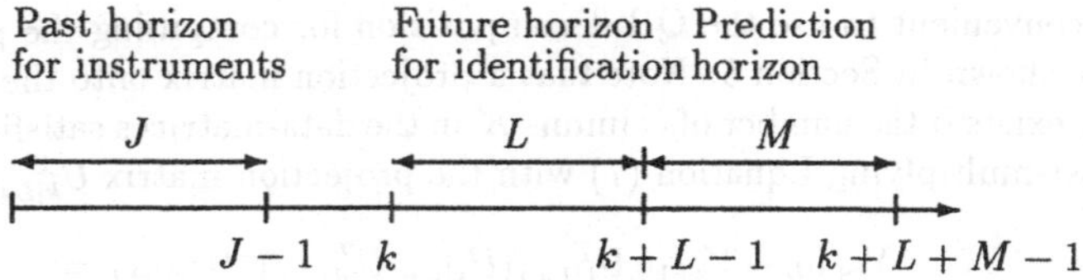

Figure 1: Illustration of horizons involved in the **DSR** algorithm. Usually $k = J$. That is, the end of the past is the beginning of the future.

The choice of instruments is unimportant in the deterministic case, i.e. when the process as well as the observation noise are identically zero which also means that $e_k = 0$ for all discrete time instants. However, the "optimal" choice of instruments is important in the combined deterministic and stochastic case.

Assume that $U_{l|J}$ with $l > 0$ is chosen as the instrumental variable matrix. This means that not only past inputs but also future inputs are used to remove future noise (first time instant in the future horizon satisfy $J \leq k$). Our experience from Monte Carlo simulations indicates that this is not an "optimal" choice. Note also that the future inputs are already used in the projection matrix $U_{k|L+1}^{\perp}$. Hence, it make sense to use only past inputs as instruments, i.e. the choice $U_{0|J}$. It can also be shown (Verhagen, 1994) that by using only past inputs as instruments, only the deterministic part of the model can be recovered.

Past outputs are assumed to be uncorrelated with future noise. This gives a first constraint on the discrete time instant k, i.e. $k \geq J$. We have

$$\lim_{K\to\infty} \frac{1}{K} E_{k|L+1} Y_{0|J}^T = 0 \quad \forall\ k \geq J \tag{25}$$

This statement can be proved from Equations (16) and (20). By incorporating past outputs as instruments we are also able to recover the stochastic part of the model. Note that the states which are exited from the known inputs are not necessarily the same as those which are exited from the unknown process noise variables. It is necessary that all states are exited from both known and unknown inputs and that they are observable from the output, in order to identify them.

Hence, the following past inputs and past outputs instrumental variable matrix is recommended to remove future noise from the model

$$W_1 = \begin{bmatrix} Y_{0|J} \\ U_{0|J} \end{bmatrix} \in \mathbb{R}^{J(m+r)\times K} \qquad \forall\ J \geq 1 \tag{26}$$

A consistent equation for $\tilde{A}$ is then given by the following autonomous matrix equation

$$Z_{k+1|L} = \tilde{A} Z_{k|L} \qquad \forall\ \ k \geq J \tag{27}$$

where

$$Z_{k+1|L} \stackrel{\text{def}}{=} \tfrac{1}{K} Y_{k+1|L} U_{k|L+1}^{\perp} W_i^T \ \in \mathbb{R}^{mL\times ni} \tag{28}$$

$$Z_{k|L} \stackrel{\text{def}}{=} \frac{1}{K} Y_{k|L} U_{k|L+1}^{\perp} W_i^T \in \mathbb{R}^{mL \times ni} \tag{29}$$

Equation (27) is consistent because W_i, given by Equations (23) and (24), satisfies Equation (22). See also Equation (25).

We can now prove that the column space of the matrix $Z_{k|L}$ coincides with the column space of the extended observability matrix O_L, when the identification (*future*) horizon parameter L is chosen great enough to observe all states and the *past* horizon parameter J is chosen adequately. Using Equations (16) and (29) with the past inputs and past outputs instrumental variable matrix gives

$$Z_{k|L} = \frac{1}{K} Y_{k|L} U_{k|L+1}^{\perp} W_1^T = O_L X_k U_{k|L+1}^{\perp} \frac{1}{K} W_1^T \quad \in \mathbb{R}^{mL \times J(m+r)} \tag{30}$$

Assume that both the row and column dimensions of $Z_{k|L}$ are greater or equal to the number of states, i.e., $Lm \geq n$ and $J(m+r) \geq n$, and that L is chosen such that the system is observable. The dimension of the column space of the left hand side matrix must be equal to the system order, i.e. $\text{rank}(X_k) = n$. Hence,

$$\text{rank}(Z_{k|L}) = \text{rank}(\frac{1}{K} Y_{k|L} U_{k|L+1}^{\perp} W_1^T) = \text{rank}(O_L X_k U_{k|L+1}^{\perp} \frac{1}{K} W_1^T) = n \tag{31}$$

The row constraints have a theoretical lower limit. From system theory we know that a number of $L \geq n - \text{rank}(D) + 1$ observations of the output is sufficient in order to observe the states of a linear system, Kalman, Falb and Arbib (1969), p. 37. However, the theoretical lower limit is the ceiling function $L = \lceil n/m \rceil$, defined as the integer ratio n/m rounded towards plus infinity.

From the column dimension we must ensure that the past horizon J for defining the instrumental variable matrix must satisfy $J(m+r) \geq n$. Hence, the theoretical lower limit is $J = \lceil n/(m+r) \rceil$.

The maximum system order which can be specified by the user for a specified choice of the parameter L is $n = Lm$. In this case the observability matrix can be estimated as the column space of $Z_{k|L}$ only if the past horizon parameter J is chosen such that $J(m+r) \geq Lm$. A reasonable choice is therefore $J = L$.

Monte Carlo simulation experiments shows a relatively constant statistical behavior of the estimates as a function of the past horizon parameter J. Hence, we simply recommend putting $J = L$.

We have the following algorithm for analysis and modeling of system dynamics.

Algorithm 4.1 (System order, n, and the pair (D, A))
Given the positive integer parameters L and J and the matrices $Z_{k+1|L}$ and $Z_{k|L}$ with $k \geq J$ which satisfy the autonomous matrix equation

$$Z_{k+1|L} = \tilde{A} Z_{k|L} \tag{32}$$

where

$$\tilde{A} = O_L A (O_L^T O_L)^{-1} O_L^T \tag{33}$$

and O_L is the extended observability matrix for the pair (A, D).

1. The system order n

Determine the Singular Value Decomposition (SVD)

$$Z_{k|L} = USV^T \tag{34}$$

where $U \in \mathbb{R}^{mL \times mL}$, $S \in \mathbb{R}^{mL \times ni}$ and $V \in \mathbb{R}^{ni \times ni}$ are given by

$$U = \begin{bmatrix} U_1 & U_2 \end{bmatrix} \quad S = \begin{bmatrix} S_n & 0 \\ 0 & S_2 \end{bmatrix} \quad V = \begin{bmatrix} V_1 & V_2 \end{bmatrix} \tag{35}$$

where $S_n \in \mathbb{R}^{n \times n}$ and n is the number of "non-zero" singular values of $Z_{k|L}$, which is equal to the system order. n is determined by inspection of the "non-zero" diagonal elements of S or SS^T. The term $U_2 S_2 V_2^T$ represents the error by estimating the system order as the n first principal singular values.

2. The extended observability matrix O_L for the pair (D, A)

The (extended) observability matrix can be taken directly as the first left part in U, i.e. U_1. We have

$$O_L = U(1 : Lm, 1 : n) = U_1 \tag{36}$$

3. The system matrix A

The system matrix A can be determined as

$$A = O_L^T Z_{k+1|L} V \begin{bmatrix} S_n^{-1} \\ 0 \end{bmatrix} = U_1^T Z_{k+1|L} V_1 S_n^{-1} \tag{37}$$

4. The system output matrix D

The matrix D can be taken as the $m \times n$ upper sub-matrix in the observability matrix O_L, i.e.

$$D = U(1 : m, 1 : n) \tag{38}$$

5. The extended system matrix $\tilde{A}$

We have

$$\tilde{A} = O_L A (O_L^T O_L)^{-1} O_L^T = Z_{k+1|L} V_1 S_n^{-1} U_1^T \tag{39}$$

$\triangle$

Note that it can be difficult to estimate the system order as the non-zero or large singular values of the matrix $Z_{k|L}$ when the signal to noise ratio is "small". In practice, the model order is chosen by a trial and error procedure combined with both model validation and physical knowledge of the data generating process. A procedure for chosing the model order from some statistical test is possibile, see e.g. Ch. 9.6 in Aoki (1990).

We have chosen $O_L = U_1$ in Step 2 for simplicity, because we have $O_L^T O_L = I_{n \times n}$ in this case. This gives an output normal realization when $L \to \infty$. The algorithm can also be formulated with the choice $O_L = U_1 S_n^{\frac{1}{2}}$ which gives a balanced realization when $L \rightarrow \infty$. $O_L^T O_L$ is equal to the observability grammian as L tends to infinity because, in this case, DA^{L-1} tends to zero. A third choice is $O_L = U_1 S_n$ which gives an input normal realization. These definitions are attributable to Moore (1981). These choices only represent different scalings of the column space and give similar state space model matrices. The scaling does not affect the statistical properties of the algorithm.

4.2 Realization of the Deterministic Sub-system

At this stage the system matrices A and D as well as the extended observability matrix O_L are known, see Section 4.1. In order to obtain a complete realization for the deterministic part of the system we need to compute the system matrices B and E.

There are many alternatives for extracting the B and E matrices. See e.g. Section 5.2 for an alternative to the method presented below.

A consistent estimate of the $\tilde{B}$ matrix can be computed from

$$\tilde{B}\frac{1}{K}U_{k|L+1}U_{k|L+1}^T = \frac{1}{K}(Y_{k+1|L} - \tilde{A}Y_{k|L})U_{k|L+1}^T \tag{40}$$

where $\tilde{A}$ is determined from Algorithm 4.1, because

$$\lim_{K\rightarrow\infty} \tilde{C}\frac{1}{K}E_{k|L+1}U_{k|L+1}^T = 0 \quad \forall\, k \geq 0 \tag{41}$$

For known data and system matrices A and D, Equation (40) can be written as an over determined set of linear equations in the unknown system matrices B and E .

We will in the rest of this section discuss the simpler solution when $U_{k|L+1}U_{k|L+1}^T$ is non-singular. The matrix $\tilde{B}$ can be computed directly from Equation (40) in this case. We refer to Section 5.2 for the case when $U_{k|L+1}U_{k|L+1}^T$ is singular. The system matrices B and E can be extracted from $\tilde{B}$. E is given directly as the lower right $m \times r$ sub-matrix of $\tilde{B}$. We have

$$E = \tilde{B}(m(L-1)+1 : mL, rL+1 : r(l+1)) \tag{42}$$

B is given as a function of the block columns in $\tilde{B}$ and the matrices $\tilde{A}$ and O_L. $\tilde{A}$ and O_L are known from Algorithm 4.1. Define

$$\tilde{B}_i = \tilde{B}(1 : mL, r(i-1)+1 : ri) \quad 1 \leq i \leq L+1 \tag{43}$$

as block column number i of $\tilde{B}$. We have

$$O_L B = \sum_{i=1}^{L+1} \tilde{A}^{i-1}\tilde{B}_i \tag{44}$$

and

$$B = \sum_{i=1}^{L+1} A^{i-1}(O_L^T O_L)^{-1} O_L^T \tilde{B}_i \tag{45}$$

One strategy for recursively extracting the B, E, H_L^d and O_LB matrices is as follows.

Algorithm 4.2 *Determination of H_L^d, O_LB, E and B from known $\tilde{B}$, O_L and A.*

$\tilde{A} = O_LA(O_L^TO_L)^{-1}O_L^T$
$E_L = \tilde{B}_{L+1}$
$E = E_L(m(L-1)+1 : mL, 1 : m)$
for $i = 1, \ldots, L$
    $E_{L-i} = \tilde{B}_{L-i+1} + \tilde{A}E_{L-i+1}$
    if $i < L$
        $E_{L-i}(1 : r(L-i-1), 1 : r) = zeros(r(L-i-1), r)$
    end
end
$O_LB = E_0$
$B = (O_L^TO_L)^{-1}O_L^TO_LB$

$\triangle$

The lower block triangular Toeplitz matrix H_L^d is given by the block columns E_1, $\cdots$, E_L according to Equations (14) and (5).

4.3 Realization of the Stochastic Sub-system

This section is mainly concerned with the problem of identifying the stochastic part of the system. However, for natural reasons most of the results in Sections 4.1 and 4.2 are extended and resolved in parallel.

The ESSM, Equation (7), gives a relation between the future data matrices and the SSM matrices. The following extension of the ESSM can be proved

$$\begin{aligned} Y_{k|L+1} &= O_{L+1}A^kO_J^\dagger Y_{0|J} \\ &+ \begin{bmatrix} O_{L+1}C_k^d & H_{L+1}^d \end{bmatrix} \begin{bmatrix} U_{0|k} \\ U_{k|L+1} \end{bmatrix} \\ &- \begin{bmatrix} O_{L+1}A^kO_J^\dagger H_J^d & 0_{Lm\times(k+L+1-J)r} \end{bmatrix} \begin{bmatrix} U_{0|J} \\ U_{J|k+L+1-J} \end{bmatrix} \\ &+ \begin{bmatrix} O_{L+1}C_k^s & H_{L+1}^s \end{bmatrix} \begin{bmatrix} E_{0|k} \\ E_{k|L+1} \end{bmatrix} \\ &- \begin{bmatrix} O_{L+1}A^kO_J^\dagger H_J^s & 0_{Lm\times(k+L+1-J)r} \end{bmatrix} \begin{bmatrix} E_{0|J} \\ E_{J|k+L+1-J} \end{bmatrix} \end{aligned} \tag{46}$$

Assume $k = J$ for the sake of simplicity. Then we have

$$\begin{aligned} Y_{J|L+1} &= O_{L+1}A^JO_J^\dagger Y_{0|J} \\ &+ \begin{bmatrix} O_{L+1}C_J^d - O_{L+1}A^JO_J^\dagger H_J^d & H_{L+1}^d \end{bmatrix} \begin{bmatrix} U_{0|J} \\ U_{J|L+1} \end{bmatrix} \end{aligned}$$

$$+ \begin{bmatrix} O_{L+1}C_J^s - O_{L+1}A^J O_J^\dagger H_J^s & H_{L+1}^s \end{bmatrix} \begin{bmatrix} E_{0|J} \\ E_{J|L+1} \end{bmatrix} \quad (47)$$

This last equation nicely shows the connection between the past and future data matrices and the unknown model matrices. As we will see, it also yields some important results. We have the following projection

$$Y_{J|L+1} \begin{bmatrix} U_{J|L+1} \\ U_{0|J} \\ Y_{0|J} \end{bmatrix}^\perp$$

$$= \begin{bmatrix} O_{L+1}C_J^s - O_{L+1}A^J O_J^\dagger H_J^s & H_{L+1}^s \end{bmatrix} \begin{bmatrix} E_{0|J} \begin{bmatrix} U_{J|L+1} \\ U_{0|J} \\ Y_{0|J} \end{bmatrix}^\perp \\ E_{J|L+1} \begin{bmatrix} U_{J|L+1} \\ U_{0|J} \\ Y_{0|J} \end{bmatrix}^\perp \end{bmatrix} \quad (48)$$

The projection of future inputs, past inputs and outputs onto the null-space of future noise is equal to $E_{J|L+1}$. We have separated the deterministic part of the system from the data by the projection in Equation (48). Hence we have the following theorem concerning the stochastic part of the system

Theorem 4.1 (Realization of the Toeplitz matrix H_{L+1}^s)
The lower triangular Toeplitz matrix H_{L+1}^s with Markov parameters for the stochastic sub-system is given by the projection of past inputs, past outputs, and future inputs onto the null-space of the future outputs, i.e.

$$Z_{J|L+1}^s \stackrel{\text{def}}{=} Y_{J|L+1} \begin{bmatrix} U_{J|L+1} \\ U_{0|J} \\ Y_{0|J} \end{bmatrix}^\perp$$

$$= P_{L+1}^s E_{0|L+1} \begin{bmatrix} U_{J|L+1} \\ U_{0|J} \\ Y_{0|J} \end{bmatrix}^\perp + H_{L+1}^s E_{J|L+1} \begin{bmatrix} U_{J|L+1} \\ U_{0|J} \\ Y_{0|J} \end{bmatrix}^\perp \quad (49)$$

where

$$P_{L+1}^s = O_{L+1}C_J^s - O_{L+1}A^J O_J^\dagger H_J^s \quad (50)$$

and when $N \to \infty$, then

$$Z_{J|L+1}^s \stackrel{\text{def}}{=} Y_{J|L+1} \begin{bmatrix} U_{J|L+1} \\ U_{0|J} \\ Y_{0|J} \end{bmatrix}^\perp = P_{L+1}^s E_{0|L+1} \begin{bmatrix} U_{0|J} \\ Y_{0|J} \end{bmatrix}^\perp + H_{L+1}^s E_{J|L+1} \quad (51)$$

which is a linear problem in H_{L+1}^s.

In order to see that this is a linear problem in the unknown matrix H^s_{L+1}, one only needs to write up the equations obtained by using the structure of H^s_{L+1}, Equation (15), and a partition of the last term on the right hand side into sub-matrices.

This result with proof is to our knowledge new. However, the same solution resulting from a QR decomposition was presented in Di Ruscio (1995b).

From Theorem 4.1 and Equation (47) we can immediately state the following important result. The extended observability matrix can be recovered from the column space of the matrix on the left hand side of the following matrix Equation (52).

Theorem 4.2 (Realization of the extended observability matrix O_{L+1})

Given the following matrix equation

$$Z_{J|L+1} \overset{\text{def}}{=} (Y_{J|L+1} / \begin{bmatrix} U_{J|L+1} \\ U_{0|J} \\ Y_{0|J} \end{bmatrix}) U^{\perp}_{J|L+1} =$$

$$O_{L+1} \begin{bmatrix} A^J O^{\dagger}_J & C^d_J - A^J O^{\dagger}_J H^d_J & C^s_J - A^J O^{\dagger}_J H^s_J \end{bmatrix} \begin{bmatrix} Y_{0|J} \\ U_{0|J} \\ E_{0|L+1} / \begin{bmatrix} U_{J|L+1} \\ U_{0|J} \\ Y_{0|J} \end{bmatrix} \end{bmatrix} U^{\perp}_{J|L+1} \tag{52}$$

then the column space of the matrix $Z_{J|L+1}$ coincides with the column space of the extended observability matrix O_{L+1} and the system order n of the SSM is given as the dimension of the column space.

The proof of Theorem 4.2 and Equation (52) is simple. From Theorem 4.1, Equation (49), we have

$$Y_{J|L+1} - Y_{J|L+1}/\tilde{\underline{Y}} = P^s_{L+1} E_{0|J} - P^s_{L+1} E_{0|J}/\tilde{\underline{Y}} + H^s_{L+1} E_{J|L+1} \tag{53}$$

where

$$\tilde{\underline{Y}} \overset{\text{def}}{=} \begin{bmatrix} U_{J|L+1} \\ U_{0|J} \\ Y_{0|J} \end{bmatrix} \tag{54}$$

and for the sake of simplicity the other matrices are defined according to Equations (49) and (50). Substituting Equation (53) into equation (47) in order to remove the stochastic term $H^s_{L+1} E_{J|L+1}$ gives

$$Y_{J|L+1}/\tilde{\underline{Y}} = O_{L+1} A^J O^{\dagger}_J Y_{0|J} + P^d_{L+1} U_{0|J} + P^s_{L+1} E_{0|J}/\tilde{\underline{Y}} + H^d_{L+1} U_{J|L+1} \tag{55}$$

Hence, an extra projection $U^{\perp}_{J|L+1}$ on the right hand side is necessary in order to remove the deterministic term $H^d_{L+1} U_{J|L+1}$ and in order to recover the extended observability matrix. The matrix on the right hand side of Equation (52) is proportional with the extended observability matrix O_{L+1}. Hence, the column space of the matrix $Z_{J|L+1}$ coincides with O_{L+1}. The dimension of the column space of the matrix $Z_{J|L+1}$ is the order n of the SSM. **QED.**

From Theorem 4.2 we immediately have the following corollary concerning the system dynamics.

Corollary 4.1 (Identification of system dynamics)
From $Z_{J|L+1}$ defined in Theorem 4.2, Equation (52) we have the following relationship

$$Z_{J+1|L} = \tilde{A} Z_{J|L} \tag{56}$$

where

$$\tilde{A} \stackrel{\text{def}}{=} O_L A (O_L^T O_L)^{-1} O_L^T$$

and where $Z_{J+1|L}$ is the L last $(m \times K)$ block rows in $Z_{J|L+1}$ and $Z_{J|L}$ is the L first $(m \times K)$ block rows in $Z_{J|L+1}$, i.e.

$$Z_{J+1|L} \stackrel{\text{def}}{=} (Y_{J+1|L} / \begin{bmatrix} U_{J|L+1} \\ U_{0|J} \\ Y_{0|J} \end{bmatrix}) U^{\perp}_{J|L+1}, \quad Z_{J|L} \stackrel{\text{def}}{=} (Y_{J|L} / \begin{bmatrix} U_{J|L+1} \\ U_{0|J} \\ Y_{0|J} \end{bmatrix}) U^{\perp}_{J|L+1}$$

Moreover, the column space of $Z_{J|L}$ coincides with the column space of the extended observability matrix O_L and the dimension of the column space is equal to the order n of the SSM. A realization of n, O_L, A, D and $\tilde{A}$ is determined by using Equation (56) in combination with Algorithm 4.1.

The matrix on the left hand side of Equation (52) can be shown to be equivalent to

$$\begin{aligned} Z_{J|L+1} &= Y_{J|L+1} W_c \\ W_c &\stackrel{\text{def}}{=} U^{\perp}_{J|L+1} W_1^T (W_1 U^{\perp}_{J|L+1} W_1^T)^{-1} W_1 U^{\perp}_{J|L+1} \end{aligned} \tag{57}$$

Comparing Equation (57) with Equation (30) shows that these matrices are related. We have shown that the column space of both equations coincides with the column space of the extended observability matrix. The difference can be viewed as a different column weighting matrix W_c, on the right hand side of Equation (57). Equation (30) can be viewed as a special case factorization of equation (57) with $k = J$ and $W_c = U^{\perp}_{J|L+1} W_1^T / K$. Later in Section 6, we will show that this last Equation (57) is extremely important and useful in order to partly compare other subspace identification methods.

From Theorem 4.1 we have the following result concerning the stochastic part of the system.

Theorem 4.3 (Realization of Δ and $\tilde{C}$)
Assume that the number of system input and output observations, $N \to \infty$. Define

$$Z^s_{J|L+1} \stackrel{\text{def}}{=} Y_{J|L+1} \begin{bmatrix} U_{J|L+1} \\ U_{0|J} \\ Y_{0|J} \end{bmatrix}^{\perp} \tag{58}$$

then we have

$$Z^s_{J+1|L} = \tilde{A} Z^s_{J|L} + \tilde{C} E_{J|L+1} \tag{59}$$

and

$$Z^s_{J+1|L}(Z^s_{J|L})^{\perp} = \tilde{C} E_{J|L+1}(Z^s_{J|L})^{\perp}$$

where

$$\tilde{C} \stackrel{\text{def}}{=} \begin{bmatrix} O_L C & H^s_L \end{bmatrix} - \tilde{A} \begin{bmatrix} H^s_L & 0_{Lm\times m} \end{bmatrix}$$

The covariance matrix of the innovations Δ is estimated directly from the column space of the left hand side matrix Equation (58), e.g.

$$Z^s_{J+L|1}(Z^s_{J|L})^{\perp} = F E_{J+L|1}(Z^s_{J|L})^{\perp} = F E_{J+L|1} \tag{60}$$

and

$$\Delta = F F^T \tag{61}$$

when

$$E_{J+L|1} E^T_{J+L|1} = I_{m\times m}$$

$\triangle$

The importance of Theorem 4.3 is that it shows that the innovations covariance matrix can be estimated directly from the column space of the data matrix, Equation (58). The factorization of the left hand side matrix of Equation (60) into the product $FE_{J+L|1}$ and $E_{J+L|1}E^T_{J+L|1} = I_{m\times m}$ can be performed by the QR decomposition. See also Section 5. The Kalman filter gain matrix C can be extracted when A and D is known.

Finally, we have the following Theorem 4.4 for the realization of the deterministic part of the system.

Theorem 4.4 (Realization of $\tilde{B}$)
Given

$$Z^d_{J|L+1} \stackrel{\text{def}}{=} Y_{J|L+1} / \begin{bmatrix} U_{J|L+1} \\ U_{0|J} \\ Y_{0|J} \end{bmatrix} \tag{62}$$

then we have

$$Z^d_{J+1|L} = \tilde{A} Z^d_{J|L} + \tilde{B} U_{J|L+1} \tag{63}$$

and

$$Z^d_{J+1|L} (Z^d_{J|L})^\perp = \tilde{B} U_{J|L+1} (Z^d_{J|L})^\perp \tag{64}$$

where

$$\tilde{B} \stackrel{\text{def}}{=} \begin{bmatrix} O_L B & H^d_L \end{bmatrix} - \tilde{A} \begin{bmatrix} H^d_L & 0_{Lm \times r} \end{bmatrix}$$

Theorem 4.4 is proved from Equation (55). Remark that the SSM matrix E can be estimated directly from Equation (64). This is so because of the structure of the ESSM matrix $\tilde{B}$.

All the projections in this section, Equations (49), (52) and (62) can be effectively computed from a QR decomposition, either directly from the projections defined in this section, or as will be shown in Section (5).

We will conclude this section by pointing out the relationship between the data matrices and the projection matrices. The data matrix with future system outputs and the projection matrices, (49) and (62) are related as

$$Y_{J|L+1} = Z^d_{J|L+1} + Z^s_{J|L+1}$$

Z^d represents the outputs from the deterministic part of the system. Z^s represents the outputs from the stochastic part of the system. $Y_{J|L+1}$ is the data matrix with future outputs from the combined deterministic and stochastic system.

5 Implementation with QR Decomposition

We will here use the QR decomposition in order to compute the column space of the projection matrices derived in Section 4. The QR decomposition is also used in the subspace identification methods by Verhagen (1994) and Van Overschee and De Moor (1994).

Define the following QR decomposition

$$\frac{1}{\sqrt{K}} \tilde{Y} = RQ =$$

$$\frac{1}{\sqrt{K}} \begin{bmatrix} U_{k|L+1} \\ W_i \\ Y_{k|L} \\ Y_{k+1|L} \end{bmatrix} = \begin{bmatrix} R_{11} & 0 & 0 & 0 \\ R_{21} & R_{22} & 0 & 0 \\ R_{31} & R_{32} & R_{33} & 0 \\ R_{41} & R_{42} & R_{43} & R_{44} \end{bmatrix} \begin{bmatrix} Q_1 \\ Q_2 \\ Q_3 \\ Q_4 \end{bmatrix} \tag{65}$$

where

$$R \in \mathbb{R}^{(r(L+1)+ni+2mL)\times(r(L+1)+ni+2mL)} \tag{66}$$

$$Q \in \mathbb{R}^{(r(L+1)+ni+2mL)\times K} \tag{67}$$

Note that this decomposition could perhaps more precisely have been defined as a lower Left, Q-orthogonal (LQ) decomposition. See Golub and Van Loan (1983) for the computation.

The QR decomposition can be viewed as a data compression step. The data matrix $\tilde{Y}$ which usually has a large number of columns is compressed to a usually much smaller lower triangular matrix R which contains all relevant information of the system for which the data was generated. As we will show, the orthogonal Q matrix is not needed in the algorithm.

Note that the first $(L-1)m$ rows in $Y_{k+1|L}$ are equal with the last $(L-1)m$ rows in $Y_{k|L}$. This means that $Y_{k+1|L}$ can be substituted for $Y_{k+L|1}$ in the QR decomposition, Equation (65). This is utilized in the efficient implementation of the **DSR** algorithm. However, for the sake of simplicity we will present the results according to Equation (65).

By definition, the instrumental variable matrix W_i is uncorrelated with $E_{k|L+1}$. We can therefore remove the noise matrix $E_{k|L+1}$ from Equation (7) by post-multiplying with $\frac{1}{K}W_i^T$. We have from (65) that

$$\lim_{K\to\infty}\frac{1}{K}E_{k|L+1}W_i^T = (\lim_{K\to\infty}\frac{1}{K}E_{k|L+1}\left[\begin{array}{cc} Q_1^T & Q_2^T \end{array}\right])\left[\begin{array}{c} R_{21}^T \\ R_{22}^T \end{array}\right] = 0 \tag{68}$$

Post-multiplying Equation (7) with $[Q_1^T Q_2^T]$, using (68) and substituting for the corresponding R_{ij} sub-matrices from (65) gives

$$\left[\begin{array}{cc} R_{41} & R_{42} \end{array}\right] = \tilde{A}\left[\begin{array}{cc} R_{31} & R_{32} \end{array}\right] + \tilde{B}\left[\begin{array}{cc} R_{11} & 0 \end{array}\right] \tag{69}$$

which gives one matrix equation for $\tilde{A}$ and one for both $\tilde{A}$ and $\tilde{B}$. We will in the next Sections 5.1 and 5.2 show how the order n and the system quadruple (A, B, D, E) are computed from (69).

The stochastic part of the system, defined by the matrices C and Δ, is computed from

$$R_{43} - \tilde{A}R_{33} = \tilde{C}E_{k|L+1}Q_3^T \tag{70}$$

$$R_{44} = \tilde{C}E_{k|L+1}Q_4^T \tag{71}$$

This will be shown in Section 5.3.

Note that the QR decomposition compresses the possible large data matrices into a number of (smaller) matrices which contain the information of the system. It is also interesting that the matrices (information) which define the deterministic part (A, B, D, E) and the stochastic part (Δ, C) are separated by the QR decomposition. The user must specify the parameter $k \geq J$ in Equation 65. See Figure 1 for a definition of the horizons involved. We recommend putting $k = J$. The matrix W_1 with $l = 0$, Equation (24), is recommended for W_i in 65.

5.1 Realization of A and D

We have from Equation (69) that

$$R_{42} = \tilde{A}R_{32} \tag{72}$$

and we choose

$$Z_{k+1|L} = R_{42} \tag{73}$$
$$Z_{k|L} = R_{32} = USV^T \tag{74}$$

in Algorithm (4.1) in order to determine A, D and the extended observability matrix O_L. The system order is determined by inspection of the dominant singular values of S or SS^T.

Note that the first $(L-1)m$ rows in R_{42} are equal to the last $(L-1)m$ rows in R_{32}. This is utilized in the efficient implementation of the **DSR** algorithm, in order to reduce the computational work.

Note also that if $\tilde{A}$ is computed as the projection of R_{32} onto R_{42} then $\tilde{A}$ takes a special canonical form. This is due to the common rows.

The A matrix may also be determined as follows. From the extended observability matrix we have

$$O_1 \stackrel{\text{def}}{=} O_L(1:(L-1)m, 1:n) \tag{75}$$
$$O_2 \stackrel{\text{def}}{=} O_L(m+1:Lm, 1:n) \tag{76}$$

then

$$A = (O_1^T O_1)^{-1} O_1^T O_2 \tag{77}$$

However, we must put $L =: L+1$ in this case if the extended observability matrix is estimated as the left singular vectors in (74). This will increase the computational work. This last method is the so called *shift invariance* method for computing the transition matrix A from the extended observability matrix, Kung (1978). The *shift invariance* method is used in the subspace algorithms (**N4SID**) by Van Overschee and De Moor (1994) and (**MOESP**) by Verhagen (1994). The parameter which defines the number of block rows in the **N4SID** and **MOESP** algorithms is denoted I. This parameter is related to the **DSR** parameter L as $I = L+1$. This is one of the differences between the **DSR** algorithm and the **N4SID** and **MOESP** algorithms.

This means that **N4SID** and **MOESP** computes a number $Im = (L+1)m$ of singular values. However, the system order can only be chosen according to Lm of these singular values, i.e. the maximum system order which can be chosen for a specified parameter $I = L+1$ is $n = Lm$. For comparison, the **DSR** algorithm computes only a number Lm of singular values and the maximum system order which can be chosen for a specified parameter L is $n = Lm$. Hence, the **DSR** algorithm seems to be more consistent with respect to choosing the system order as the number of non-zero singular values.

The *shift invariance* method can be included in the **DSR** algorithm, but at a higher computational expense. This strategy can be described as follows. The extended observability matrix can be estimated from the column space of the matrix formed from R_{32} and R_{42}. Compute the SVD

$$\begin{bmatrix} R_{32} \\ R_{42} \end{bmatrix} = \begin{bmatrix} U_{11} & U_{12} \\ U_{21} & U_{22} \end{bmatrix} \begin{bmatrix} S_n & 0 \\ 0 & 0 \end{bmatrix} \begin{bmatrix} V_1^T \\ V_2^T \end{bmatrix} = \begin{bmatrix} U_{11} S_n V_1^T \\ U_{21} S_n V_1^T \end{bmatrix}$$

We then have that A and D are determined from the *shift invariance* method, e.g. from the left singular vectors as follows

$$\begin{aligned} O_L &= U_{11} && (78)\\ O_L A &= U_{21} && (79)\\ A &= (U_{11}^T U_{11})^{-1} U_{11}^T U_{21} && (80)\\ D &= U_{11}(1:m, 1:n) && (81) \end{aligned}$$

5.2 Realization of B and E

We have from Equation (69) that

$$\tilde{B} R_{11} = R_{41} - \tilde{A} R_{31} \tag{82}$$

$\tilde{B}$ can be determined directly from (82) if the input u is persistently exiting of order $L+1$. R_{11} is non singular in this case. We have

$$\tilde{B} = (R_{41} - \tilde{A} R_{31}) R_{11}^T (R_{11} R_{11}^T)^{-1} \tag{83}$$

The B and E matrices are then extracted from $\tilde{B}$ as pointed out in Section 4.2.

At this stage, the system order is identified (Algorithm 4.1). It is possible to determine B and E if the input is only persistently exiting of order $p+1$ where $L_{\min} \leq p \leq L$, directly without recomputing the algorithm with $L = L_{\min}$ or $L = p$. The minimal observability index, for a given system order, is $L_{\min} = n - \text{rank}(D) + 1$ when $n \geq \text{rank}(D)$ and $L_{\min} = 1$ when $n \leq \text{rank}(D)$.

Define

$$\begin{aligned} O_p &= O_L(1:mp, 1:n) && (84)\\ \tilde{A}_p &= O_p A (O_p^T O_p)^{-1} O_p^T && (85)\\ R_{11}^p &= R_{11}(1:r(p+1), 1:r(p+1)) && (86)\\ R_{31}^p &= R_{31}(1:mp, 1:r(p+1)) && (87)\\ R_{41}^p &= R_{41}(1:mp, 1:r(p+1)) && (88) \end{aligned}$$

We then have

$$\tilde{B}_p R_{11}^p = R_{41}^p - \tilde{A}_p R_{31}^p \tag{89}$$

This result is a consequence of Equation (7) with L substituted with p. Note that the minimal observability index, for a given system order, is $p = n - \text{rank}(D) + 1$ when $n \geq \text{rank}(D)$. The B and E matrices are then extracted from $\tilde{B}_p$ as shown in Section 4.2 and Algorithm 4.2 with L substituted with p.

5.3 Realization of C and Δ

Corollary 5.1 (Realization of C and Δ)
Consider the lower left triangular matrix R, determined by the QR decomposition

in Equation (65). An estimate of the square root of the innovations covariance matrix is given by the $m \times m$ lower right sub-matrix of R, i.e.

$$F = R_{44}(m(L-1)+1 : mL, m(L-1)+1 : mL) \tag{90}$$

and the estimate of the innovations covariance matrix is

$$\Delta = FF^T \tag{91}$$

Furthermore, when $J > 1$, then an estimate of the Kalman filter gain matrix C can be computed from

$$O_L CF = R_{43}(1 : mL, 1 : m) \tag{92}$$

If F is non-singular, then we have

$$C = (O_L^T O_L)^{-1} O_L^T R_{43}(1 : mL, 1 : m) F^{-1} \tag{93}$$

An estimate of the lower left block triangular Toeplitz matrix H_L^s for the stochastic subsystem (D, A, CF, F) is given by

$$H_L^s = R_{43}(1 : mL, m+1 : m(L+1)) \tag{94}$$

The estimate of the lower left block triangular Toeplitz matrix for the stochastic subsystem (D, A, C, I), according to Equation (15), can be formed from the block columns $F_1, \cdots, F_L$ which can be computed from

$$\begin{bmatrix} F_1F & F_2F & \cdots & F_LF \end{bmatrix} = R_{43}(1 : mL, m+1 : m(L+1)) \tag{95}$$

$\triangle$

The stochastic subsystem is identified separately from the deterministic subsystem. The necessary separation into deterministic and stochastic subsystems are implicitly done by the QR decomposition.

The first $(L-1)m$ rows in $Y_{k+1|L}$ are equal to the $(L-1)m$ last rows in $Y_{k|L}$. Q_4 is uncorrelated with $U_{k|L+1}$, W_i, $Y_{k|L}$ and with the $(L-1)m$ first rows in $Y_{k+1|L}$. The first $(L-1)m$ rows in $Y_{k+1|L}Q_4^T$ and $E_{k|L+1}Q_4^T$ are therefore zero. We then have from (65) and the structure of $\tilde{C}$, given by (13) and (15), that

$$Y_{k+1|L}Q_4^T = \tilde{C}E_{k|L+1}Q_4^T = R_{44} = \begin{bmatrix} 0 & 0 & \cdots & 0 \\ 0 & 0 & & 0 \\ \vdots & \vdots & \ddots & \vdots \\ 0 & 0 & \cdots & F \end{bmatrix} \tag{96}$$

where $R_{44} \in \mathbb{R}^{Lm \times Lm}$. Hence, the square root of the innovations noise process covariance matrix is estimated directly as the $m \times m$ lower left matrix, denoted F, in the lower triangular matrix R from the QR decomposition, Equation (65). Note that F also is lower left triangular and can be compared to a Cholesky factorization of Δ. This result is believed to be of some importance. The result (91) is then clarified.

The matrices Q_i, $i = 1, 2, 3, 4$, are orthogonal matrices and we have

$$\frac{1}{K} E_{k|L+1} Q_3^T = \begin{bmatrix} F & 0 & \cdots & 0 \\ 0 & F & & 0 \\ \vdots & & \ddots & \vdots \\ 0 & 0 & \cdots & F \\ 0 & 0 & \cdots & 0 \end{bmatrix} \in \mathbb{R}^{(L+1)m \times Lm} \tag{97}$$

Another strategy is then to compute OCF from Equations (70) and (97), and an algorithm similar to Algorithm 4.2. This is formulated in the following Corollary 5.2.

Corollary 5.2 (Realization of C)
Given the sub-matrices R_{43} and R_{33} from the QR decomposition in Equation (67) and the ESSM transition matrix $\tilde{A}$. Define according to Equation (70)

$$\tilde{CF} \stackrel{\text{def}}{=} R_{43} - \tilde{A} R_{33} \tag{98}$$

The matrix $O_L CF$ can then be computed from $\tilde{CF}$, e.g. by a procedure similar to Algorithm 4.2.
$\triangle$

This strategy is found from Monte Carlo simulations to be the best one when the past horizon parameter is $J = 1$, but no significant difference is found when $J > 1$.

5.4 Special Remarks

One advantage of the QR implementation of the algorithm is that potential ill-conditioning of the covariance matrices are concentrated in a certain triangular matrix. This ill-conditioning usually results from ill-conditioned noise processes (process noise and measurements noise) and from rounding-off errors. Note that the triangular matrix R is the square root of the covariance matrix ($H = \frac{1}{K}\tilde{Y}\tilde{Y}^T$ where $\tilde{Y}$ is defined in (65)) and that the triangular matrix is computed without ever computing the covariance matrix. The method can therefore be defined as numerically stable.

The QR decomposition is not unique. The R matrix is post-multiplied by a diagonal permutation matrix E such that $R := RE$ has positive diagonal elements. The diagonal elements of E are equal to the sign of the corresponding diagonal elements of R which was the result from the QR decomposition. Note also that $Q := EQ$ and $EE = I$. This gives a more unique coordinate system for the estimated (A, B, D, E) quadruple. This scaling is also one of the reasons for the simple solutions for the C and Δ matrices in Section 5.3. The scaling ensures that the diagonal blocks of (97) gets the same sign.

Note that common rows in the data matrices $Y_{k|L}$ and $Y_{k+1|L}$ can be removed prior to obtaining the QR decomposition in equation (67). It is also clear from the above that the orthogonal Q matrix ($QQ^T = I$) is not needed in the algorithm. This will reduce the computational effort considerably. In fact, the QR factorization works on a data matrix of size only $r(2L+1)+m(2L+1) \times K$ and not of size $r(2L+1)+3mL \times K$ as indicated in (65).

Another strategy for determining R is to first compute $H = \frac{1}{K}\tilde{Y}\tilde{Y}^T$ where $\tilde{Y}$ is defined in (65) and then the SVD, $H = USV^T$, followed by a QR decomposition of $US^{\frac{1}{2}}$ in order to obtain the lower triangular matrix, R. This strategy reduced the number of flops and increased the accuracy of the R matrix when MATLAB was used for the computations. However, no significant difference in the estimated models was observed. This strategy can be numerically ill-conditioned due to possible rounding errors when forming the product of rows in $\tilde{Y}$ with columns in $\tilde{Y}^T$ in order to compute the correlation matrix $\tilde{Y}\tilde{Y}^T$. This strategy is therefore not recommended.

Equation (82) is defined by the triangular factors for Equation (40) with $k \geq J$.

$$\tilde{B}U_{k|L+1}U_{k|L+1}^T = Y_{k+1|L}U_{k|L+1}^T - \tilde{A}Y_{k|L}U_{k|L+1} \tag{99}$$

It is also possible to extract the triangular factors for

$$\tilde{B}U_{0|L+1}U_{0|L+1}^T = Y_{1|L}U_{0|L+1}^T - \tilde{A}Y_{0|L}U_{0|L+1} \tag{100}$$

directly from the QR decomposition (65) when $k = J = L$. The first block row in $U_{L|L+1}$ is equal to the last block row in $U_{0|L+1}$. Hence, the matrices in Equation (100) can be defined from the lower triangular R matrix, equation (65). It is therefore natural to choose

$$\tilde{B}\left[\begin{array}{cc} R_{11} & R_{11}^0 \end{array}\right] = \left[\begin{array}{cc} R_{41} & R_{41}^0 \end{array}\right] - \tilde{A}\left[\begin{array}{cc} R_{31} & R_{31}^0 \end{array}\right] \tag{101}$$

for defining $\tilde{B}$ and an equation for computing the B and E system matrices. Equation (101) consists of the triangular factors for both $\frac{1}{K}U_{k|L+1}U_{k|L+1}^T$ and $\frac{1}{K}U_{0|L+1}U_{0|L+1}^T$. Equation (101) has effect for systems where the input signal is poor with frequencies, but gives no additional effect compared to (82) for e.g. white noise inputs.

Note that the stochastic part of the model is determined from QR and SV Decompositions only. The Markov parameters and the square root of the innovations covariance matrix are determined from a QR decomposition only. The Kalman filter gain matrix is determined from the Markov parameters and the extended observability matrix. No matrix Lyapunov or non-linear matrix Riccati equations have to be solved.

The method has in this work been illustrated for systems which are not strictly proper (only proper), i.e. the case when $E \neq 0$ in the underlying model (2). The method can also be implemented to handle proper systems, i.e. systems where E is known to be zero. This can be done by deleting the last block row in $U_{k|L+1}$ and the last block column in $\tilde{B}$, see Equation (7).

6 Comparison with Existing Algorithms

A comparison with the **DSR** algorithm and three different subspace algorithms will be given, namely **N4SID**, Van Overschee and De Moor (1994), **CVA**, Larimore (1983), (1990) and **PO-MOESP**, Verhagen (1994).

The first and common step in subspace identification algorithms is to estimate the extended observability matrix from the column space of a known data matrix. We will therefore concentrate our discussion on the similarities and differences in which these methods estimate the extended observability matrix. We will only briefly discuss how the system matrices are estimated by the different methods.

It is shown in Van Overschee and De Moor (1995) that the different methods are related through certain row and column weightings with the **N4SID** data matrix as the key matrix.

We will below present a different approach, with the matrix $Z_{J|L+1}$ defined in Section 4.3, Theorem 4.2 and Equation (57), as the key matrix.

Multiplying Equation (57) from left with an extra row weighting matrix W_r and using the SVD as discussed in Algorithm 4.1 gives

$$W_r Z_{J|L+g} = W_r Y_{J|L+g} W_c = U_1 S_n V_1^T + U_2 S_2 V_2^T \tag{102}$$

where g is an integer parameter and

$$W_c \stackrel{\text{def}}{=} U_{J|L+1}^{\perp} W_1^T (W_1 U_{J|L+1}^{\perp} W_1^T)^{-1} W_1 U_{J|L+1}^{\perp}$$

The extended observability matrix can be estimated as, e.g.

$$O_{L+g} = W_r^{-1} U_1 \tag{103}$$

where W_r is a non-singular matrix. The matrix W_c is symmetric and can be written as

$$W_c \stackrel{\text{def}}{=} \underbrace{U_{J|L+1}^{\perp} W_1^T (W_1 U_{J|L+1}^{\perp} W_1^T)^{-\frac{1}{2}}}_{W_c^2} \underbrace{(W_1 U_{J|L+1}^{\perp} W_1^T)^{-\frac{1}{2}} W_1 U_{J|L+1}^{\perp}}_{(W_c^2)^T} \tag{104}$$

From the above factorization of the matrix W_c we have at least four matrices (W_c^i, $i = 1, \cdots, 4$) which all have essentially the same column space as W_c, i.e.

$$\begin{aligned}
W_c^1 &= U_{J|L+1}^{\perp} W_1^T (W_1 U_{J|L+1}^{\perp} W_1^T)^{-1} W_1 U_{J|L+1}^{\perp} \\
W_c^2 &= U_{J|L+1}^{\perp} W_1^T (W_1 U_{J|L+1}^{\perp} W_1^T)^{-\frac{1}{2}} \\
W_c^3 &= U_{J|L+1}^{\perp} W_1^T (W_1 U_{J|L+1}^{\perp} W_1^T)^{-1} K \\
W_c^4 &= U_{J|L+1}^{\perp} W_1^T \frac{1}{K} \\
W_c^5 &= U_{J|L+1}^{\perp}
\end{aligned}$$

The matrix W_c^5 is sufficient only for purely deterministic systems and is shown for the sake of completeness.

These column weighting matrices are used in the **DSR** algorithm which is presented in this work. The past horizon parameter J is usually chosen as

$J \geq L$. The parameter $g = 0$ in Equation (102) and the row weighting matrix is the $Lm \times Lm$ identity matrix, denoted $W_r = I_{Lm}$. Algorithm 4.1 is used in order to identify the extended observability matrix O_L from the column space of the matrix $Z_{J|L}$.

We will now illustrate the **DSR** algorithm's similarity and difference to two published algorithms, **CVA** by Larimore (1990) and **PO-MOESP** by Verhagen (1994).

6.1 PO-MOESP

The **PO-MOESP** algorithm in Verhagen (1994) estimates the extended observability matrix O_{L+1} from Equations (102) and (103) with the following matrices

$$\left.\begin{aligned} W_c &= U^{\perp}_{L+1|L+1} W_1^T (W_1 U^{\perp}_{L+1|L+1} W_1^T)^{-1} W_1 U^{\perp}_{L+1|L+1} \\ W_r &= I_{(L+1)m} \\ g &= 1 \end{aligned}\right\} \textbf{PO-MOESP} \qquad (105)$$

From Theorem 4.2 and the factorization in Equation (104) we conclude that the two algorithms **PO-MOESP** and **DSR** estimate the extended observability matrix from a known data matrix which essentially has the same column space. The only difference is that **PO-MOESP** estimates the extended observability matrix O_{L+1} (of larger size than **DSR** does) in order to use the shift invariance method for extracting the system matrix A as explained in Section 5.1.

Using the triangular factors from the QR decomposition in Section 5, we have

$$R_{32}R_{22}^T(R_{22}R_{22}^T)^{-1}R_{22}Q_2 = (U_1 S_n V_1^T + U_2 S_2 V_2^T)Q_2 \qquad (106)$$

where the orthogonal matrix Q_2 is not needed because the column space can be estimated as the matrix U_1.

A major difference is that the **PO-MOESP** algorithm does not estimate the stochastic part of the model. **DSR** estimates the Kalman gain and innovations covariance matrix directly from the data as shown in Sections 4.3 and 5.3.

6.2 Canonical Variate Analysis (CVA)

The **CVA** algorithm in Larimore (1990) estimates the extended observability matrix O_{L+1} from Equations (102) and (103) with the following matrices

$$\left.\begin{aligned} W_c &= U^{\perp}_{L+1|L+1} W_1^T (W_1 U^{\perp}_{L+1|L+1} W_1^T)^{-\frac{1}{2}} \\ W_r &= (Y_{L+1|L+1} U^{\perp}_{L+1|L+1} Y^T_{L+1|L+1})^{-\frac{1}{2}} \\ g &= 1 \end{aligned}\right\} \textbf{CVA} \qquad (107)$$

As we can see, the column weighting matrix W_c used by the **CVA** algorithm fits into the factorization in Equation (104). A difference is that the **CVA** algorithm uses a row weighting matrix W_r.

The only difference is that the **DSR** algorithm takes the SVD of a matrix of size only $Lm \times J(r + m)$ where usually $J = L$, in order to identify O_L. The other methods take the SVD of a matrix of size $(L+1)m \times J(r+m)$, in order to identify O_{L+1}. This is to separate out the sub-matrices O_L and $O_L A$ from O_{L+1}. See also Section 5.1 for a discussion.

From Theorem 4.2 and the factorization in Equation (104) we conclude that the two algorithms **CVA** and **DSR** essentially have the same column space.

An interpretation of the **CVA** algorithm is that the system order is estimated as the number of principal angles between the matrix $Y_{L+1|L+1}U^{\perp}_{L+1|L+1}$ and $W_1 U^{\perp}_{L+1|L+1}$ different from $\pi/2$. The principal angles can be effectively computed using the SVD, see e.g. Van Overschee (1995), p. 29 and Golub and Van Loan (1989), p. 428.

By using the triangular factors as shown in Section 5 we get the following method for computing the principal angles

$$(R_{32}R_{32}^T + R_{33}R_{33}^T)^{-\frac{1}{2}} R_{32}R_{22}^T (R_{22}R_{22}^T)^{-\frac{1}{2}} = U_1 S_n V_1^T + U_2 S_2 V_2^T \tag{108}$$

The system order is here identified as the number of singular values equal to one.

The next step in the **CVA** algorithm is to define a memory which defines a valid sequence of system states. The system matrices can then be estimated from a least squares problem.

6.3 N4SID

The **N4SID** algorithm is different. The following weighting matrices are used

$$\left.\begin{aligned} W_c &= (W_1 U^{\perp}_{L+1|L+1} W_1^T)^{-1} W_1 \\ W_r &= I_{(L+1)m} \\ g &= 1 \end{aligned}\right\} \textbf{N4SID} \tag{109}$$

The column weighting matrix used in the **N4SID** algorithm does not generally have the same column space as W_c or any of the column weighting matrices which result from Equation (104) and Theorem 4.2. This is possibly the reason for why **N4SID** gives bad results for deterministic input signals.

In Viberg (1995) it is pointed out that only the difference between the **PO-MOESP** and **N4SID** algorithms is the extra projection $U^{\perp}_{L+1|L+1}$. However, it is also claimed that *the resulting subspace estimates should therefore have very similar properties.* From Theorem 4.2 and the above discussion we conclude that this conclusion in Viberg (1995) is wrong. This is illustrated in example 2, Section 7.2.

From the above discussion we have the following relationship between the column weighting matrix W_c in Equation (104) and the matrix W_c in (109) used by **N4SID**.

$$W_c = W_c^{\text{N4sid}} U^{\perp}_{L+1|L+1} \tag{110}$$

In Theorem 4.2, Equations (52) and (55), it is proved that the extra projection $U^{\perp}_{L+1|L+1}$ is necessary in order to remove the deterministic term $H^d_{L+1} U_{L+1|L+1}$ and establish the data matrix $Z_{L+1|L+1}$ which has the same column space as the extended observability matrix. See also Example 2, Section 7.2, for an illustration.

The **N4SID** method computes the SVD of the data matrix defined in Equation (102) with the above matrices W_r and W_c, Equation (109). If the triangular factors as shown in Section 5 are used then we have

$$R_{32}R_{22}^T(R_{22}R_{22}^T)^{-1}[R_{21}R_{22}]\begin{bmatrix} Q_1 \\ Q_2 \end{bmatrix} = (U_1S_nV_1^T + U_2S_2V_2^T)\begin{bmatrix} Q_1 \\ Q_2 \end{bmatrix} \tag{111}$$

The orthogonal matrices Q_1 and Q_2 are not used. The system order is identified as the number of non-zero singular values and the extended observability matrix O_{L+1} is estimated from the column space. The rest of the **N4SID** algorithm can briefly be described as follows. From the extended observability matrix O_{L+1} and the system data, a valid sequence of system states are estimated and a least squares problem is solved in order to construct the system matrices A, B, D, E. The covariance matrices for the process and measurements noise are then identified from a residual, and a Riccati equation is solved in order to construct the Kalman filter gain matrix C and the innovations covariance matrix Δ.

The **DSR** method does not use state sequences and the Kalman gain matrix C and the innovations covariance matrix Δ are constructed directly from the data, without recursions of non-linear matrix Riccati equations.

6.4 Main Differences and Similarities

Both of the algorithms **N4SID** and **CVA** estimate in the first instance a sequence of states. When the states are known, the state space model matrices can be determined by simple linear regression. Both methods must solve a matrix Riccati equation in order to identify the Kalman filter gain and the innovations covariance matrices.

The **DSR** algorithm is based on first writing up an extended state space model (ESSM) where the unknown states are eliminated from the problem. The ESSM shows us the relationship between the known data matrices and the SSM matrices. Hence, the **DSR** algorithm does not have any problems with unknown states; unknown initial values etc. The state space model matrices are then extracted from the ESSM. **DSR** estimates the Kalman gain and innovations covariance matrices (stochastic part of the model) directly from the data, without recursions of non-linear matrix equations, e.g. the Riccati equation.

The **PO-MOESP** algorithm does not estimate the stochastic part of the model.

We have shown that the **CVA**, **PO-MOESP** and **DSR** algorithms gives consistent estimates of the extended observability matrix. The algorithms fit into the same Theorem 4.2. We have shown that the **N4SID** algorithm in general does not give consistent estimates of the extended observability matrix. However, it will give consistent results if an extra projection of future inputs is included. From the above discussion we have the following relationship between the column weighting matrix W_c in Equation (104) and the matrices used by **PO-MOESP**, **CVA** and **N4SID**. See (105), (107) and (109).

$$W_c = W_c^{\text{PO-MOESP}} = W_c^{\text{CVA}}(W_c^{\text{CVA}})^T = W_c^{\text{N4sid}} U_{L+1|L+1}^{\perp}$$

These are the most important similarities and differences between the method presented in this work and previously published methods.

7 Numerical Examples

7.1 Example 1: Monte Carlo Simulation

A single input single output (SISO) system with one state is chosen to compare the algorithm presented in this paper, entitled **DSR** (**D**eterministic and **S**tochastic System Identification and **R**ealization), with two other algorithms, **CVA** (which stands for Canonical Variate Analysis, Larimore (1983), (1990)) and the prediction error method implemented in the MATLAB function **ARMAX** (i.e., in the system identification toolbox, Ljung (1991)).

$$x_{k+1} = 0.9x_k + 0.5u_k + 0.6e_k \tag{112}$$

$$y_k = 1.0x_k - 1.0u_k + e_k \tag{113}$$

Three types of input signals were used. One input equal to a sum of four sinusoid signals, u^1, one input equal to a white noise signal with unit covariance, u^2, and one equal to a sine, u^3.

u^1 $\quad u_k = 0.2(\sin(\frac{k}{25}) + \sin(\frac{k}{10}) + \sin(\frac{k}{5}) + \sin(k))$

u^2 $\quad$ White noise, unit covariance

u^3 $\quad u_k = \sin(k)$

For each input the time series (y_k, u_k) was generated by simulating the model with 100 different white noise sequences e_k, also with unit variance.

The **DSR** algorithm parameter L was changed from 1 to 5 and the **CVA** parameter I from 2 to 6. For each L and I, the mean and standard deviation of the parameters of the 100 different estimated models are presented in Tables 1 to 6. The results obtained by the **ARMAX** algorithm are also shown in the tables. See Ljung (1991) for the description of the parameters $nn = [1, 2, 1, 0]$ which is used as arguments for **ARMAX**.

The true deterministic system quadruple is denoted $(a, b, d, e) := (0.9, 0.5, 1, -1)$ and the deterministic steady state gain and deterministic zero are denoted $H^d(1) = 4.0$ and $p_d(1) = 1.4$, respectively. The parameters in the stochastic part of the model are $(c, \Delta) := (0.6, 1)$. The stochastic steady state gain and stochastic zero are denoted $H^s(1) = 7.0$ and $p_s(1) = 0.3$, respectively. The signal to noise ratio is approximately 0.4, hence, the identification problem is not simple.

The **CVA** algorithm sometimes estimated systems with negative (b,d) parameters, i.e. sometimes an estimated quadruple (a,b,d,e) and sometimes (a,-b,-d,e). This happened with the algorithm parameter $I = 3$ and with a random input signal. It is believed that this can be avoided by using a scaling similar to that presented in Section 5.4.

The results are very good for both the **DSR** and the **CVA** algorithms, see Tables 1 and 2. There are small differences in the estimated models for both methods when N is large, see Tables 1 and 2. This indicates that the asymptotic statistical distribution of the parameter estimates is the same. The example indicates that both the **DSR** and **CVA** algorithms are insensitive to variation in the number of block rows.

The **DSR** algorithm is found to be marginally better than the **CVA** when the number of samples is small (for this example and with $N = 200$ and $N = 500$ samples) see Tables 3 to 6.

It is also interesting to observe that the results from **DSR** are as good as the results from the **ARMAX** function, even for a simple SISO system. This indicates that **DSR** gives asymptotically statistical optimal results for this example, both for purely deterministic inputs (u^1 and u^3) and stochastic input sequences (u^2). Note that the prediction error method (**ARMAX** function) is based on iterative optimization but that the **DSR** algorithm is based on SVD and QR decompositions only. Prediction error methods are rather complicated for MIMO systems while the **DSR** algorithm is very simple.

Figures 2 to 8 are included in order to illustrate the asymptotic mean and variance properties of the **DSR** algorithm for varying horizon parameters L and J. The figures illustrate that the algorithm is numerically robust and that the estimates are consistent and fairly insensitive for the parameters L and J.

Table 1: Mean of the parameters of 100 different estimated models: input type u^1, $N = 10000$. The models was scaled so that $d = 1$.

Alg., Par.	a	b	e	$p_d(1)$	$H^d(1)$	c	Δ	$p_s(1)$	$H^s(1)$
DSR, L=1	0.8998	0.497	-1.002	1.396	3.9598	0.6088	1.0045	0.2910	7.0910
DSR, L=2	0.8995	0.497	-1.002	1.398	3.9516	0.6003	0.9989	0.2992	6.9867
DSR, L=3	0.8998	0.497	-1.002	1.398	3.9604	0.5992	0.9985	0.3006	6.9940
DSR, L=4	0.8996	0.497	-1.002	1.398	3.9572	0.5992	0.9981	0.3004	6.9875
DSR, L=5	0.8998	0.496	-1.002	1.398	3.9644	0.5989	0.9977	0.3009	6.9976
CVA, I=2	0.8996	0.497	-1.002	1.398	3.9545	0.6008	1.0046	0.2988	7.0015
CVA, I=3	0.8996	0.497	-1.003	1.398	3.9572	0.5995	1.0000	0.3001	6.9849
CVA, I=4	0.8998	0.495	-1.003	1.396	3.9471	0.5990	0.9994	0.3008	6.9930
CVA, I=5	0.8996	0.498	-1.001	1.400	3.9714	0.5992	0.9993	0.3003	6.9830
CVA, I=6	0.8996	0.498	-1.002	1.399	3.9659	0.5993	0.9993	0.3003	6.9872
ARMAX	0.8995	0.498	-1.002	1.399	3.9604	0.5995		0.3000	6.9769

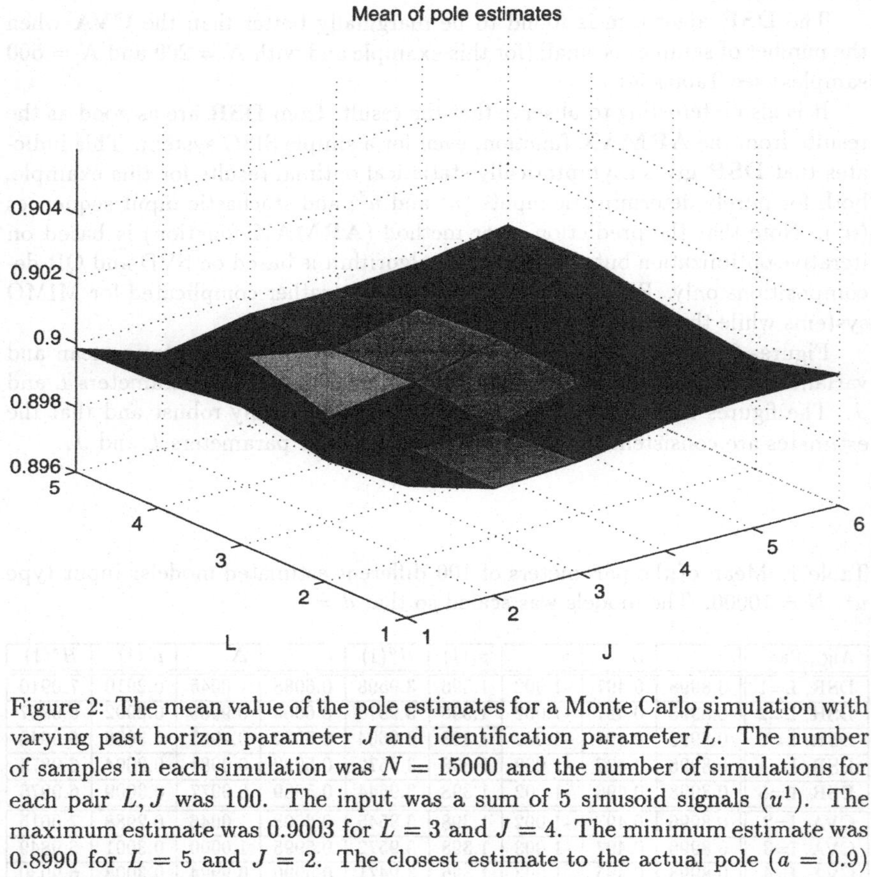

Figure 2: The mean value of the pole estimates for a Monte Carlo simulation with varying past horizon parameter J and identification parameter L. The number of samples in each simulation was $N = 15000$ and the number of simulations for each pair L, J was 100. The input was a sum of 5 sinusoid signals (u^1). The maximum estimate was 0.9003 for $L = 3$ and $J = 4$. The minimum estimate was 0.8990 for $L = 5$ and $J = 2$. The closest estimate to the actual pole ($a = 0.9$) was 0.89997 for $L = 3$ and $J = 5$.

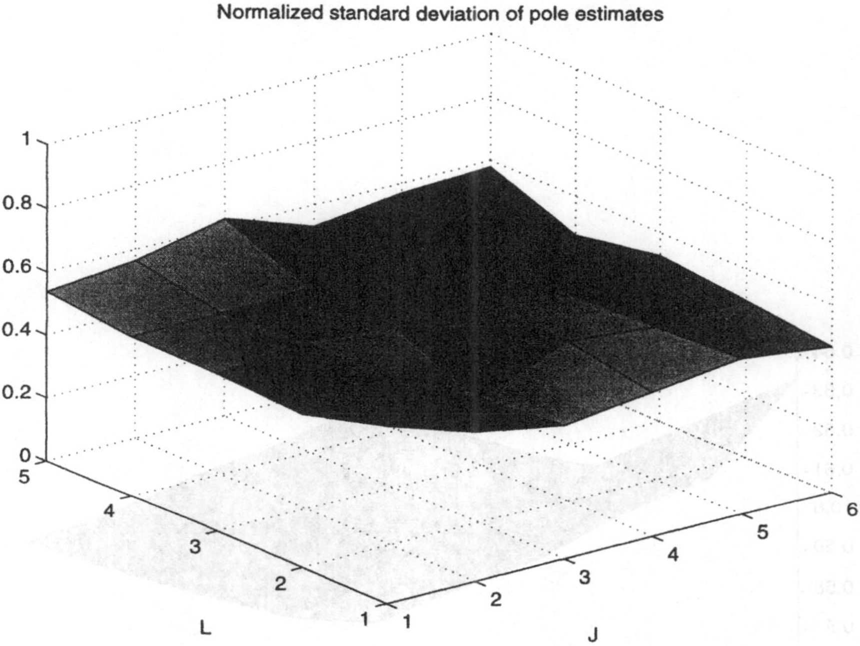

Figure 3: The standard deviation multiplied (normalized) with $N^{1/2}$ of the pole estimates for a Monte Carlo simulation with varying past horizon parameter J and varying identification parameter L. The number of samples in each simulation was $N = 15000$ and the number of simulations for each pair L, J was 100. The input was a sum of 5 sinusoid signals (u^1). The maximum standard deviation of the pole estimates was $0.6265/N^{1/2}$ at $L = 5$ and $J = 3$. The minimum standard deviation was $0.4005/N^{1/2}$ at $L = 2$ and $J = 3$.

Table 2: Standard deviation of the parameters of 100 different estimated models: input type u^1, $N = 10000$.

Alg., Par.	a	b	e	$p_d(1)$	$H^d(1)$	c	Δ	$p_s(1)$	$H^s(1)$
DSR, L=1	0.0054	0.027	0.068	0.045	0.3020	0.0101	0.0153	0.0130	0.3170
DSR, L=2	0.0052	0.026	0.068	0.043	0.3063	0.0099	0.0152	0.0121	0.3021
DSR, L=3	0.0054	0.026	0.068	0.043	0.3190	0.0102	0.0152	0.0128	0.3211
DSR, L=4	0.0056	0.026	0.068	0.044	0.3236	0.0105	0.0151	0.0135	0.3309
DSR, L=5	0.0059	0.026	0.068	0.043	0.3328	0.0111	0.0151	0.0145	0.3447
CVA, I=2	0.0052	0.027	0.068	0.044	0.2959	0.0101	0.0153	0.0124	0.2994
CVA, I=3	0.0051	0.029	0.069	0.047	0.3423	0.0097	0.0152	0.0118	0.3009
CVA, I=4	0.0053	0.033	0.070	0.052	0.3920	0.0096	0.0152	0.0118	0.3155
CVA, I=5	0.0055	0.038	0.070	0.057	0.4456	0.0096	0.0151	0.0119	0.3236
CVA, I=6	0.0058	0.037	0.07	0.056	0.4554	0.0096	0.0152	0.0122	0.3391
ARMAX	0.0051	0.027	0.066	0.042	0.2925	0.0095		0.0115	0.2983

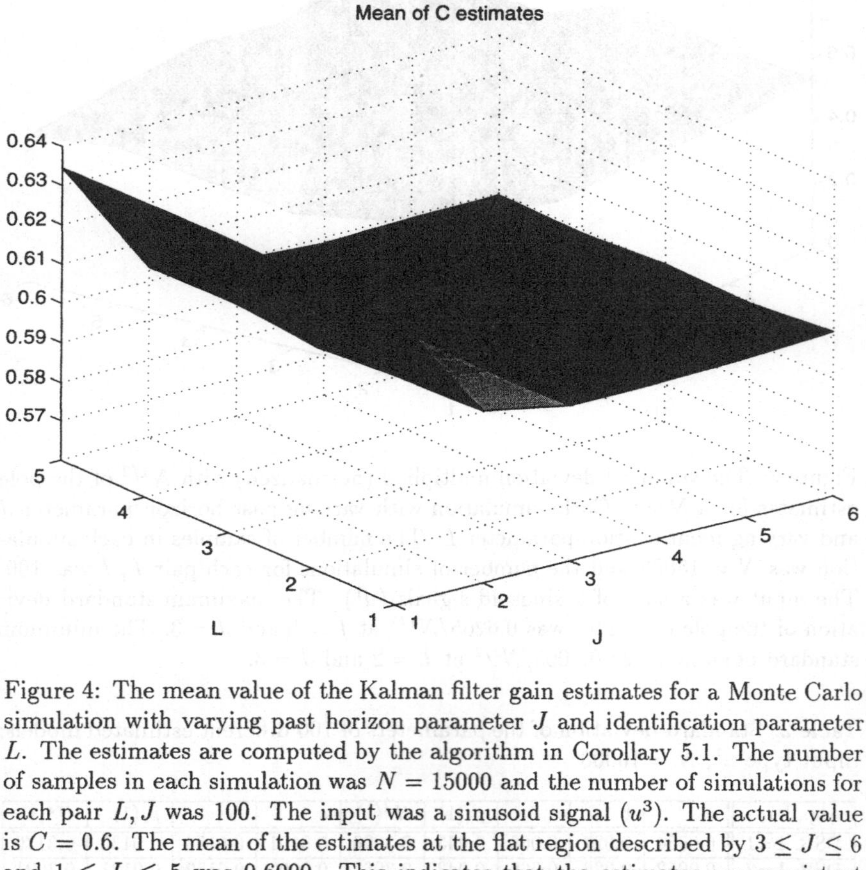

Figure 4: The mean value of the Kalman filter gain estimates for a Monte Carlo simulation with varying past horizon parameter J and identification parameter L. The estimates are computed by the algorithm in Corollary 5.1. The number of samples in each simulation was $N = 15000$ and the number of simulations for each pair L, J was 100. The input was a sinusoid signal (u^3). The actual value is $C = 0.6$. The mean of the estimates at the flat region described by $3 \leq J \leq 6$ and $1 \leq L \leq 5$ was 0.6000. This indicates that the estimates are consistent for past horizon parameters $J > 2$ independently of the choice of L, but clearly biased for $J = 1$. The estimates for $J = 2$ should be further investigated.

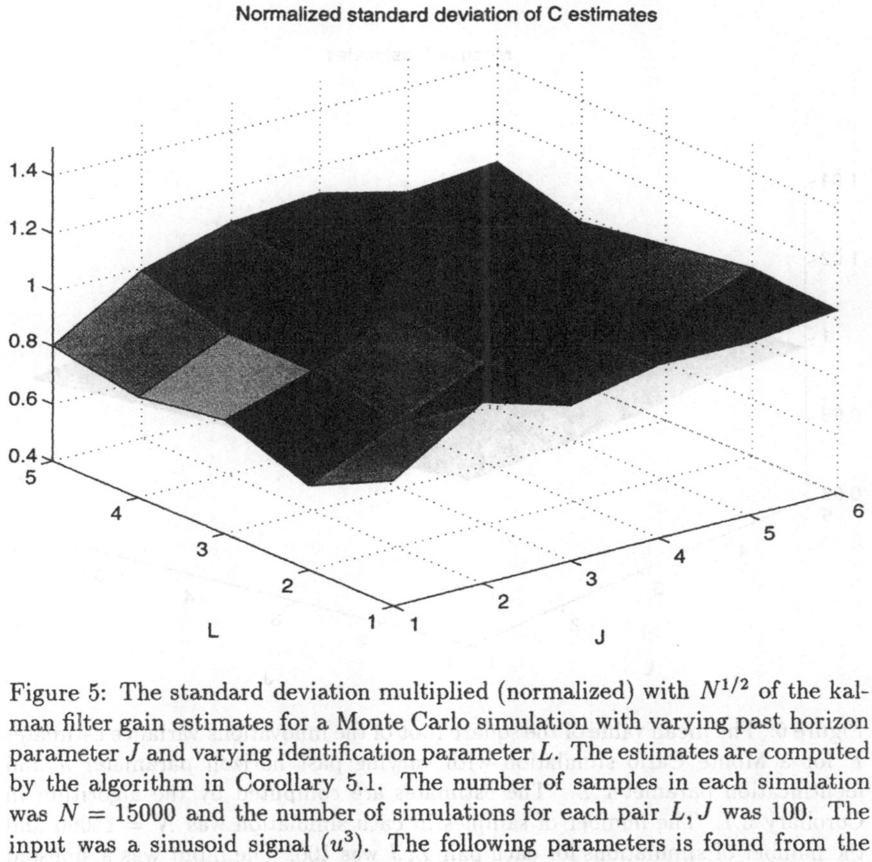

Figure 5: The standard deviation multiplied (normalized) with $N^{1/2}$ of the kalman filter gain estimates for a Monte Carlo simulation with varying past horizon parameter J and varying identification parameter L. The estimates are computed by the algorithm in Corollary 5.1. The number of samples in each simulation was $N = 15000$ and the number of simulations for each pair L, J was 100. The input was a sinusoid signal (u^3). The following parameters is found from the region with consistent estimates. The maximum standard deviation of the pole estimates was $1.2135/N^{1/2}$ at $L = 4$ and $J = 5$. The minimum standard deviation was $0.9344/N^{1/2}$ at $L = 2$ and $J = 5$. The mean of all standard deviations was $1.0315/N^{1/2}$.

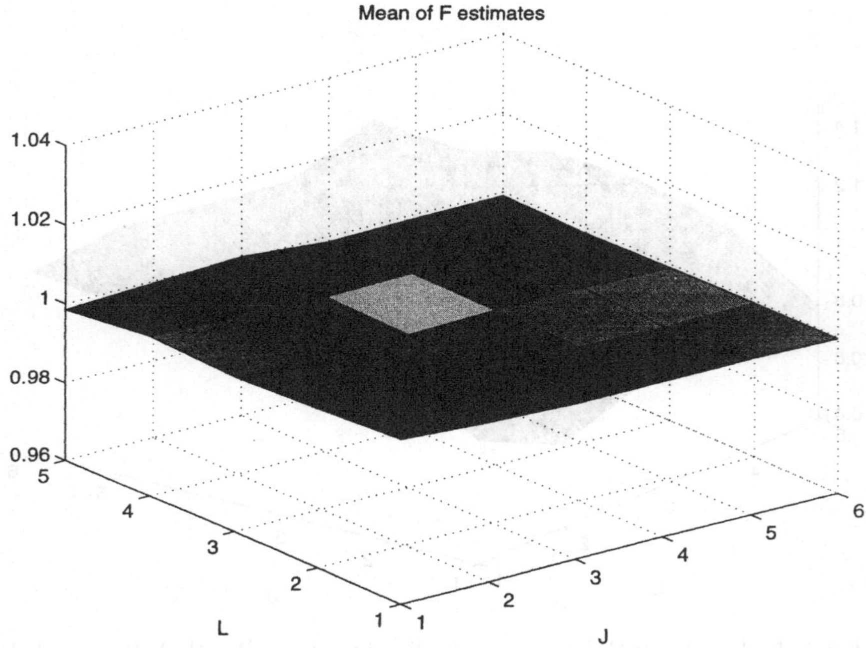

Figure 6: The mean value of the square root of the innovations variance estimates F for a Monte Carlo simulation with varying past horizon parameter J and identification parameter L. The estimates are computed by the algorithm in Corollary 5.1. The number of samples in each simulation was $N = 15000$ and the number of simulations for each pair L, J was 100. The input was a sinusoid signal (u^3). The actual parameter value is $F = 1$. The mean of all the estimates in the figure is 0.9996 with a standard deviation of $7.4 \cdot 10^{-4}$. This indicates that the estimates are consistent.

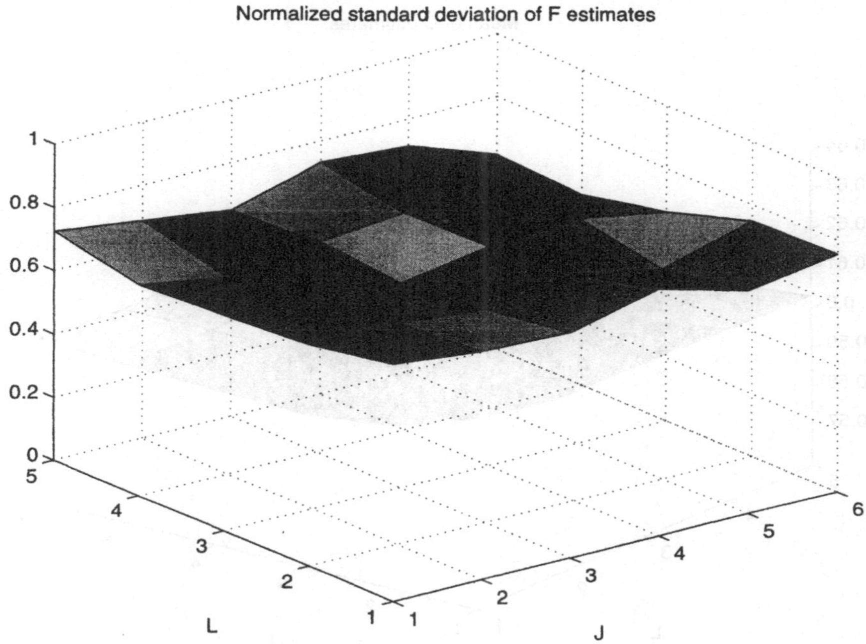

Figure 7: The standard deviation multiplied (normalized) with $N^{1/2}$ of the innovations variance estimates F for a Monte Carlo simulation with varying past horizon parameter J and varying identification parameter L. The estimates are computed by the algorithm in Corollary 5.1. The number of samples in each simulation was $N = 15000$ and the number of simulations for each pair L, J was 100. The input was a sinusoid signal (u^1). The minimum standard deviation was $0.6016/N^{1/2}$ at $L = 4$ and $J = 6$.

Table 3: Mean of the parameters of 100 different estimated models: input type u^2, $N = 200$. The models was scaled sa that $d = 1$.

Alg., Par.	a	b	e	$p_d(1)$	$H^d(1)$	c	Δ	$p_s(1)$	$H^s(1)$
DSR, L=1	0.8861	0.507	-0.992	1.405	4.0874	0.6093	0.9961	0.2768	7.0385
DSR, L=2	0.8884	0.502	-0.995	1.399	3.9833	0.6045	0.9699	0.2839	6.9509
DSR, L=3	0.8886	0.502	-0.997	1.398	3.9781	0.6094	0.9476	0.2792	7.0266
DSR, L=4	0.8891	0.5	-0.993	1.399	4.0110	0.6128	0.9218	0.2763	7.1207
DSR, L=5	0.8902	0.5	-0.994	1.399	4.0741	0.6139	0.8991	0.2763	7.2495
CVA, I=2	0.8865	0.508	-0.992	1.406	3.9629	0.6027	1.0102	0.2838	6.8462
CVA, I=3	0.8865	0.503	-0.995	1.399	3.8955	0.6072	0.9953	0.2793	6.8840
CVA, I=4	0.8859	0.500	-0.997	1.396	3.8534	0.6111	0.9916	0.2749	6.9022
CVA, I=5	0.8852	0.499	-0.997	1.394	3.8399	0.6122	0.9844	0.2730	6.9141
CVA, I=6	0.8840	0.501	-0.931	1.396	3.7969	0.6145	0.9801	0.2695	6.8643
ARMAX	0.8864	0.504	-0.994	1.399	3.8848	0.5987	0.9971	0.2877	6.7799

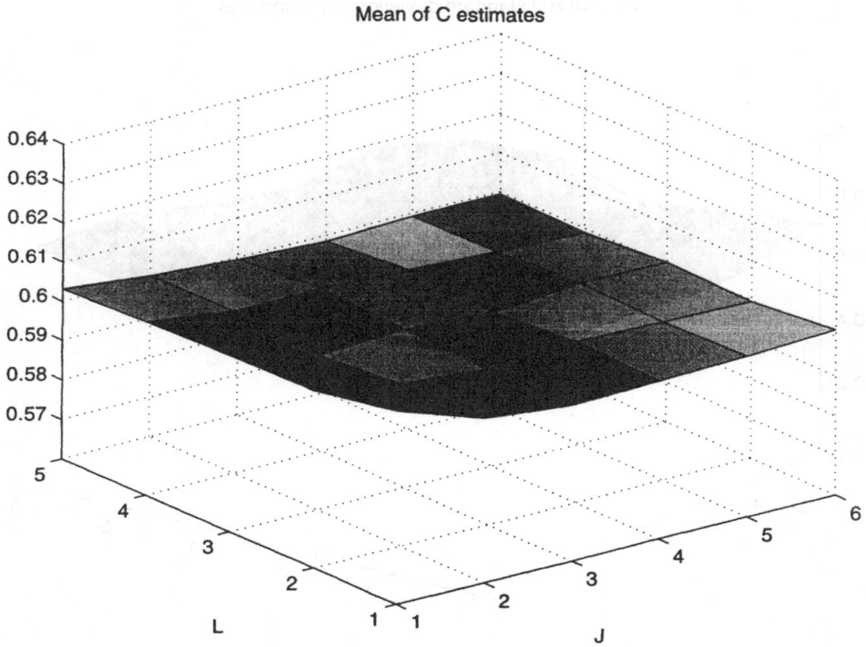

Figure 8: The mean value of the Kalman filter gain estimates for a Monte Carlo simulation with varying past horizon parameter J and identification parameter L. The estimates are computed by the algorithm in Corollary 5.2. The number of samples in each simulation was $N = 15000$ and the number of simulations for each pair L, J was 100. The input was a sinusoid signal (u^3). The actual value is $C = 0.6$. The bias for $J = 1$ as shown in Figure 4 when the estimates are computed as in Corollary 5.2 is approximately eliminated.

Table 4: Standard deviation of the parameters of 100 different estimated models: input type u^2, $N = 200$.

Alg., Par.	a	b	e	$p_d(1)$	$H^d(1)$	c	Δ	$p_s(1)$	$H^s(1)$
DSR, L=1	0.0379	0.096	0.073	0.132	2.3261	0.0101	0.0907	0.0884	2.4984
DSR, L=2	0.0351	0.091	0.072	0.122	1.8766	0.0131	0.0889	0.0915	1.9967
DSR, L=3	0.0357	0.093	0.079	0.123	1.8585	0.0158	0.0887	0.0964	2.0583
DSR, L=4	0.0363	0.095	0.084	0.123	1.9096	0.0183	0.0872	0.1046	2.1884
DSR, L=5	0.0361	0.098	0.089	0.121	1.9666	0.0210	0.0865	0.1095	2.4102
CVA, I=2	0.0354	0.096	0.073	0.133	1.9081	0.0104	0.0936	0.0844	1.9693
CVA, I=3	0.0351	0.099	0.074	0.137	1.8680	0.0106	0.0959	0.0839	2.0161
CVA, I=4	0.0357	0.099	0.077	0.138	1.8757	0.0115	0.0931	0.0871	2.0303
CVA, I=5	0.0367	0.099	0.078	0.135	1.9511	0.0119	0.0951	0.0904	2.1405
CVA, I=6	0.0376	0.098	0.079	0.133	1.8513	0.0123	0.0955	0.0974	2.0864
ARMAX	0.0348	0.092	0.068	0.123	1.7760	0.0723	0.0933	0.0867	1.8969

Table 5: Mean of the parameters of 100 different estimated models: input type u^3, $N = 500$. The models was scaled so that $d = 1$.

Alg., Par.	a	b	e	$p_d(1)$	$H^d(1)$	c	Δ	$p_s(1)$	$H^s(1)$
DSR, L=1	0.8952	0.493	-1.001	1.392	4.0591	0.6013	1.0006	0.2939	7.0975
DSR, L=2	0.8938	0.494	-1.002	1.390	3.9881	0.5961	0.9879	0.2977	6.9587
DSR, L=3	0.8934	0.493	-1.003	1.389	3.9382	0.6000	0.9784	0.2934	6.9510
DSR, L=4	0.8931	0.493	-1.002	1.389	3.9376	0.6018	0.9695	0.2914	6.9572
DSR, L=5	0.8936	0.493	-1.002	1.390	3.9738	0.6030	0.9612	0.2906	7.0060
CVA, I=2	0.8944	0.493	-1.002	1.391	3.9908	0.5944	1.0047	0.2999	6.9579
CVA, I=3	0.8933	0.492	-1.001	1.389	3.9558	0.5959	0.9998	0.2974	6.9273
CVA, I=4	0.8931	0.492	-1.000	1.389	3.9105	0.5966	0.9992	0.2965	6.8970
CVA, I=5	0.8930	0.492	-1.000	1.389	3.9064	0.5967	0.9986	0.2963	6.8926
CVA, I=6	0.8931	0.491	-1.999	1.389	3.8941	0.5972	0.9977	0.2960	6.8930
ARMAX	0.8936	0.493	-1.002	1.39	3.9354	0.5952	0.9980	0.2983	6.8954

Table 6: Standard deviation of the parameters of 100 different estimated models: input type u^3, $N = 500$.

Alg., Par.	a	b	e	$p_d(1)$	$H^d(1)$	c	Δ	$p_s(1)$	$H^s(1)$
DSR, L=1	0.0277	0.066	0.077	0.102	1.6168	0.0409	0.0644	0.0542	1.4790
DSR, L=2	0.0274	0.066	0.075	0.099	1.6257	0.0442	0.0637	0.0580	1.4551
DSR, L=3	0.0268	0.066	0.077	0.099	1.5491	0.0456	0.0638	0.0583	1.4129
DSR, L=4	0.0272	0.066	0.076	0.100	1.5699	0.0470	0.0626	0.0600	1.4067
DSR, L=5	0.0275	0.067	0.076	0.100	1.6054	0.0479	0.0640	0.0609	1.4341
CVA, I=2	0.0269	0.066	0.076	0.100	1.5547	0.0417	0.0653	0.0542	1.4149
CVA, I=3	0.0275	0.066	0.076	0.099	1.5903	0.0431	0.0642	0.0560	1.4462
CVA, I=4	0.0268	0.067	0.075	0.099	1.5164	0.0415	0.0645	0.0532	1.3767
CVA, I=5	0.0267	0.066	0.075	0.099	1.5091	0.0420	0.0645	0.0539	1.3780
CVA, I=6	0.0262	0.067	0.076	0.101	1.5003	0.0433	0.0660	0.0539	1.3668
ARMAX	0.0260	0.066	0.075	0.099	1.5207	0.0421	0.0643	0.0535	1.3581

7.2 Example 2

We will in this example investigate the problem with colored input signals and the **N4SID** algorithm.

Consider the same SISO one state example as in Example 1, Equations (112) and (113). Two different input signals were chosen, one equal to a pure sinusoid signal, input type u^3, and one equal to a white noise sequence with unit variance, input type u^2. The inputs are the same as defined in Example 1. The number of samples was fixed to $N = 500$. The standard deviation of the innovation was varied from $\Delta^{0.5} = 0$ to $\Delta^{0.5} = 0.01$ in order to investigate the sensitivity for noise. The number of block rows in the data matrices was chosen as $L = 2$.

The extended observability matrix O_3 was estimated from the column space of the matrix $Z_{L+1|L+1}$, Equation (102). The dimension of the column space is estimated as the number of "non zero" singular values, see Figure 9. We have also introduced the normalized singular value $(s_1 - s_2)/s_1$ as shown in Figure 10. This means that when $(s_1 - s_2)/s_1 = 1$ then the number of states is $n = 1$.

The conclusions to be drawn from Figures 9 and 10 are that the **DSR** algorithm gives reasonable estimates for both the system order and the actual pole (the **CVA** and **PO-MOESP** give essentially the same results for this example) and that the **N4SID** algorithm does not work at all for this system with a pure deterministic sinusoid input signal ($u_k = sin(k)$). However, note that when the input was changed to a white noise sequence (input type u^3) then the two algorithms gave essentially the same singular values as well as pole estimates.

7.3 Example 3

A two input two output system with the following model matrices is considered.

$$A = \begin{bmatrix} 1.5 & 1 & 0.1 \\ -0.7 & 0 & 0.1 \\ 0 & 0 & 0.85 \end{bmatrix} \quad B = \begin{bmatrix} 0 & 0 \\ 0 & 1 \\ 1 & 0 \end{bmatrix} \tag{114}$$

$$D = \begin{bmatrix} 3 & 0 & -0.6 \\ 0 & 1 & 1 \end{bmatrix} \quad E = \begin{bmatrix} 0 & 0 \\ 0 & 0 \end{bmatrix} \tag{115}$$

$$C = \begin{bmatrix} 0 & 0.1 \\ 0.1 & 0 \\ 0 & 0.2 \end{bmatrix} \quad \Delta = \begin{bmatrix} 1 & 0 \\ 0 & 1 \end{bmatrix} \tag{116}$$

The system have a deterministic zero at -3, i.e. the system is non-minimum phase. The eigenvalues of A are $0.85, 0.75 \pm 0.3708i$.

The deterministic and stochastic gain matrices are given by

$$H^d(1) = \begin{bmatrix} 16 & 15 \\ 2.6667 & -2.5 \end{bmatrix} \quad H^s(1) = \begin{bmatrix} 2.5 & 4.7 \\ -0.25 & 1.1833 \end{bmatrix}$$

The algorithm gives exact results when $\Delta = 0$. Hence, this result is not presented. The time series y_k, u_k was generated by simulating the model with one particular random noise process e_k with covariance Δ. The input was $u = [u^2 u^1]^T$.

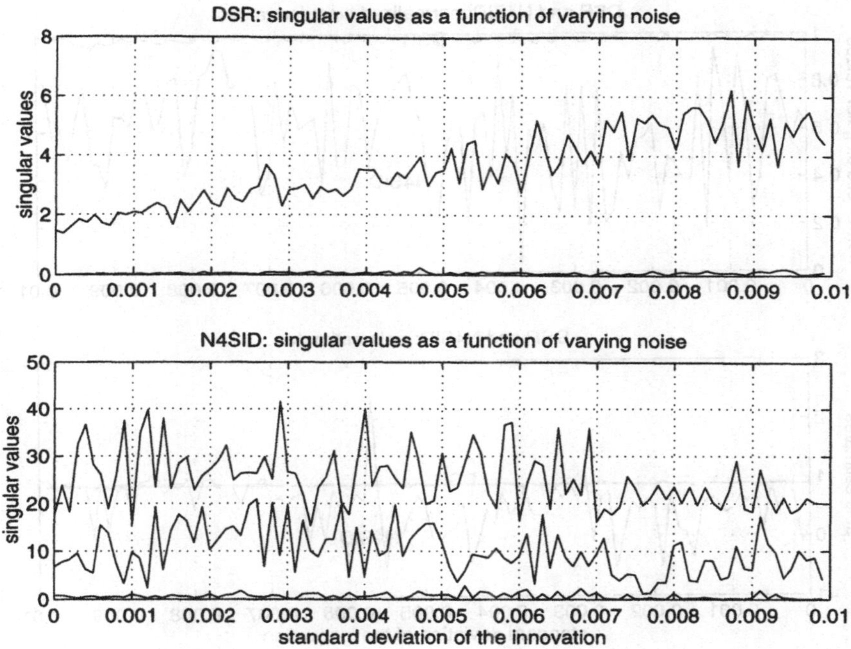

Figure 9: The singular values for model order selection as a function of varying innovations noise level. The input to the system was a pure sinusoid signal (input type u^3). The other parameters are as described in Example 2. The singular values from the **DSR** algorithm are shown in the upper figure and for the **N4SID** algorithm in the lower. The actual system order is $n = 1$. As we can see, the system order is fairly well detected by the **DSR** algorithm, while the **N4SID** algorithm does not work at all for this system with a sinusoid input signal.

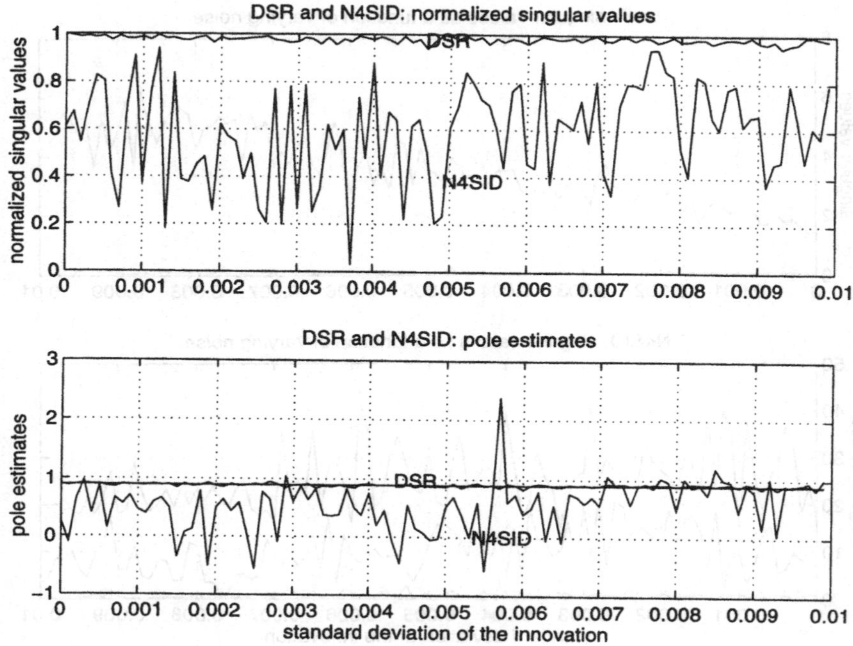

Figure 10: This figure shows the normalized singular values $(s_1 - s_2)/s_1$ and the pole estimates as a function of varying innovations noise level for the system in Example 2. The input to the system was a pure sinusoid signal (input type u^3). The other parameters are as described in Example 2. The normalized singular values from both the **DSR** and **N4SID** algorithms are shown in the upper figure. The pole estimates are shown in the lower figure. The actual system order is $n = 1$ and the actual pole is 0.9. As we can see, the system order and the pole is fairly well estimated by the **DSR** algorithm, while the **N4SID** algorithm does not work at all for this system with a sinusoid input signal.

The **DSR** parameter was fixed to $L = 6$. The following estimates are obtained by **DSR**.

$$\hat{H}^d(1) = \begin{bmatrix} 15.6575 & 14.6168 \\ 2.5576 & -2.4188 \end{bmatrix} \quad \hat{H}^s(1) = \begin{bmatrix} 2.3625 & 4.5583 \\ -0.2436 & 1.1275 \end{bmatrix}$$

$$\hat{\Delta} = \begin{bmatrix} 1.0531 & -0.0244 \\ -0.0244 & 0.9859 \end{bmatrix}$$

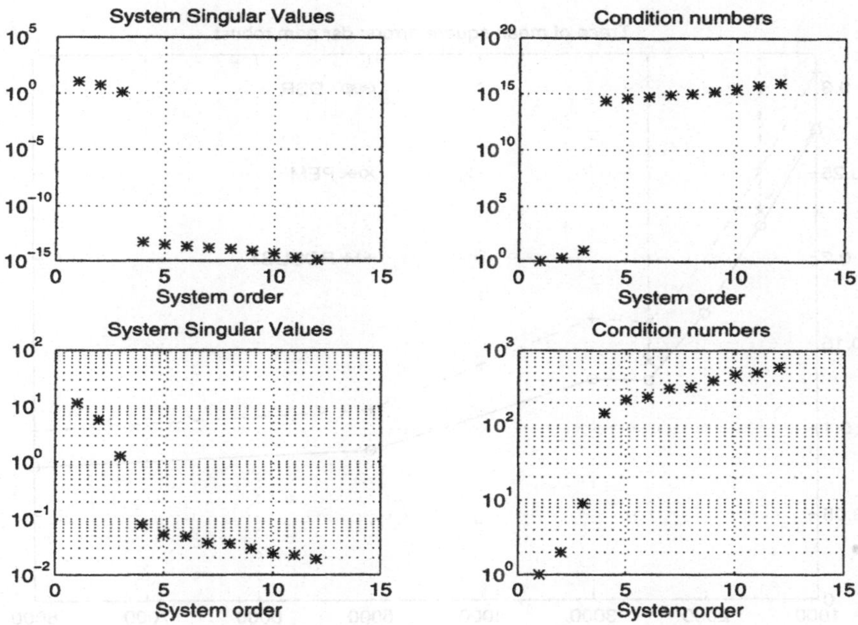

Figure 11: Singular values ($s_i, i = 1, ..., 12$) and condition numbers ($\frac{s_1}{s_i}, i = 1, ..., 12$) for the system order selection in Example 3. The noise free deterministic case with $\Delta = 0$, upper figure. The case with $\Delta = I$, lower figure.

7.4 Example 4: Modeling the crude oil prize

The idea of this example is to investigate if the crude oil prize can be modeled by a combined deterministic and stochastic model with the USD vs. NKR exchange rate treated as an exogenous input variable. It is assumed that the exchange rate is managed and exactly known. There are of course other variables than the exchange rate which influences upon the crude oil prize. These variables are in this example assumed to be unknown stochastic variables.

A number of 504 observations of the crude oil prize ($[\frac{NKR}{barrel}]$) in the period from 1993 to 1995 is shown in Figure 13. The data is from Norges Bank (the

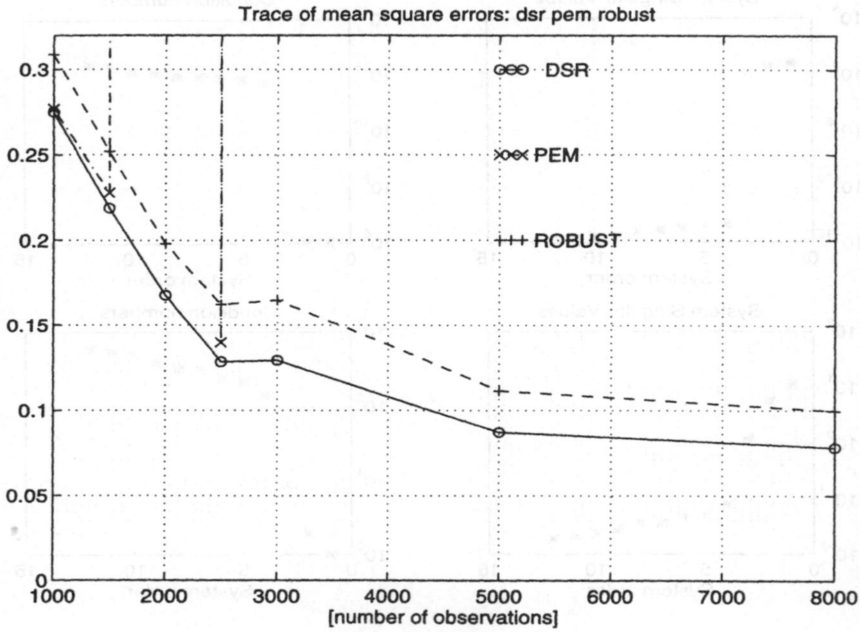

Figure 12: Results from a Monte Carlo simulation in order to investigate the accuracy of the innovation estimates. The model in Example 3 is simulated 100 times with varying number of samples. The **DSR** estimates are computed from Theorem 4.3. The **PEM** algorithm converged only for number of samples $N < 3000$.

Bank of Norway). The crude oil prize at observation k is defined as y_k and also defined as an output variable. At the same time instants the USD vs. NKR exchange rate ($[\frac{USD}{NKR}]$) in the period from 1993 to 1995 is shown in Figure 14. The exchange rate at day k is defined as u_k and also defined as an input variable.

The time series are separated into a trend and a series of variations. Define

$$dy_k = y_k - y_0(k) \quad (117)$$

$$du_k = u_k - u_0(k) \quad (118)$$

where dy_k is the variation in the crude oil prize around the trend $y_0(k)$ and du_k is the variation in the exchange rate around the trend $u_0(k)$. The exchange rate trend is simply chosen as the first observation, i.e. $u_0(1)$. Inspection of the series shows that there is an immediately fall in the crude oil prize at observation number $k = 368$ of about 18 $[\frac{NKR}{barrel}]$. This drop in the crude oil prize can hardly be explained by changes in the exchange rate (u_k). The explain-nation is more likely an economical decision (from a OPEC meeting etc.). The trend is therefore chosen as shown in Figure 13. However, note that it is not necessary to remove trends from the time series when the algorithm for Subspace Identification (4SID), (Di Ruscio (1995) algorithm name DSR) is used.

The 200 first observations in the time series dy_k and du_k was used as inputs to the DSR algorithm. The algorithm was also executed with the parameter $L = 5$ (number of block rows in the extended observability matrix). The singular values and condition numbers used for estimating the order of the state space model is shown in Figure 15. It is clear from the figure that a first order model ($n = 1$) is a good choice. The estimated model parameters are $A = 0.9669$, $B = 160.6463$, $C = 1.8302$, $D = 0.4759$, $E = -719.6302$ and $\Delta = 2.9844$ and the model

$$x_{k+1} = 0.9669 x_k + 160.6463 du_k + 1.8302 e_k \quad (119)$$

$$\hat{dy}_k = 0.4759 x_k - 719.6302 du_k + e_k \quad (120)$$

where e_k is a zero mean, Gaussian distributed stochastic sequence with covariance

$$\Delta = \mathrm{E}(e_k e_k^T) = 2.9844 \quad (121)$$

and unknown realization.

First, the deterministic part of the estimated model was simulated with all 504 exchange rate observations as inputs. The simulated deterministic model output (dy_k^d) and the actual observations (dy_k) are illustrated in Figure 16. The error between the actual and simulated time series can also be measured as (the covariance estimate)

$$J_1 = \frac{1}{200}\sum_{i=1}^{200}(dy_k - dy_k^d)^2 = 23.996 \quad (122)$$

$$J_2 = \frac{1}{504}\sum_{i=1}^{504}(dy_k - dy_k^d)^2 = 38.88 \quad (123)$$

The total combined deterministic and stochastic model was simulated. The stochastic input (innovations process) was generated with the MATLAB function $e = randn(501,1) * sqrt(\Delta)$ also with seed zero. Note that a different noise realization (different seed) will give a slightly different result. However, the results are illustrated in Figure 17.

The innovations model can be written as an optimal prediction model for the output (the crude oil prize). Define the innovations process as $e_k = y_k - \bar{y}_k$ where $\bar{y}_k$ is the optimal prediction of the output y_k. Then we have

$$x_{k+1} = Ax_k + Bu_k + C(y_k - Dx_k - Eu_k) \quad (124)$$

$$\bar{y}_k = Dx_k + Eu_k \quad (125)$$

The optimal prediction of the crude oil prize variations is illustrated in Figure 18.

Even if the model is generated only by the 200 first observations it is capable of roughly predicting the slowly variations in all the 504 crude oil prize observations.

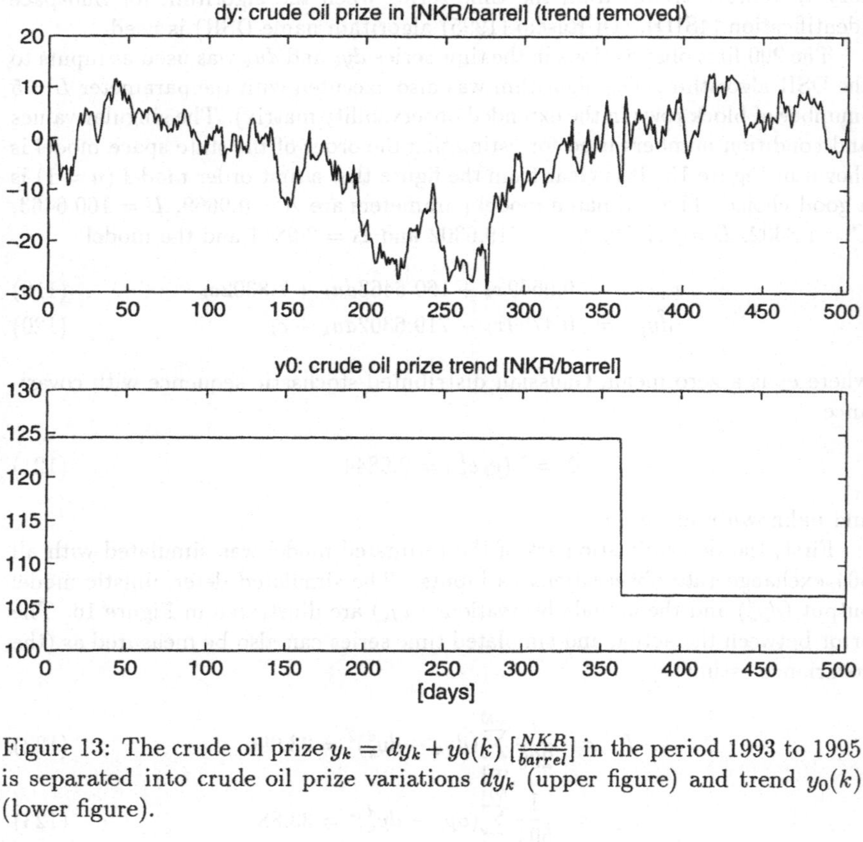

Figure 13: The crude oil prize $y_k = dy_k + y_0(k)$ $[\frac{NKR}{barrel}]$ in the period 1993 to 1995 is separated into crude oil prize variations dy_k (upper figure) and trend $y_0(k)$ (lower figure).

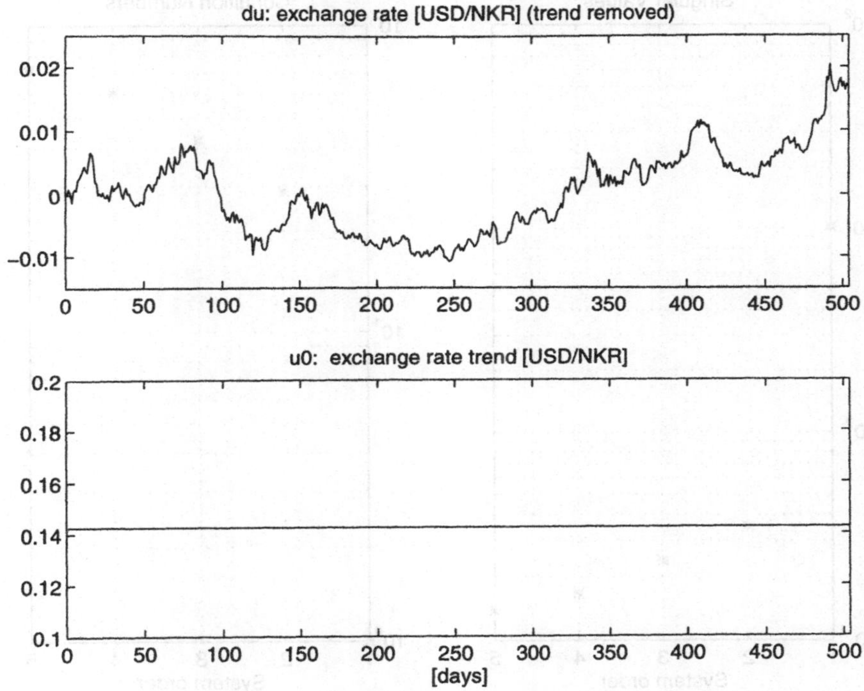

Figure 14: The exchange rate $u_k = du_k + u_0(k)$ $[\frac{USD}{NKR}]$ in the period 1993 to 1995 is separated into rate variations du_k (upper figure) and trend $u_0(k)$ (lower figure).

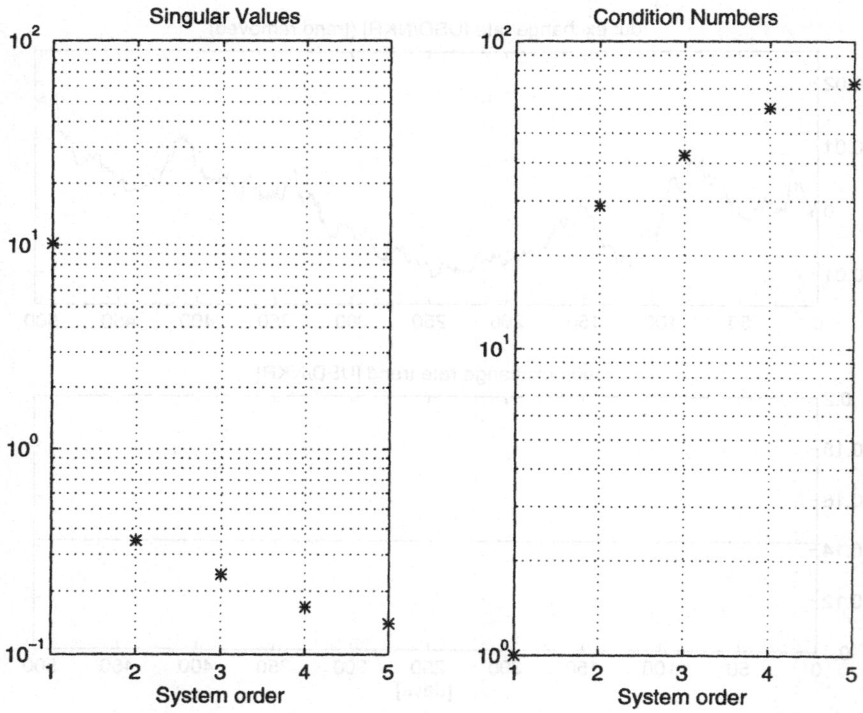

Figure 15: The singular values ($s_i,\ \ i = 1, ..., 5$) and condition numbers ($\frac{s_1}{s_i},\ \ i = 1, ..., 5$) used to investigate the order of the state space model in Example 4. The model order is estimated as the number of large singular values or the number of small condition numbers.

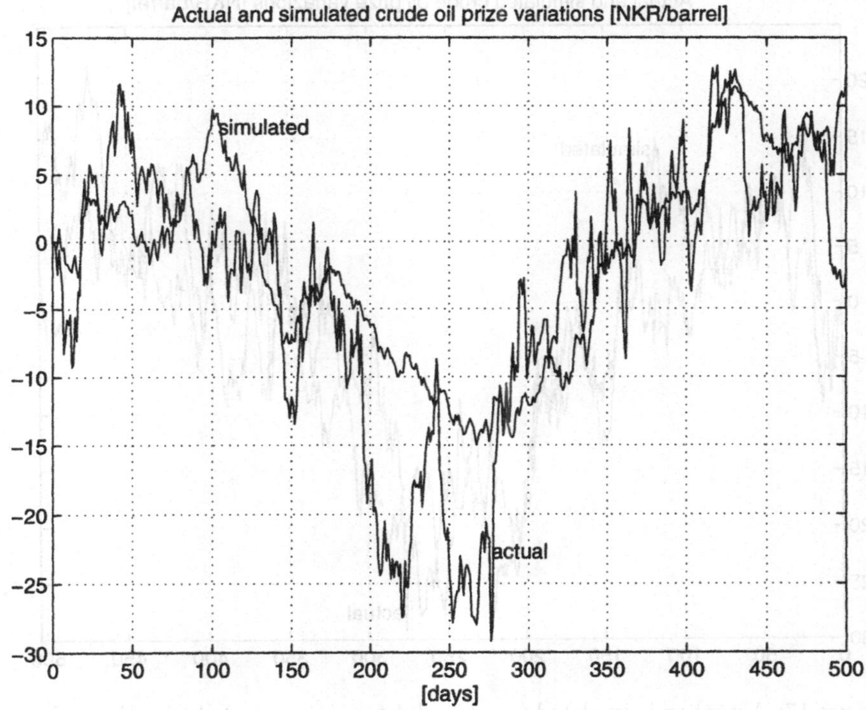

Figure 16: Actual and simulated (deterministic part of model) crude oil prize.

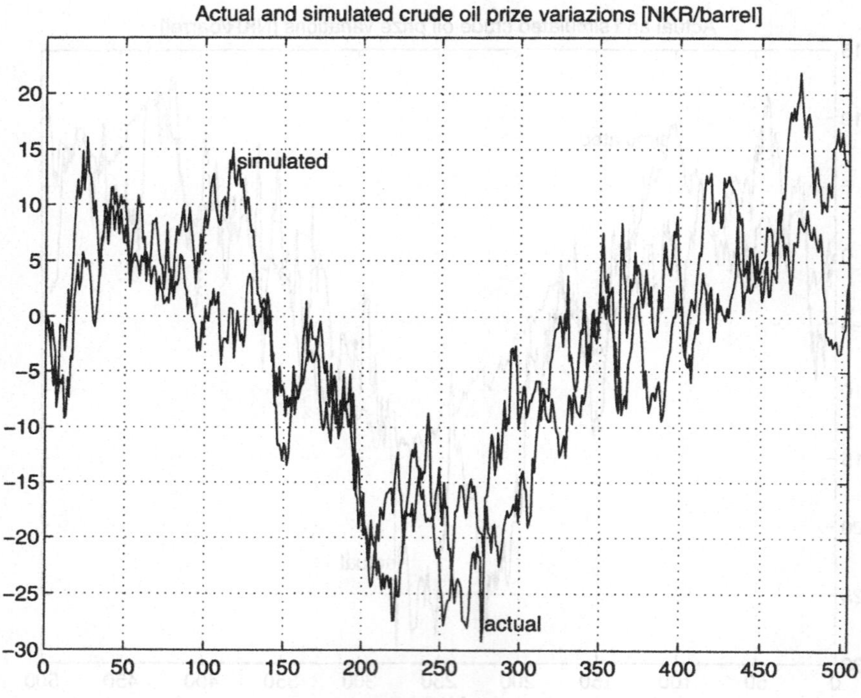

Figure 17: Actual and simulated (combined deterministic and stochastic model) crude oil prize. This model was generated from the 200 first samples of the input output samples, only. The algorithm parameter was $L = 5$.

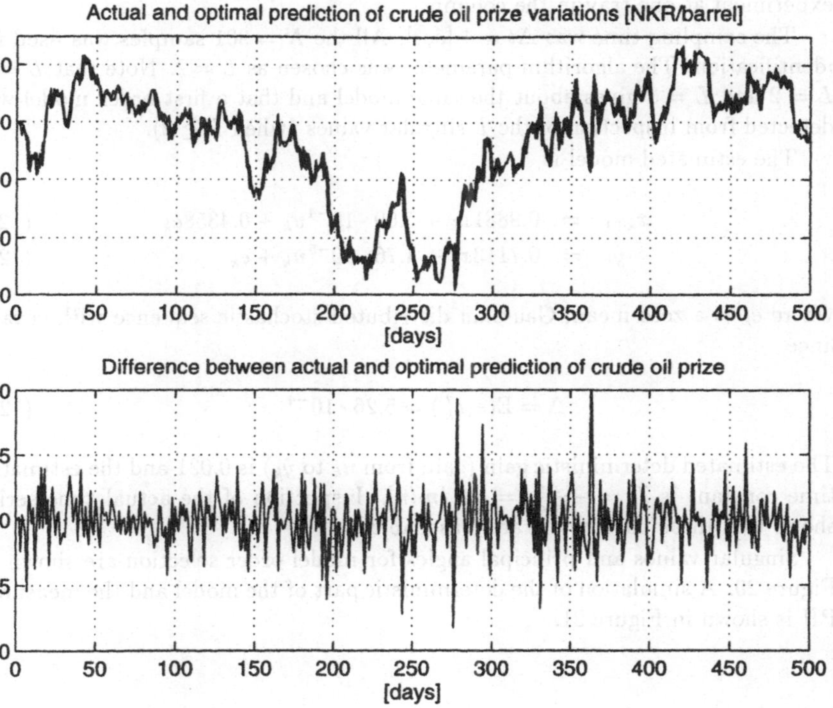

Figure 18: Actual and simulated optimal prediction of the crude oil prize is shown in the upper part of the figure. The difference between the actual and the optimal prediction of the crude oil prize is shown in the lower part of the figure. This model was generated from the 200 first samples of the input output samples, only. The algorithm parameter was $L = 5$.

7.5 Example 5: Modeling PH

The time series in this example is from a tray (plate) column scrubber at Union Co., Skien, Norway (printed with permission). Liquid travels down (vertical) the column over a series of trays. Gas travels up through contacting devices in each tray. There is a PH control loop at each tray. A suspension of slurry is used as input variable (here defined as u_k) to control the PH (output variable here defined as y_k). The time series shown in Figure 19 are from an open loop experiment at one tray in the column.

The sampling time was $\Delta t = 4$ $[sec]$. All the $N = 901$ samples was used for identification. The algorithm parameter was chosen as $L = 2$. Note that $L = 1$, $L = 2$ and $L = 3$ gives about the same model and that a first order model was detected from inspection of the L singular values (when $L \geq 2$).

The estimated model is

$$x_{k+1} = 0.9881 x_k + 5.09 \cdot 10^{-4} u_k + 0.4358 e_k \tag{126}$$
$$\hat{y}_k = 0.7113 x_k + 4.76 \cdot 10^{-5} u_k + e_k \tag{127}$$

where e_k is a zero mean, Gaussian distributed stochastic sequence with covariance

$$\Delta = \mathrm{E}(e_k e_k^T) = 5.26 \cdot 10^{-4} \tag{128}$$

The estimated deterministic gain (gain from u_k to $\hat{y}_k$) is 0.031 and the estimated time constant is $T = -\frac{\Delta t}{ln(a)} = 5.6$ $[min]$. Inspection of the actual time series shows that these parameters are reasonable.

Singular values and principal angles for model order selection are shown in Figure 20. A simulation of the deterministic part of the model and the measured PH is shown in Figure 21.

8 Conclusions

A method for subspace identification and realization of state space models on innovations form directly from given input and output data is presented.

The method determines both the deterministic part and the stochastic part of the model.

The algorithm gives exact results in the deterministic case and consistent results when the system is influenced by noise.

The stochastic part of the model is computed from standard linear algebra decomposition methods and no matrix equations (e.g. Riccati or Lyapunov equations) need to be solved.

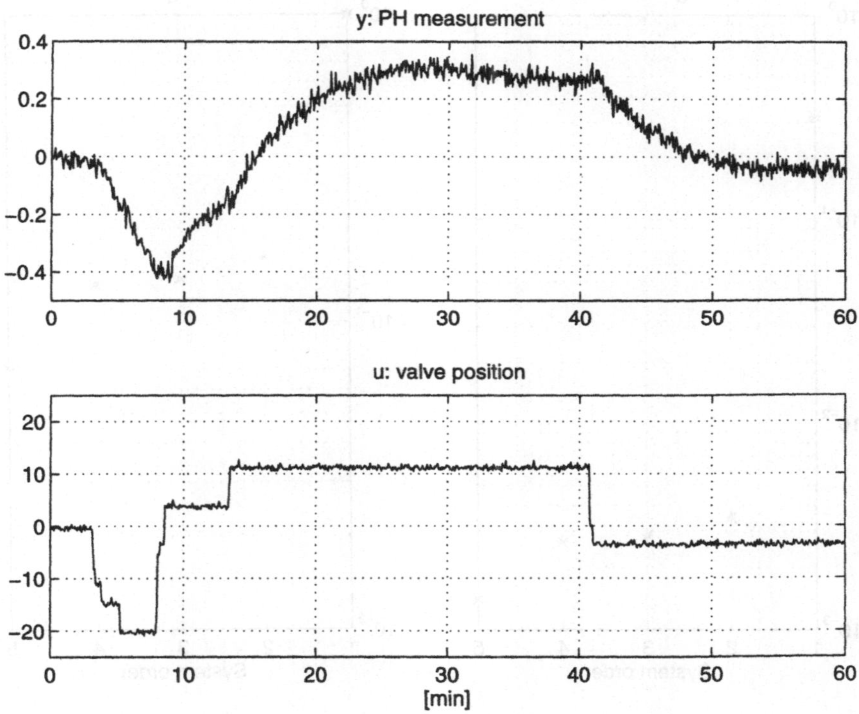

Figure 19: The measured PH (output from the process) upper figure. The manipulated input to the process (valve position) is shown in the lower figure.

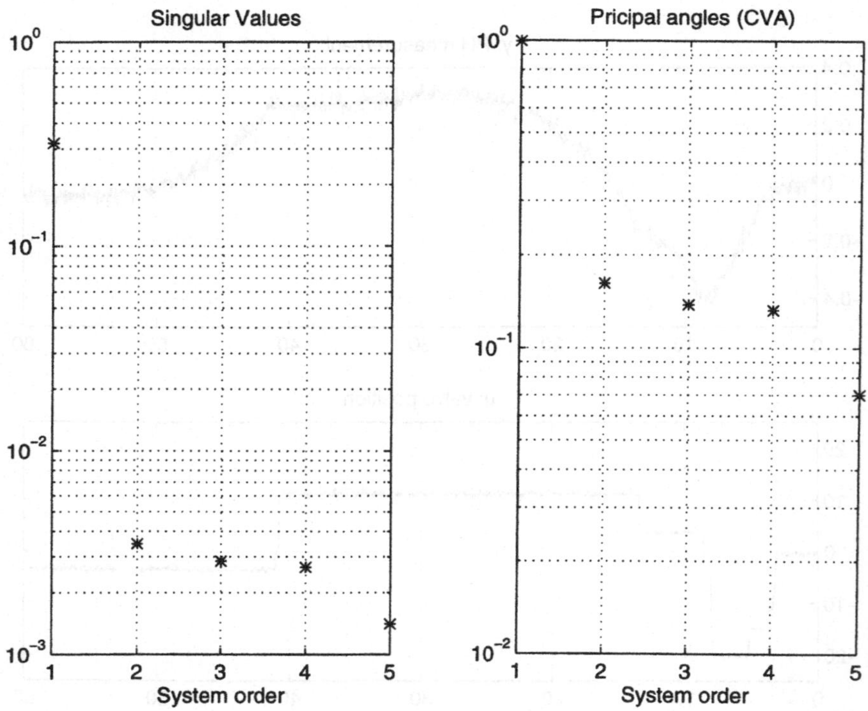

Figure 20: Singular values and principal angles for inspection of model order. The principal angles are computed by the **CVA** algorithm. The **DSR** parameter $L = 5$ was chosen. The figure shows that a reasonable model order is $n = 1$. See Example 5 for details.

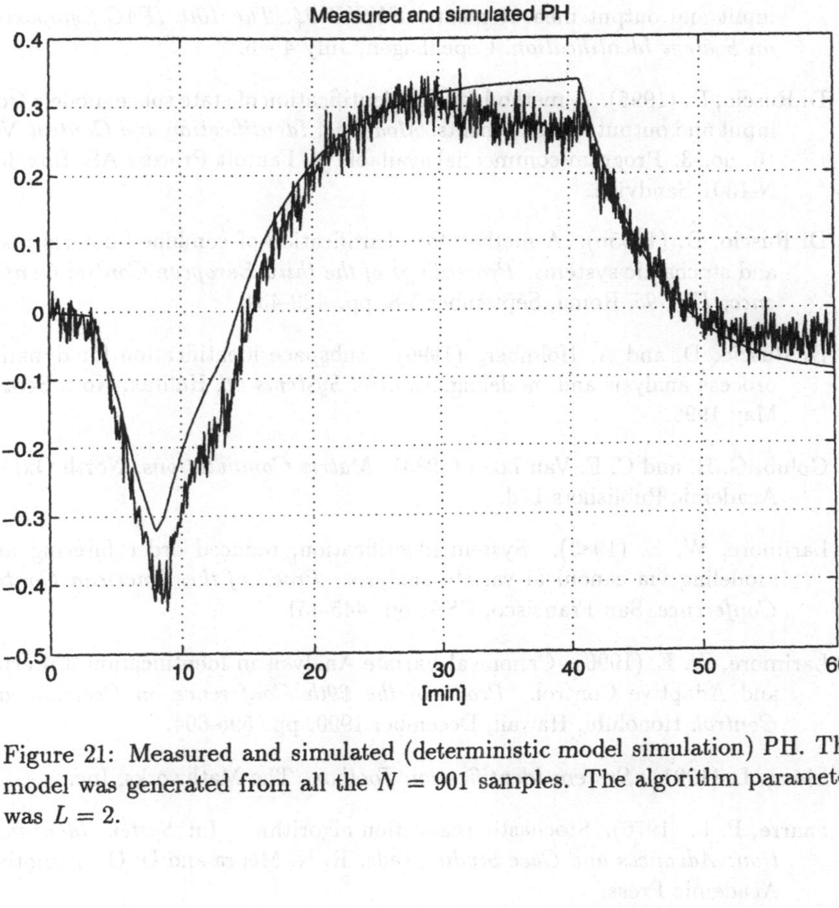

Figure 21: Measured and simulated (deterministic model simulation) PH. The model was generated from all the $N = 901$ samples. The algorithm parameter was $L = 2$.

References

Aoki, M. (1990). *State Space Modeling of Time Series.* Second, Revised and Enlarged Edition. Springer-Verlag Berlin, Heidelberg.

Aoki, M. (1994). Two Complementary Representations of Multiple Time Series in State-Space Innovation Forms. *Journal of Forecasting*, Vol. 13, pp. 69-90.

Di Ruscio, D. (1994). Methods for the identification of state space models from input and output measurements. *SYSID 94, The 10th IFAC Symposium on System Identification*, Copenhagen, July 4 - 6.

Di Ruscio, D. (1995). A method for the identification of state space models from input and output measurements. *Modeling, Identification and Control*, Vol. 16, no. 3. Program commercial available by Fantoft Process AS, Box 306, N-1301 Sandvika.

Di Ruscio, D. (1995b). A method for identification of combined deterministic and stochastic systems. *Proceedings of the third European Control Conference, ECC95*, Roma, September 5-8, pp. 429-434.

Di Ruscio, D. and A. Holmberg (1996). Subspace identification for dynamic process analysis and modeling. *Control Systems 96*, Halifax, Nova Scotia, May 1996.

Golub, G. H. and C. F. Van Loan (1983). *Matrix Computations.* North Oxford Academic Publishers Ltd.

Larimore, W. E. (1983). System identification, reduced order filtering and modeling via canonical variate analysis. *Proc. of the American Control Conference*, San Francisco, USA, pp. 445-451.

Larimore, W. E. (1990). Canonical Variate Analysis in Identification, Filtering and Adaptive Control. *Proc. of the 29th Conference on Decision and Control*, Honolulu, Hawaii, December 1990, pp. 596-604.

Ljung, L. (1991). *System Identification Toolbox.* The Mathworks, Inc.

Faurre, P. L. (1976). Stochastic realization algorithms. In: *System Identification: Advances and Case Studies*, (eds. R. K. Mehra and D. G. Lainiotis), Academic Press.

Kalman, R. E., P. L. Falb and M. A. Arbib (1969). *Topics in mathematical system theory.* McGraw-Hill Book Company.

Kung, S. Y. (1978). A new identification and Model Reduction Algorithm via Singular Value Decomposition. *Conf. on Circuits, Systems and Computers*, Pacific Grove, CA, November 1978, pp. 705-714.

Moore, B. C. (1981). Principal Component Analysis in Linear Systems: Controllability, Observability, and Model Reduction. *IEEE Trans. on Automatic Control*, Vol. AC-26, pp. 17-31.

Van Overschee, P. and B. De Moor (1994). N4SID: Subspace Algorithms for the Identification of Combined Deterministic Stochastic Systems. *Automatica*, vol. 30, No. 1, pp.75-94.

Van Overschee, P. (1995). *Subspace Identification: theory-implementation-application.* PhD thesis, Katholieke Universiteit Leuven, Belgium.

Van Overschee, P. and B. De Moor (1995). A Unifying Theorem for Three Subspace System Identification Algorithms. *Automatica*, vol. 31, No. 12, pp. 1853-1864.

Verhagen, M. (1994). Identification of the deterministic part of MIMO state space models given on innovations form from input output data. *Automatica*, vol. 30, No. 1, pp. 61-74.

Viberg, M. (1995). Subspace-Based Methods for the Identification of Linear Time-invariant Systems. *Automatica*, vol. 31, No. 12, pp. 1835-1851.

Competing Exchange Rate Models: A State Space Model vs Structural and Time Series Alternatives

Jeffrey H. Dorfman
University of Georgia
Athens, Georgia 30602-7509

Exchange rate modeling has been popular among economists interested in the fields of international trade, finance, monetary policy, and time series analysis for the past twenty years, beginning shortly after the current regime of floating rates. Articles have been written to propose structural models [e.g., Bilson (1979), Dornbusch and Fischer (1980), Edwards (1983), Frankel (1979), Kiguel (1987), Mussa (1976), and Stulz (1987)], to propose time series models [Ahking and Miller (1987), Baillie and Bollerslev (1989), and Havenner and Modjtahedi (1988)], to reject previously proposed models [Meese and Rogoff (1983)], and even to examine the underlying statistical properties of the exchange rates that all the others had been modeling [Boothe and Glassman (1987a) and (1987b)]. This paper will take a somewhat different approach by attempting to provide a fair testing ground for several of the previously proposed models along with one new model, a state space model based on results from linear systems theory. An evaluation of the models will be made with a set of nonnested hypothesis tests. While articles have described the empirical failure of some exchange rate models advanced in out-of-sample forecasting [cf. Meese and Rogoff (1983)], no direct tests of competing paradigms have been done giving equal treatment to all competing hypotheses while including both in-sample and out of-sample tests.

Five models of exchange rates will be evaluated. Two behavioral (or structural) models, based on economic theory, are included in the study, one based on monetarist assumptions and one from a finance vein employing a version of the capital asset pricing model. The monetarist model follows those developed by Mussa (1976) and Bilson (1978), although a more general form suggested by Ahking and Miller (1987) will be employed to avoid any implication that the particular monetarist model chosen was an inferior one. It is derived from a stable money demand function with several standard monetarist assumptions imposed. The second behavioral model is taken directly from Stulz (1987). This model has a finance flavor to it, with a base of households maximizing time separable logarithmic utility functions that lead to well-established portfolio and consumption results. The important feature of the model as it is compared to the usual monetarist models is that Stulz' model assumes imperfect information concerning the growth rates of the money stocks. Three time series models are examined. First, a random walk model is included because it is a common benchmark model for hard to predict series such as exchange rates. Second, a balanced representation state-space model is estimated; this particular type of state space model was developed primarily by Aoki (1987). The third time series model is a traditional ARMA model.

If either the state space or traditional time series model proves superior that will imply that the process that drives the exchange rate is not a random walk, but that the two behavioral explanations do not capture this process correctly. Such a result would be support for the use of time series models rather than structural models in the modeling of hard-to-predict series. Alternatively, it is possible that one of the behavioral models will prove itself superior to the time series models. This would support the advantage of including relevant variables in a structure suggested by economic theory. If the random walk model proves superior, it must be concluded that the existing economic theory of exchange rates and the current econometric techniques available for time series analysis are both unable to provide any useful information about the future path of the exchange rate.

Several tests will be used to evaluate the models. Two standard nonnested hypothesis tests, the J-test and the Cox test, will be performed. These each test one model against another, so a whole series of pairings are needed and inconclusive results often occur. Out-of-sample prediction tests will also be performed. This will allow all five models to be compared simultaneously by such statistics as MSE, the difference in absolute forecast errors, and a nonparametric test of direction suggested by Henriksson and Merton (1981).

The paper is organized into four main sections. Section I presents the models to be evaluated and a brief development of the two behavioral models. The second section describes the data. The third section contains the discussion of the nonnested testing procedures. Finally, the results of the model estimation and hypothesis tests are in the fourth section.

1 The Models

Because so many exchange rate models have been developed with the exchange rates in logarithmic form (to avoid Seigel's paradox), all five models will have the natural logarithm of the spot exchange rate as the dependent variable. This is the standard presentation for the monetarist model and the time series models. However, Stulz' imperfect information equilibrium model results in an expression for the spot rate in levels. The model can be adapted to have the common dependent variable, although some thought must then be given to the stochastic specifications. The five models can be written in the following manner, defining e_t as the exchange rate, s_t as the natural logarithm of the exchange rate, and u_t as a serially uncorrelated stochastic error term.

<u>The random walk model:</u>

(1) $s_t = s_{t-1} + u_t$

<u>The state space model:</u>

(2) $z_{t+1} = Az_t + Bu_t$

$s_t = Cz_t + \kappa + u_t$,

where z is a vector of state variables that represent a set of minimally sufficient statistics for the series s_t, κ is a constant representing the mean of the series s_t, and the subscripts on the state variables indicate that the variables enter the equation in conditional means.

The traditional time series model:

(3) $\phi(B)s_t = \eta + \Theta(B)u_t$

where $\phi(B)$ and $\Theta(B)$ are polynomials in the backshift operator B and η is a constant.

The monetarist model:

(4) $s_t = \mu + \beta_1(m_{1t} - m_{2t}) + \beta_2(i_{1t} - i_{2t}) + \beta_3(i_{1t} - i_{2t}) + \beta_4(\pi^e_{1t} - \pi^e_{2t}) + u_t$

where μ is a constant and the right hand side variables in equation (4) are the natural logarithm of the each country's money supply (m_j), the natural logarithm of each country's real income level (y_j), the natural logarithm of one plus each country's nominal interest rate (i_j), and the natural logarithm of one plus each country's expected inflation rate (π^e_j). The exchange rate is assumed to be in units of country one's currency per unit of country two's currency. This model varies slightly from the more general model presented in Ahking and Miller (1987) by restricting the variables to those that appear in the more traditional versions of the monetarist models. The model in (4) encompasses both the sticky-price monetarist model of Dornbusch (1976) and Frankel (1979) and the flexible price monetarist model of Mussa (1976), Bilson (1979), and Edwards (1983). The sticky price model would set β_4 equal to zero. The model above does not encompass the rational expectations formulation of the monetarist model directly, but is closely related; see Bilson (1978) for a discussion of various expectational formulations. The model in (4) is derived by assuming the two countries' money demand functions are of double logarithmic form and imposing purchasing power parity on the countries. See Mussa (1976) for a good development of this model.

The Stulz imperfect information equilibrium model:

(5) $s_t = \alpha + (m_{1t} - m_{2t}) + (\tilde{r}_{1t} - \tilde{r}_{2t}) + u_t$

where m_j is as before, the $\tilde{r}_j$ are the natural logarithm of a particular form of expected nominal interest rate (not one plus the rate which was denoted by i_j in the monetarist model), and α is a constant that comes from the proportion of household wealth held as money balances. Stulz' model is built from a two country world with stochastic investment following a Markov process, two goods produced in each country (only one of which is traded internationally), a portfolio of risky tradeable assets and one safe asset, and identical households (within a country) that maximize time separable logarithmic utility functions over their consumption and real balance choices. These assumptions lead to a version of the capital asset pricing model with the usual result that consumption and asset holding choices are linear functions of wealth. The model developed by Stulz, however, treats stochastic processes and uncertainty explicitly, leading to some new results, particularly in the exchange rate determination equation. Stulz introduces the assumption that households do not know the true rate of growth of the money stock in the domestic country (country 1). This leads to the use of the expected interest rate for this country, rather than the actual rate, and the use of a related expression for country 2 in place of its interest rate. Stulz assumes that households form their expectations of the growth rate of the money stock only from the time series of the money stock. Given the logarithmic utility functions and other assumptions, this results in the relation for the natural logarithm of the expected interest rates being

(6a) $\tilde{r}_{1t} = \ln[\mu_t(M_1) + \mu_t(R_1) + \delta - \sigma_t^2(M_1) - \sigma_t^2(R_1) - \sigma_t(M_1,R_1)]$

(6b) $\tilde{r}_{2t} = \ln[\mu_t(M_2) + \delta - \sigma_t^2(M_2)]$.

In equation (6) $\mu_t(M_1)$ is the estimate at time t of the growth rate of the money stock of country 1, $\mu_t(M_2)$ is the growth rate of the money stock in country 2, the $\mu_t(R_1)$ are the growth rates of the nominal interest rates of the countries, δ is the rate of time preference, and the σ's represent variances and covariances of the relevant growth rates. There are no coefficients to be estimated in equations in (5) and (6) except for the intercept. However, the growth rate of the domestic money stocks and the nominal interest rates will need to be calculated from the time series of these variables in the manner suggested by Stulz. Assumptions will have to be made concerning the time rate of preference, as well. More details on the estimation of (5) and (6) will be given in the data section below.

2 The Data

The exchange rate modeled is the dollar-deustchemark ($/DM) rate, using the spot price at the end of the period. All the data for the U.S. and West Germany were collected from the IMF's International Financial Statistics (IFS). The data used to estimate the models are monthly observations for the period 1978:1 to 1988:12. Of these 132 observations, the first twelve are used for generating any necessary initial conditions and lagged values [see equations (26) through (28) below]. The next 84 observations, from 1979:1 to 1985:12 are the period employed for the estimation of the models' parameters. Finally, the thirty-six observations from 1986:1 to 1988:12 are reserved for out-of-sample forecasting and tests. The following variables were collected directly from the IFS, with the number in parentheses being its exact IFS code: the variable for money supply is a money supply index (34..i), the variable for real income is an industrial production index (66..c), the interest rates are each country's Treasury bill rate (60c), and the prices are measured by each country's consumer price index(64). All other variables needed to estimate the model were created from these variables.

The expected inflation rate for the monetarist model was assumed to be a weighted average of the past twelve months' inflation rates. Recalling that the variable π_j^e was defined to be the logarithm of one plus the expected inflation rate, it is approximated by

(7) $\pi_{jt}^e = \ln[1 + (12\pi_{jt-1} + 11\pi_{jt-2} + 10\pi_{jt-2} + \ldots + \pi_{jt-12})]$,

where the π_{jt} are the actual inflation rate measured by country j's CPI for period t. Because of the twelve observations saved for generating initial conditions, this expected inflation rate can be calculated for all the periods used for the estimation of the parameters without any values needing to be set to zero.

The expected interest rates necessary for the implementation of Stulz' model were calculated as described in equation (6). The assumption was made to treat the information set as if it started with the period 1978:1. Thus, the variables in equation (6) are defined as follows:

(8a) $\mu_t(M_1) = (m_{1t} - m_{10})/t + (1/2)\sigma_t^2(M_1)$

(8b) $\mu_t(M_2) = (m_{2t} - m_{20})/t$

(9) $\mu_t(R_1) = (r_{1t} - r_{10})/t$

The m_{jt} and r_{jt} have already been defined as the natural logarithms of the money supply and interest rate. The variance and covariance variables are computed by the standard sample variance and covariance estimates of the growth rates using only the values from t = 0 (1978:1) to t = t. Thus agents' estimates of the growth rates, their variances, and their covariance all evolve throughout the sample period as the information set is updated. The difference between (8a) and (8b) is due to the assumed uncertainty concerning country 1's money growth rate. Given the assumption made by Stulz that the money stock follows a lognormal diffusion process, equation (8a) is the expected value of the growth rate of the money stock. The only variable in equation (6) remaining is the rate of time preference δ. This parameter was estimated by a grid-search (due to the nonlinearity involved).

3 The Nonnested Tests

Five nonnested tests will performed on the five exchange rate models. The first test will be a mean squared error (MSE) test of the five models for their predictions of the in-sample and out-of-sample data. The measure of out-of-sample performance is important in light of the argument advanced by Meese and Rogoff (1983) that many exchange rate models fit well in- sample and then perform very poorly in out-of-sample tests. Although the MSE's are directly comparable across models and can be easily ordered, a test due to Granger and Newbold (1986) for a statistically significant difference between MSE's of two models will be performed. This will allow a hypothesis of equal MSE's to be tested across models. The second and third tests described below are the Cox test and Davidson and McKinnon's J-test. These tests are more structural in nature. They formally test the null hypothesis that a single model is "true" against the hypothesis of an alternative model. The fourth test is a nonparametric test for information in a model's forecasts, due to Henriksson and Merton (1981). This test is performed on both the in-sample and out-of sample data. The last test was suggested by Wecker (1988). This test measures the paired differences in absolute forecast errors and tests the differences for significance. It is also performed on both the in-sample and out-of-sample data.

The first nonnested test is the Granger and Newbold MSE test. The five models will be compared on their in-sample and out-of-sample mean squared errors, with the tests performed on successive pairings of the models. The null hypothesis is that the MSE's of the two models are equal. The test statistic is

$$(10) \quad r = \frac{\sum_{t=1}^{T} (e_{1t} + e_{2t})(e_{1t} - e_{2t})}{\sqrt{\sum_{t=1}^{T} (e_{1t} + e_{2t})^2 \sum_{t=1}^{T} (e_{1t} - e_{2t})^2}}$$

calculated as where the subscripts 1 and 2 identify the two models being tested and T is the sample size. This statistic is asymptotically normal with a variance of (1/T).

A Cox test is a variation of the more usual Neyman-Pearson likelihood ratio test. It tests two models against each other with the null hypothesis that one is the true model. Because null and alternate hypotheses are not treated equally in classical statistics, a Cox test between two models must be performed once with each model as the null hypothesis. Let the first model be represented by $y = X\beta + \varepsilon_1$ and the second model by $y = Z\gamma + \varepsilon_2$. To test the null hypothesis that the first model is the true model, the test statistic is

$$(11) \qquad C = \frac{c_{12}}{\sqrt{var(c_{12})}}$$

where C_{12} and $var(c_{12})$ are defined by

$$(12) \quad c_{12} = \frac{T}{2}\ln\left[\frac{\hat{\sigma}_2^2}{\hat{\sigma}_1^2 + (1/T)\hat{\beta}'X'[I-Z(Z'Z)^{-1}Z']X\hat{\beta}}\right] = \frac{T}{2}\ln\frac{\hat{\sigma}_2^2}{\hat{\sigma}_{21}^2}$$

$$(13) \quad var(c_{12}) = \frac{\hat{\sigma}_2^2}{\hat{\sigma}_{21}^2}\,\hat{\beta}'X'[I-Z(Z'Z)^{-1}Z'][I-X(X'X)^{-1}X'][I-Z(Z'Z)^{-1}Z']X\hat{\beta}$$

and $\hat{\sigma}_1^2$ and $\hat{\sigma}_2^2$ are the usual estimates of the variances of ε_1 and ε_2. The resulting test statistic is asymptotically distributed as a standard normal. In this application, the Cox test statistic will be computed for twenty pairings of null and alternate hypotheses. For more details see Cox (1962).

The J-test, proposed by Davidson and MacKinnon (1981), is a linearized version of the Cox test. The motivation for a J-test is to examine whether the predicted values from one model would have a significant coefficient if included as an additional regressor in another model. If the coefficient of the predicted values is significantly different from zero, as measured by its t-value, the other model is rejected. Again, the test must cover all possible pairings. To test the null hypothesis that model 1 is the true model, the equation

(14) $s_t = x_t\beta + \hat{y}_2\lambda + \varepsilon^*_t$

is estimated, where $\hat{y}_2$ is the vector of predictions from the second model, λ is the coefficient to be tested, s_t is the logarithm of the exchange rate, and x_t is regressors of model 1 in time period t. For the state space model, the estimated z_t serve as the x's.

The fourth nonnested test performed is a nonparametric test due to Henriksson and Merton (1981). A Henriksson-Merton test for the information value of each model's forecasts allows the null hypothesis of no information to be rejected at any desired confidence level. This test measures the conditional probabilities of a model correctly predicting the direction of revision of the series (i.e., predicting turning points). For a series y_t with a zero mean, this is equivalent to the conditional probabilities of the actual value being negative or nonnegative given the sign of the prediction. For a particular series, let $\phi_t = 1$ if the actual y_t is nonnegative, and let $\phi_t = 0$ otherwise. The null hypothesis of the Henriksson-Merton test is

(15) $\text{Prob}(\phi_t = 1|\hat{y}_t \geq 0) + \text{Prob}(\phi_t = 0|\hat{y}_t < 0) = 1.$

For the exchange rate models above with exogenous variables and nonzero means, the correct test concerns the conditional probabilities of the sign of the change in s_t given the sign of the forecasted change. For this application, the null hypothesis that there is no information in the forecasts is written

(16) $\text{Prob}(s_t - s_{t-1} \geq 0|\hat{s}_t - s_{t-1} \geq 0) + \text{Prob}(s_t - s_{t-1} < 0|\hat{s}_t - s_{t-1} < 0) = 1.$

Note that the Henriksson-Merton test would be degenerate for the random walk model since that model always predicts no change in s_t. Therefore, this test will only be performed on the other four models.

To calculate the confidence level, *c*, of rejecting the null hypothesis of no information, define for each series the following values over the set of observations to be tested:

N_1 = the number of observations with negative sales revisions,
N_2 = the number of observations with nonnegative sales revisions,
$N = N_1 + N_2$,
n_1 = the number of successful predictions of negative sales revisions
n_2 = the number of successful predictions of nonnegative sales revisions
$n = n_1 + n_2$

The confidence level, *c*, is given by

(17) $$c = 1 - \sum_{x=n_1}^{\min(N_1,n)} \binom{N_1}{x}\binom{N_2}{n-x} \Big/ \binom{N}{n}.$$

It is the researcher's choice what level of c will be considered an acceptable level for rejecting the null hypothesis of no information in a model's forecasts. For more details, see Henriksson and Merton (1981).

The final test, suggested by Wecker (1988), tests for significant differences in forecast errors. By exploiting the knowledge that the forecasts are for the same actual values, they can be paired and all the difficulties concerning covariances of the forecasts are avoided. Define $d_t = |s_t - f_{1t}| - |s_t - f_{2t}|$ as the difference in absolute forecast errors between the forecast of model 1 (f_{1t}) and the forecast of model 2 (f_{2t}) in time period t. Letting σ_d be the standard deviation of the d_t's, the test statistic for period t is

$$(18) \qquad W_1 = \frac{d_t}{\sigma_d}.$$

The test statistic W_t is distributed as a t-statistic. To make this test an explicit test for one model's forecast errors being smaller than another's, a slight modification is made. Instead of the null hypothesis being that the absolute forecast errors are equal, let the null be that d_t is less than or equal to zero, i.e., that model 1's absolute forecast error is less than model 2's absolute forecast error. The test statistic remains the same, but the test is now a one-tailed t-test. It will be performed with each model as the null (model 1 in the formula) and against each alternative model in turn, just as the Cox and J-tests must be performed with each model as the null. An estimate of σ_d can be calculated for both the in-sample and out-of-sample data, allowing the test to be done on both sets of observations without any assumption that σ_d remains constant across the two sub-samples. The assumption that σ_d is constant within the set of in-sample data and within the set of out-of-sample data is necessary to perform the test. This test determines the number of times one model's absolute forecast error proves to be significantly larger than each alternative model's comparable error.

4 Empirical Results

Estimation

The models were all estimated using the data and variables described above without any reference to the out-of-sample data. Only after each model's estimated parameters had been finalized were the out-of-sample forecasts calculated. The parameter estimates for each model are presented in Table 1. While the parameter estimates themselves are of little interest here, a few comments are in order. The ARMA(1,1) model was chosen from a set of fifteen models considered by three selection criteria: AIC, FPE, and the Schwarz criterion. Not only did the ARMA(1,1) prove superior in these respects, but it closely matched the patterns of the autocorrelation and partial autocorrelation profiles. The state space model shows both trend and cycle components, with the dynamics matrix A having one real and one pair of complex conjugate eigenvalues. The monetarist model's coefficients for the money stock, income, and the expected inflation rate were all highly significant (t-values of approximately 5, 7, and 8, respectively), but the coefficient on the interest rate term had a t-value of only 1.52. The elasticities of the exchange rate with respect to both interest rates and the money stock were less than one in magnitude, while the elasticity of the exchange rate with respect to income was greater than one.

Table 1. Model Parameter Estimates

1. Random Walk Model

$$s_t = s_{t-1} + u_t$$

No parameters.

2. State Space Model

$$z_{t+1} = Az_t + Bu_t$$
$$s_t = Cz_t + \kappa + u_t$$

$$\hat{A} = \begin{bmatrix} .95535 & .06748 & -.00495 \\ -.06748 & 1.0344 & -.27668 \\ -.00495 & .27668 & .55563 \end{bmatrix} \quad \hat{B} = \begin{bmatrix} -6.470 \\ -4.4493 \\ 5.8399 \end{bmatrix} \quad \hat{C} = \begin{bmatrix} -.18262 \\ .02188 \\ .00429 \end{bmatrix}$$

$$\hat{\kappa} = -.84821$$

3. ARMA Model

$$\phi(B)s_t = \eta + \Theta(B)u_t$$

$$\hat{\phi}(B) = (1 - .98629B) \quad \hat{\Theta}(B) = (1 - .18512B) \quad \hat{\eta} = -.010997$$

4. Monetarist Model

$$s_t = \mu + \beta_1(m_{1t} - m_{2t}) + \beta_2(i_{1t} - i_{2t}) + \beta_3(i_{1t} - i_{2t}) + \beta_4(\pi^e_{1t} - \pi^e_{2t}) + u_t$$

$$\hat{\mu} = -.97446 \quad \hat{\beta}_1 = .84713 \quad \hat{\beta}_2 = -.1.2383 \quad \hat{\beta}_3 = .88584 \quad \hat{\beta}_4 = 3.2338$$

5. Stulz Imperfect Information Model

$$s_t = \alpha + (m_{1t} - m_{2t}) + (\tilde{r}_{1t} - \tilde{r}_{2t}) + u_t$$

$$\tilde{r}_{1t} = \ln[\mu_t(M_1) + \mu_t(R_1) + \delta - \sigma^2_t(M_1) - \sigma^2_t(R_1) - \sigma_t(M_1,R_1)]$$

$$\tilde{r}_{2t} = \ln[\mu_t(M_2) + \delta - \sigma^2_t(M_2)].$$

$$\hat{\alpha} = 0.97299 \quad \hat{\delta} = .0460$$

The elasticity of the exchange rate with respect to the expected inflation rate was enormous, with an estimated value of 3.23. This suggests that for the U.S. or Germany to maintain a strong exchange rate it is essential to keep any changes in expected inflation in line with the other country's. The estimation of the Stulz imperfect information model was minimal, as discussed above. The grid search for the rate of time preference yielded the surprisingly large value of .0460, equivalent to approximately 4.7% per month when converted to a discrete discount factor (the model uses $e^{-\delta t}$ as a discount factor). Clearly, this is a very high rate of time preference. It implies that agents would be willing to pay approximately 1.73 more forsomething that they can have today than for the delivery of the same good one year from today. While this seems to be an excessively large premium perhaps it reflects the characteristics of international foreign exchange markets where traders and multinational corporations cannot always afford to wait for delivery of goods (or foreign currency).

<u>Hypothesis Testing</u>

The first test performed is the test for equality of MSE's using the test statistic described in equation (29) above. The results of the test and the MSE's of the five models, both in-sample and out-of sample, are in Table 2. The tabled MSE's have been multiplied by a factor of 1000 for ease of presentation. For convenience, the models are referred to by labels as follows: the random walk, RW; the state space model, SS; the ARMA model, ARMA; the monetarist model, M; and the Stulz model, SII. Given the 84 in-sample observations and the 36 out-of-sample observations, the critical values of the test can be calculated. For the in-sample tests, the critical value is 0.2138 for a five percent significance level (α = .05) and 0.2811 for a one percent level (α = .01). For the out-of sample tests, the critical values are 0.3267 and 0.4293, respectively. In Table 2, a negative value for r implies that the model in that column had a larger MSE than the model in that row of the table. For example, the value in the second row and third column of Table 2 is 0.2267. This test statistic is for the state space model versus the ARMA model. The negative sign shows that the state space model had a smaller in-sample MSE, a fact confirmed by the last row of the in-sample part of the table. The size of the test statistic allows the hypothesis of equal MSE's between the state space and ARMA models to be rejected at the 95% confidence level.

The general conclusion of the MSE test is that the time series methods dominate the structural models in terms of MSE. The MSE's of both structural models are statistically significantly larger than the MSE of any of the three time series models at the one percent level. Among the time series models, the ARMA model is judged to have a significantly larger MSE than both the random walk and state space models in-sample, but the three time series models have out-of-sample MSE's that cannot be differentiated statistically. The MSE's of the state space and random walk models are not significantly different in either sample.

Table 2. Mean Squared Error Test Values of the Correlation Test Statistic r[1]

In-sample

Model:	RW	SS	ARMA	M	SII
RW	--	.1155	-.2436*	-.6066†	-.9418†
SS	--	--	-.2267*	-.6290†	-.9473†
ARMA	--	--	--	-.5812†	.9372†
M	--	--	--	--	-.8046†
MSE	.9101	.8515	1.0129	3.4941	30.0238

Out-of-Sample

Model:	RW	SS	ARMA	M	SII
RW	--	-.0654	-.1515	-.9860†	-.9605†
SS	--	--	-.0412	-.9863†	-.9620†
ARMA	--	--	--	-.9861†	-.9603†
M	--	--	--	--	.9157†
MSE	1.4276	1.4801	1.5254	194.5896	68.2792

* significant at $\alpha = .05$ † significant at $\alpha = .01$

[1] Large values of r indicate that the model with the smaller MSE is significantly more accurate, as measured by MSE.

The next set of tests performed were the Cox tests. These test the null hypothesis that one model is the true model against the alternative hypothesis that it is not. The Cox test is only performed on the in-sample data since it involves parameters estimated in the regressions performed on the in-sample observations. The results are presented in Table 3 with the model that represents the null hypothesis along the top and the alternative model down the side; thus a significant entry rejects the top model in favor of the side model. The clear outcome of the Cox tests is again a very strong rejection of the structural models, although this time it is accompanied by a fairly strong rejection of the time series models, as well. Both of the structural models are rejected by all four alternatives at the 99% confidence level. The ARMA model is rejected by all four alternatives at the 95% confidence level, and by three of the four at the 99% level. The random walk model is rejected by three of the four models at the 99% confidence level, but is not rejected when the monetarist model is the alternative. The state space model provides the best showing in the Cox tests; it is only rejected by two alternative models--the ARMA model and Stulz' model. Further, the model with the next best showing, the random walk model, can be rejected by the state space model, but the random walk model cannot reject the state space model. The extremely high values of the Cox test for the pairings of the random walk and ARMA model are caused by the high multicollinearity across the two data sets. The appearance of s_{t-1} in both data matrices leads to a very small value of $var(c_{12})$ and a correspondingly large test statistic.

Table 3. Cox Test: Values of C

	Null Model				
Alternate Model	RW	SS	ARMA	M	SII
RW	--	-.9546	$-6.x10^{8}$†	-18.90†	-.83.74†
SS	-2.634t	--	-11.07†	-19.91†	-184.7†
ARMA	$2x10^{13}$†	8.153†	--	-18.57†	-158.7†
M	- 1.503	-1.613	-2.446*	--	-36.98†
SII	19.80†	26.40†	22.75†	61.28†	--

* significant at $\alpha = .05$ † significant at $\alpha = .01$

The results of the J-tests generally support the time series models over the structural models with the random walk model providing the best overall showing. The results are presented in Table 4. None of the four alternative model's predictions had a t-value with a magnitude greater than one when included as an additional regressor in the random walk model. For the state space model, only the monetarist model's forecasts had a t-value greater in magnitude than one, but that one t-value (2.023) is significant at the $\alpha = .05$ level. The ARMA model J-tests produce no significant t-values for included forecasts, but the forecasts from the state space and monetarist models are both close to significant with t values of 1.932 and 1.881, respectively. The monetarist model is rejected by all three time series models, with all three t-values being significant at the $\alpha = .05$ level. The Stulz model is rejected by two models, the random walk and the ARMA, both at the $\alpha = .01$ level. While one conclusion of the test is that the random walk model seems to include all significant information, there is perhaps another conclusion. The state space model as the null hypothesis has the three lowest t-values in the entire set oftests, with only the monetarist model's forecasts containing any new information. Further, the monetarist model has a considerably higher t-value than the Stulz model when the ARMA model is the null, and the state space model's forecasts are almost significant in the ARMA model. Thus, if one wanted to combine structural and time series methods, the results of the J-tests imply that the ideal pair might be a monetarist structural model and a state space time series model.

Table 4. J-Test: Values for $\hat{\lambda}$

	Null Model				
Alternate Model	RW	SS	ARMA	M	SII
RW	--	.2033	-.3704	2.212*	3.029†
SS	.9911	--	1.932	2.588*	.4034
ARMA	-.4358	.0972	--	2.238*	3.061†
M	.9367	2.023*	1.881	--	.6031
SII	.9042	.0807	.1560	1.641	--

* significant at $\alpha = .05$ † significant at $\alpha = .01$

The results of the non-parametric Henriksson-Merton test of directions show clearly why in-sample and out-of-sample testing is important for model selection. The Henriksson-Merton test is not performed on the random walk model because, by definition, the confidence level in the random walk model's ability to predict turning points is zero. The results for the other four models are in Table 5. In sample, the state space and ARMA models clearly exceed the ability of the two structural models to predict turning points, with the two time series models having very high confidence levels. This indicates the presence of significant information in these two model's forecasts. Out of sample, however, the results change dramatically. None of the models have out-of-sample confidence levels that are particularly impressive, and the two time series models are out performed by the structural models (although the ARMA model ties the Stulz model). The state space model appears to perform especially poorly. In fact, the difference in out-of sample performance more difficult to measure with the test statistic due to the smaller sample involved inthe out-of-sample data set. For the state space model to have achieved the same confidence level as the monetarist model (which had the highest out-of-samplec), the state space model would only have had to predict one additional turning point correctly. If the Stulz or ARMA models had correctly forecast a single additional turning point both would have had confidence levels of 0.67. Because of the smaller number of turning points to predict in the 36 out-of-sample observations, a single correct forecast can add considerably to the confidence level that the Henriksson-Merton test calculates for a model. Thus, while this does not negate the fact that the two structural models do better retain their power to predict turning points in the out-of-sample data than the time series models (which suffered large drops in their confidence levels), the differences in the out-of-sample test statistics are heavily dependent on the relatively small number of observations involved in the test.

Finally, let us examine the results of the paired difference test. These calculations support the advantage of the time series methods in forecast accuracy that was found in the Granger and Newbold MSE tests. The null hypothesis is that model i's absolute forecast error is less than model j's. The number of times

Table 5. Results of the Henriksson-Merton Tests Confidence Levels

Model	RW	SS	ARMA	M	SII
In-Sample:	--	.9814	.9987	.4080	.6193
Out-of-Sample:	--	.2674	.4041	.5283	.4041

this null hypothesis could be rejected at the 95% confidence level is presented in Table 6. The models across the top are the null hypothesis, with the numbers in each row representing the rejections of the null hypothesis for each alterative model. For example, the number nine in the third column of the part of the table for the in-sample data means that nine of the absolute forecast errors for the ARMA model proved to be significantly larger than the corresponding period's absolute forecast error of the state space model. The general conclusion of this test is very similar to that of the MSE test. The state space and random walk models suffer the fewest rejections, followed by the ARMA model, and, finally, the two structural models suffer far more rejections than the time series models. In fact, ninety percent of the monetarist model's out-of-sample absolute forecast errors prove to be significantly larger than the alternative models', and the Stulz model suffers rejection fifty percent of the time for the out-of sample data The three time series models are tightly bunched out-of-sample; but for the in sample data, the state space model shows itself to be the most accurate. Although it suffers only three fewer rejections than the random walk model (out of 336 possible in-sample rejections), the difference is caused by the state space model rejecting the random walk model seven times to only four times for the reverse.

Table 6. Absolute Forecast Error Test

Number of Rejections of Null Hypothesis In-Sample

Alternate Model	Null Model RW	SS	ARMA	M	SII
RW	--	4	8	11	16
SS	7	--	9	12	16
ARMA	2	2	--	11	15
M	1	1	1	--	9
SII	0	0	0	0	--
SUM	10	7	18	34	56

Number of Rejections of Null Hypothesis Out-of-Sample

Alternate Model	Null Model RW	SS	ARMA	M	SII
RW	--	2	3	32	24
SS	1	--	2	32	24
ARMA	2	2	--	32	24
M	0	0	0	--	0
SII	0	0	0	33	--
SUM	3	4	5	129	72

5 Conclusions

Several general trends to the results of the nonnested tests stand out. First, if forecast accuracy is desired, time series are clearly necessary for modeling exchange rates. The MSE and absolute forecast error tests showed wide disparities in forecast accuracy between the three time series models and the two structural models. The Cox and l-tests also served to indicate a superiority of the time series models over the structural ones. However, the Stulz model did reject the null hypothesis for all three time series models in the Cox test, and the monetarist model rejected the ARMA model in the Cox test and the state space model in the J-test. Further, the two structural models had higher confidence levels in the out-of-sample Henriksson-Merton test than the state space model, while the ARMA model tied the Stulz model but lagged behind the monetarist model. These results suggest that while the structural models cannot achieve the desired level of forecast accuracy, they do indeed contain at least some valuable information about the direction of the exchange rate. Because the monetarist model rejected the state space model in the J-test and had the highest out-of-sample Henriksson-Merton confidence level, it appears to be the better candidate for a structural model. The results of the MSE test, the absolute forecast error test, and the J-tests between the ARMA and the state space model indicate that the state space model has at least a slight advantage over the ARMA model, although it is certainly not overwhelming.

Clearly, the time series models have proved to be best in terms of accuracy in the forecasting of the dollar-deutschemark exchange rate. While the monetarist model showed out-of-sample ability to predict the direction of revision, it would be very difficult to trust a model that was so inaccurate (as demonstrated by the plots). The Stulz imperfect information model appears to be completely rejected as an empirical model, leading to a suspect value of the discount parameter and wildly inaccurate forecasts which are apparent in Figure 5. In fact, except for the fact that it can offer no information concerning the direction of revision in the exchange rate, the random walk model remains at the end with no alternative model demonstrably superior.

References

Ahking, Francis W. and Stephen M. Miller. "A Comparison of the Stochastic Processes of Structural Exchange-Rate Models," *Review of Economics and Statistics* 69 (1987) 496-502.

Aoki, M. *State Space Modeling of Time Series.* (Berlin: Springer-Verlag, 1987).

Aoki, M., and A. Havenner. "A Method for Approximate Representation of Vector-valued Time Series and its Relation to Two Alternatives," *Journal of Econometrics,* October 1989.

Bilson, John F.O. "Rational Expectations and the Exchange Rate," in Jacob A. Frenkel and Harry G. Johnson eds. *The Economics of Exchange Rates* (London: Addison Wesley, 1978) 75-96.

Bilson, John F.O. "The Deutsche Mark/Dollar Rate: A Monetarist Analysis," in Karl Brunner and Allan H. Meltzer eds. *Policies for Employment, Prices, and Exchange Rates* (Amsterdam: North-Holland, 1979) 59-101.

Baillie, R. T. and T. Bollerslev. "Common Stochastic Trends in a System of Exchange Rates," *Journal of Finance 44* (March 1989) 167-181.

Boothe, P. and D. Glassman. "The Statistical Distribution of Exchange Rates," *Journal of International Economics 22* (1987a) 297-319.

Boothe, P. and D. Glassman. "Off the Mark: Lessons for Exchange Rate Modeling," *Oxford Economic Papers 39* (1987b) 443-457.

Cox, D.R. "Further Results on Tests of Separate Families of Hypotheses," *Journal of the Royal Statistical Society, Series B* 24 (1962).

Davidson, R. and J.G. MacKinnon. "Several Tests for Model Specification in the Presence of Alternative Hypotheses," *Econometrica* 49 (1981) 781.

Dornbusch, Rudiger. "Expectations and Exchange Rate Dynamics," *Journal of Political Economy* 84 (Dec. 1976) 1161-1176.

Dornbusch, Rudiger and Stanley Fischer. "Exchange Rates and the Current Account," *American Economic Review* 70:5 (Dec. 1980) 960-971.

Edwards, Sebastian. "Floating Exchange Rates in Less-developed Countries: A Monetarist Analysis of the Peruvian Experience, 1950-54," *Journal of Money, Credit, and Banking* 15 (Feb. 1983) 73-81.

Frankel, Jeffrey. "On the Mark: A Theory of Floating Exchange Rates Based on Real Interest Differentials," *American Economic Review* 69 (Sept. 1979) 610-622.

Granger, C. W. J., and Paul Newbold. *Forecasting Economic Time Series* ~New York: Academic Press. 1986).

Havenner, A., and M. Aoki. "State Space Modeling of Stationary Time Series: Theory and Applications," forthcoming, *Proceedings of the 1988 American Statistical Assoc. Business and Economic Statistics Section.* (1988a).

Havenner, A., and M. Aoki. "An Instrumental Variables Interpretation of Linear Systems Theory Estimation," *Journal of Economic Dynamics and Control, 12* (1988b).

Havenner, A., and K. Criddle. "System Theoretic Time Series: An Application to Inventories and Prices of California Range Cattle," *Computers and Mathematics with Applications 17* (1989) 1177-1188.

Havenner, Arthur and Bagher Modjtahedi. "Foreign Exchange Rates: A Multiple Currency and Maturity Analysis," *Journal of Econometrics* 37 (1988) 251-264.

Henriksson, R.D. and R.C. Merton. "On Market Timing and Investment Performance II: Statistical Procedures for Evaluating Forecasting Skills," *Journal of Business* 54 (1981) 513-533.

Kailath, T. *Linear Systems* (New Jersey: Prentice Hall, 1980).

Kiguel, M. "The Non-Dynamic Equivalence of Monetary and Exchange Rate Rules under Imperfect Capital Mobility and Rational Expectations," *Journal of International Money and Finance 6* (1987) 207-214.

Meese, R. and K. Rogoff. "Empirical Exchange Rate Models of the Seventies: Do They Fit Out of Sample," *Journal of International Economics 14* (Feb. 1983) 3-24.

Mussa, Michael. "The Exchange Rate, the Balance of Payments, and Monetarist and Fiscal Policy under a Regime of Controlled Floating," *Scandinavian Journal of Economics* 78 (May 1976) 229-248.

Stulz, Rene M. "An Equilibrium Model of Exchange Rate Determination and Asset Pricing with Nontraded Goods and Imperfect Information," *Journal of Political Economy* 95 (1987) 1024-1040.

Wecker, William. Personal communication, 1988.

Application of State-Space Models to Ocean Climate Variability in the Northeast Pacific Ocean

Roy Mendelssohn
Franklin B. Schwing
Pacific Fisheries Environmental Group
Pacific Grove, CA 93950

Abstract

State-space statistical models are used to decompose time series of oceanic parameters from the coastal Northeast Pacific Ocean into a non-parametric trend term, a non-stationary seasonal component, and a stationary autoregressive component. Climate scale variations in upwelling, sea surface temperature, and wind stress are examined, and regional differences in the component series on decadal scales are explored.

1 Introduction

In eastern boundary current (EBC) systems, the physical environment is rarely uniform in time. In addition to seasonal and higher frequency variations, El Nino-Southern Oscillation (ENSO) and other perturbations produce profound anomalies in the atmosphere and ocean on interannual to decadal and century time scales. Analogously, EBCs appear to be spatially heterogeneous. Each system can be separated into several discrete regions, dominated by different physical processes, and presumably different biological structure. These regions may be separated by sharp gradients in physical forcing and characteristics, or by broad transition zones that extend over several degrees of latitude. It is expected that environmental variability in an EBC will impact its ecosystem, and may lead to perturbations in plankton and fish abundance, biomass and distribution. The timing of seasonal cycles in each region, as well as the timing and intensity of large-scale events (*e.g.*, ENSO), may not be coherent throughout an ecosystem.

To better understand how EBC ecosystems might respond to climate variability, it is crucial to describe their primary scales of spatial and temporal variability, and discern the dynamics responsible for this variance, rather than treat EBCs as spatially homogeneous systems or use seasonally-averaged data to describe their climatology. (See also *Mann and Park* [1993] who show that interdecadal atmospheric temperature signals exhibit spatial variability that leads to considerable cancellation in a global average). Analogously, because of its seasonal nature, annual averages or low-pass "smoothed" time series will not reveal potential changes in upwelling. Monthly-averaged time series include the upwelling signal, but temporal changes in upwelling on decadal to century scales could be concealed by long-term trends related to other factors (i.e., cooler sea surface temperature (SST) in spring-summer due to increased upwelling may be masked by a general warming pattern).This background is essential if scientists are to address the likely impact of climate change scenarios on ecosystem structure and distribution.

Empirical studies of upwelling and its effects on biological production suggest that optimal fisheries production occurs within a limited range of wind speeds; at speeds greater than about 5–7 m/s the biomass of small pelagic fish decreases [*Cury and Roy,* 1989]. This has resulted in an ecosystem that is tuned to these variations. Any long-term changes in the seasonal patterns of upwelling, their intensity or the duration of upwelling events could have dramatic implications to EBC ecosystems and their living marine resources. The objective of this paper is to describe the temporal variability in the spatial texture of the California Current System (CCS), a major EBC system. We take advantage of a unique data set of long (multi-decadal) coastal environmental time series from along the west coast of North America, and we apply state-space models [cf. *Shumway,* 1988, Chapter 3 or *Harvey,* 1989] to separate the seasonal component from the long-term trend in a variety of California Current System (CCS) environmental time series, and examine the variability in the both the seasonal and trend components on climate (decadal and longer) scales. We examine the trend components for commonality across the region, to see to what extent the long-term changes are global or more regional in nature. More complete discussions of the results presented here can be found in *Schwing and Mendelssohn,* [1996] and *Schwing, Mendelssohn and Parrish,* [1996], and are part of an ongoing cooperative effort, the Climate and Eastern Ocean Systems (CEOS) Program, that is studying the effect of climate change on the major upwelling regions of the world.

2 Methods

2.1 Time series

The environmental data used to generate the monthly-averaged time series described here were obtained from a variety of sources. The primary data base was the Comprehensive Ocean-Atmosphere Data Set (COADS). The COADS contains almost 100 million reports of ocean surface conditions, mostly taken by ships-of-opportunity. The data have been collected, quality-controlled and put into common formats and units [*Slutz et al.,* 1985; *Woodruff et al.,* 1987]. Data were extracted using the CD–ROM-based version of COADS and the CODE extraction program described in *Mendelssohn and Roy* [1996]. The CD–ROM version contains Release 1 of COADS for the period 1854–1979 and the Interim release for 1980–1990, in CMR5 format. The wind data are marked as estimated, for winds approximated using the Beaufort scale; as measured, for winds measured using an anemometer or buoy; or as unknown, when the measurement method and device were not known. Only wind data marked as estimated or unknown were used in forming the mean series, to avoid as much as possible the known bias in the data due to an increase over time in the use of anemometers to measure the wind [*Cardone et al.,* 1990; *Isemer,* 1992; *Wu and Newell,* 1992]. Based on statistics from the data set, each observation for each parameter has been flagged as to the "quality" of the observation [see *Slutz et al.,* 1985, page D6]. To exclude possibly erroneous observations, all data outside of their "wide interval" (roughly equal to three standard deviations), for both the wind observations and for SST, were excluded from the summaries as well. An index of poleward (northward) pseudo-stress, henceforth referred to as wind stress, was derived by squaring the northward wind component from each record extracted. Spatial regions approximately two–degree latitude by four–degree longitude were defined based on a combination of ecological and oceanographic features as well as data density, and time series of poleward wind stress and SST were calculated for each region (Figure 1) from the monthly means of each variable. These geographic boxes are referred to in terms of their central latitude (e.g., 23N refers to the 22–24°N COADS box). The time period of extraction is 1946–90.

The shore-based monthly SST and salinity (SSS) time series were averaged from daily observations made by volunteers, which were sent to the Marine Life Research Group, Scripps Institution of Oceanography [cf. *Walker et al.*, 1993]. SSTs were reported to the nearest 0.1^oC at most sites. The observations are considered accurate to about $\pm 0.2^oC$ in most cases. Salinities were determined at Scripps from daily sea water samples using an inductive salinometer. Prior to the monthly-averaging, the daily observations were quality controlled. The number of daily samples per month varied from $15–20/month$ (and as little as $10/month$ during winter months) at some of the northern stations (e.g., Farallon, Crescent City), to nearly complete coverage (e.g., La Jolla) [*Skillman*, 1993]. Since 1979, NOAA/NOS has measured density with a hydrometer at Neah Bay and Crescent City. Salinity was back-calculated from these monthly-averaged densities. The locations of these stations are shown in Figure 1.

Monthly-averaged time series of sea level were supplied by NOAA/NOS, some through the Pacific Climate (PACLIM) data base [*Cayan et al.*, 1988]. Long sea level series are available from four sites (Figure 1). For all coastal series, months with no data were included in the analysis; the model fits through periods of missing data. With a few exceptions, missing values were sparse and of only one to a few months duration. The time- and space-averaging for the COADS series were selected to ensure no missing observations.

2.2 State-space statistical model

There are several approaches to using state-space methodologies in time series analysis. One approach estimates the state vector and the appropriate system matrices based on approximations of appropriate dynamical matrices, such as the Hankel matrix. An extensive discussion of this approach is given in *Aoki* [1990a]. Most of these methods assume stationary series, though these methods have been extended to non-stationary series and common trends in *Aoki* [1988, 1990b]. A second approach, designed specifically for dealing with non-stationary series with trends and seasonal components, is related to smoothing techniques for estimating unknown functionals observed in noise. This approach assumes the time series is additively composed of several components that have a known structural form, combines the components into a state-space model, and then estimates the parameters using a combination of Kalman filtering and smoothing and maximum likelihood estimation. This is the approach used in our analysis.

To estimate a trend as well as a time-varying (i.e., nonstationary) seasonal component for each observed time series, we assume that each observation $y(t)$ is the sum

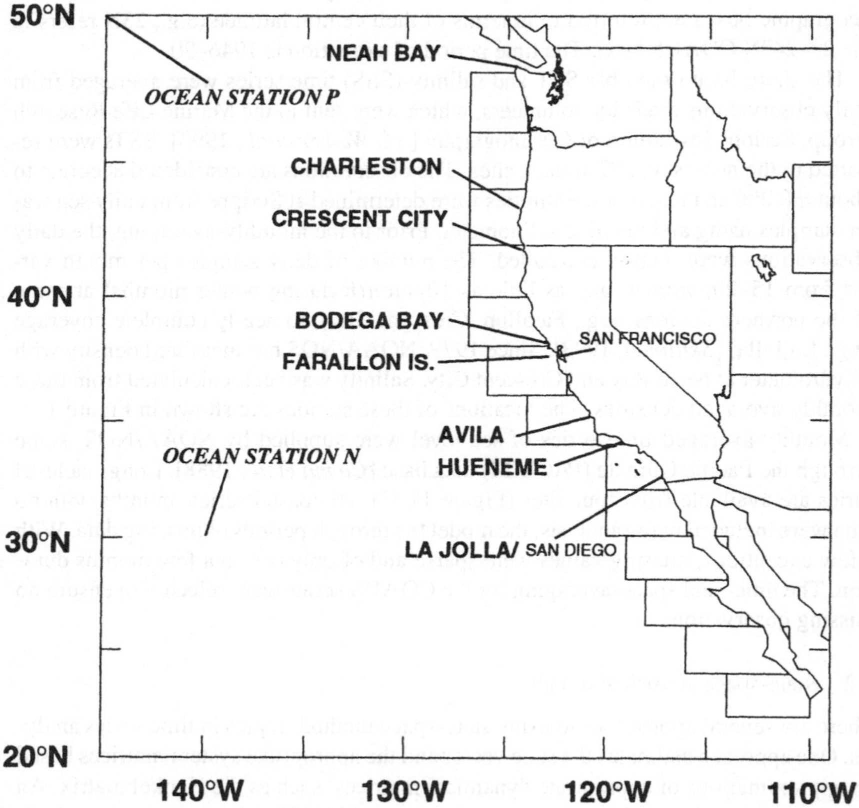

FIGURE 1. Locations of COADS 2° boxes, shown as shaded boxes, and coastal stations from which monthly time series were generated. Locations of Ocean Stations P and N also are shown.

of four components

$$y(t) = T(t) + S(t) + I(t) + e(t), \qquad t = 1, T \tag{1}$$

where, at time t, $T(t)$ is the unobserved time-dependent mean-level (trend), $S(t)$ is the seasonal component, $I(t)$ is the irregular term (stationary but autocorrelated), and $e(t)$ is the stationary, uncorrelated component which here can be viewed as "observation" or "measurement" error.

The model (Equation 1) must be more fully specified to be meaningful. The use of piecewise continuous "smoothing splines" to estimate the unobserved components dates back to *Thiele* [1880] [see *Lauritzen,* 1981], and in the more modern era to a paper by *Whittaker* [1923]. (Lauritzen provides an excellent summary of how more modern state-space models relate to these pioneering works). *Shiller* [1973] modeled distributed lag (impulse response) relationship between the input and output of time series under difference equation "smoothness" constraints on the distributed lags. He termed these constraints "smoothness priors", but did not offer an objective method of choosing the smoothing parameter. *Akaike* [1980] developed a Bayesian interpretation of the model and used maximum likelihood methods to estimate the smoothness parameter. *Brotherton and Gersch* [1981] showed how the Kalman filter and maximum likelihood estimation could be used to solve the smoothing problem—the first attempt to our knowledge to combine the techniques applied here. *Kitagawa and Gersch* [1984, 1985, 1988], in a series of papers, used "smoothness priors" to model a variety of nonstationary problems, and *Harvey* [1989] developed similar models under the title "structural time series models".

The methodology used to decompose a time series can be most clearly understood in terms of estimating a "smooth" but unknown function which has been observed with "noise", that is data of the form

$$y(t) = f(t) + e(t) \tag{2}$$

where the $y(t)$ are the observed data, $f(t)$ is an unknown smooth function and $e(t)$ are independent gaussian errors. *Whittaker* [1923] suggested that the solution should balance fidelity to the data with fidelity to a smoothness constraint

$$\left[\sum_{t=1}^{T} (y(t) - f(t))^2 + \mu^2 \sum_{t=1}^{T} (\nabla^k f(t))^2 \right]. \tag{3}$$

The first term in Equation 3 is the usual sum-of-squares criterion, while the second term constrains the k-th order finite differences of the unknown function, the discrete equivalent of splines where the k-th order derivatives are constrained. The two parts are balanced by the "smoothness parameter" μ. (Equation 3 is equivalent to a Lagrangean multiplier formulation of a sum-of squares problem constrained by the kth order differences constrained to be zero). As μ goes to zero, the smoothness constraint becomes negligible, and the solution exactly interpolates the data. As μ approaches infinity, the sum-of-squares term becomes negligible, and the solution is the appropriate k-th order polynomial (e.g., linear for $k = 1$, quadratic for $k = 2$ etc.)

Akaike [1980] gave the problem a Bayesian interpretation, where the smoothness constraint is viewed as a stochastic pertubed, zero mean difference equation, that is

$$\nabla^k f(t) \sim N(0, \sigma^2) \tag{4}$$

where $\mu^2 = \frac{1}{\sigma^2}$, which can be interpreted as a signal-to-noise ratio, is a hyperparameter that can be estimated using maximum likelihood methods. *Ansley and Kohn* [1986] and *Kohn and Ansley* [1988] have shown that for a diffuse initial prior, the solution produced by this problem is the discrete equivalent of the "smoothing spline" discussed extensively by Wahba and co-workers for nonparametric function estimation (see *Wahba* [1990] and references therein), except that the smoothness parameter is estimated by maximum likelihood instead of generalized cross-validation (see also *Wecker and Ansley* [1983]).

Returning to Equation (1), the trend term can be viewed as an unknown function of time, and parameterized as

$$\nabla^k T(t) \sim N(0, \sigma_T^2). \tag{5}$$

The seasonal component can be constrained in several fashions. In the first, the running sum of the seasonal component is constrained (assuming s periods in a season; e.g., $s = 12$ for monthly data, $s = 4$ for quarterly data)

$$\sum_{i=0}^{s-1} S(t-i) \sim N(0, \sigma_S^2). \qquad t = 1, T. \tag{6}$$

A second approach constrains the seasonal differences

$$S(t) - S(t-s) \sim N(0, \sigma_S^2). \qquad t = 1, T. \tag{7}$$

Another approach is to model the seasonal component using a second-order autoregressive (AR(2)) term with the autoregressive parameters fixed but the innovation variance to be estimated. The proper choice of the AR(2) parameters will create a component at a given frequency but with a changing phase and amplitude. *Harvey* [1989] describes a related approach using trigonometric functions. In this paper, we use the first approach as it would appear to provide the most flexibility in describing a seasonal component that could be a sum of annual and higher frequencies, as some of the time series suggest.

The irregular term $I(t)$ is assumed to be a p-th order autoregression, that is

$$I(t) = \sum_{i=1}^{p} \phi_i I(t-p) + \epsilon(t) \tag{8}$$

$$\epsilon(t) \sim N(0, \sigma_I^2), \qquad t = 1, T, \tag{9}$$

and the observation errors are assumed to be zero mean, independent, identically distributed as

$$e(t) \sim N(0, \sigma_e^2), \qquad t = 1, T. \tag{10}$$

Kitagawa and Gersch [1984, 1988] show how these assumptions can be put into a state-space model and solved by using a combination of the Kalman filter and maximum likelihood estimation. *Shumway and Stoffer* [1982] and *Shumway* [1988, Chapter 3] show how the expectation-maximization (EM) algorithm [*Dempster et al.*, 1977] can be used to calculate the maximum likelihood estimates. The minimum mean-squared error estimates of each of the components are given by the Kalman smoother calculated at the final maximum likelihood estimates of the hyperparameters. Details of the likelihood equations and the necessary computations are given in the Appendix. A similar model was used by *Young et al.* [1991] to model the Mauna Loa atmospheric CO_2 time series and a time series composed of the first principal component of Pacific ocean SST anomalies.

In all of our analyses, we use $k = 1$ in the trend (Equation 5), so that the first-order differences are assumed to be random, $p = 1$ in Equation 9 for a first-order autoregressive (AR(1)) term, and the seasonal prior term given in Equation (6) with $s = 12$, as we are analyzing monthly data.

The flexibility of this parameterization can be understood best by examining the limits of the trend and seasonal components at the extreme values of their variances (zero and infinity) when the other components have been removed (the partial residual series). If the seasonal and irregular terms were somehow known, then the algorithm would estimate a smoothed version of the observed series minus the seasonal and irregular components, in this case the discrete equivalent of a k-th order smoothing spline. When the trend variance (σ_T^2) is zero, this smoother is simply a linear least-squares fit to the partial residual series. When the trend variance approaches infinity, then the smoother simply interpolates the partial residual series.

If the trend and irregular were removed from the data, then the algorithm calculates for the seasonal component given by Equation (6), a smoothed version of the s-period running sums of the partial residual series, where the amount of smoothing applied is the same throughout the series. (This implicitly will smooth the s-period differences also.) When the seasonal variance (σ_S^2) is zero, the result is the monthly means of the partial residual series. When the seasonal variance approaches infinity, then the result again interpolates the partial residual series.

Likewise, if the trend and seasonal components were known, then the irregular term estimates a p-th order autoregressive model to the partial residual series after the trend and the seasonal were removed. While the algorithm used in this paper estimates the components simultaneously, this "backfitting" type approach of recursively smoothing the partial residual series could be used with other smoothing algorithms.

The means of the monthly seasonal model time series for April–July (the upwelling "season") were calculated in each year to produce the time series analyzed and described below (i.e., each annual value represents the average of the April–July period in that year). These series will henceforth be referred to as the upwelling time series.

3 Results

3.1 Trends in SST and Wind Stress

Time series of poleward wind stress trends for the COADS 2° boxes (Figure 2) display the spatial and temporal variability of the CCS wind field. The wind separates into three distinct regions; 22–32°N (south), 32–40°N (central), and 40–48°N (northern). Wind stress in the southern region (dashed-dotted lines) became increasingly equatorward (negative) over time in a relatively monotonic pattern. Stress also became increasingly equatorward in space between 22-30°, then less strong north and into the Southern California Bight; local maximum equatorward stress was seen over 26–30°N. The Bight featured weaker stress and more variation on 5–10-year scales, relative to adjacent boxes.

The central region (solid lines) displayed the strongest equatorward stress in the CCS. Stress became increasingly equatorward over time, but exhibited much more interannual variation compared to the southern region. A period of stronger than normal stress in the 1950s changed to decreasing equatorward stress in mid-1960s. The center of this region (39N, bold dashed line) featured the greatest negative linear tendency, shifting from the series of weakest to strongest stress.

In contrast to areas south of 44°N, the northern region (bold dotted lines) featured a mean poleward stress that became increasingly poleward over time. 40–44°N was a transition area between the central and northern regions; equatorward stress decreased rapidly with distance north. In about 1975 the 43N COADS series shifts from the pattern seen in the northern region to that of the central region.

Two temporal phenomena are notable for their absence in the stress trend series. ENSO events (*e.g.*, 1957, 1983) are not apparent in the series. Instead the model allocated their variance into the AR series (not shown), and model error (not shown) presumably because the wind field responds rapidly to ENSO conditions. The well-documented "regime shift" in the mid–1970s [*Trenberth* 1990, *Ebbesmeyer et al.* 1991, *Miller et al.* 1994, *Trenberth and Hurrell* 1994] is not seen in the wind trends either, despite its clear presence in north Pacific atmospheric pressure indices. However a substantial increase in equatorward stress occurred about 1983 in the central region and north to about 44°N. One interpretation is that the transition zone between the central and northern regions has broadened south; another is that the area off northern California and Oregon developed its own distinct wind regime after 1983. In either case, the net effect is that the zonal gradient in poleward stress has strengthened greatly over the last 45 years.

The COADS SST trends are visually more correlated in space on the interannual (1-5 year) scale than wind stress (Figure 3). As with the wind fields, SST series separate into distinct regions. SST decreased consistently with latitude south of about 40°N (fine lines), coincident with the regions of equatorward wind stress. SSTs north of 40°N (bold lines), an area where stress was either poleward or changing rapidly from strongly equatorward to poleward, were nearly uniform with latitude. While ENSO wind events are relatively ephemeral, and thus are relegated to the AR term of the model, warmer SSTs associated with ENSOs are more apparent in the SST trend series, and appear to diffuse away over time. However some ENSO signal appears in the SST AR series as well. Interannual variations in the series (*e.g.*, the 1957 ENSO event) are most evident off central and southern California (Figure 3). In contrast to the 1957 event, the 1982 ENSO shows up as a smaller local maximum in SST relative to adjacent periods.

Shore-based SST trend time series along the western North American coast, derived with the state-space model, allows an independent examination of SST variability in the CCS region, extended in space and time. These series also provide a finer spatial (across- and alongshore) scale look at decadal variability. As with the

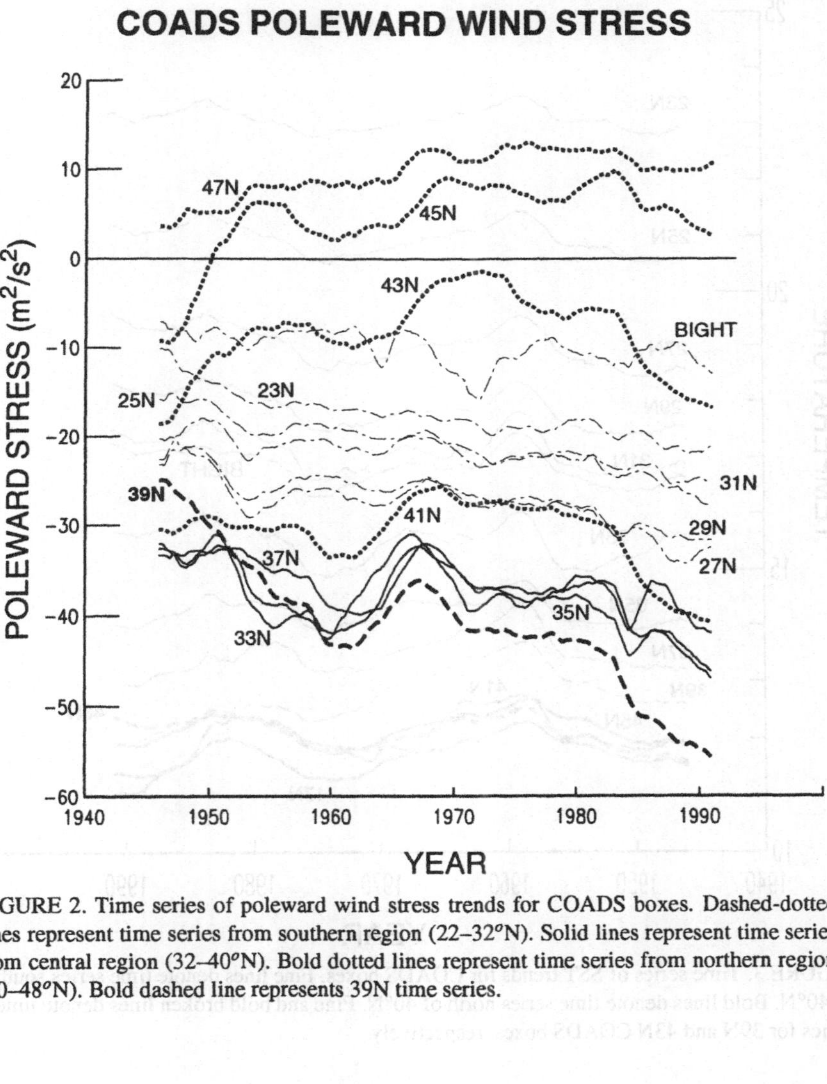

FIGURE 2. Time series of poleward wind stress trends for COADS boxes. Dashed-dotted lines represent time series from southern region (22–32°N). Solid lines represent time series from central region (32–40°N). Bold dotted lines represent time series from northern region (40–48°N). Bold dashed line represents 39N time series.

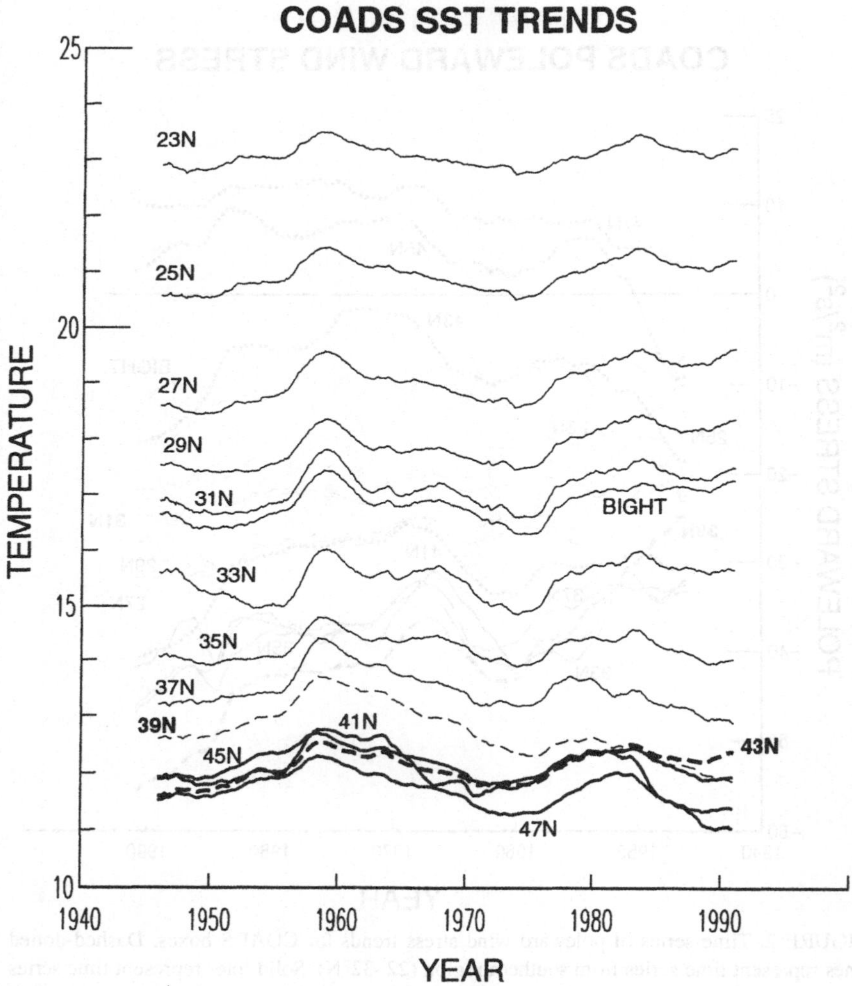

FIGURE 3. Time series of SST trends for COADS boxes. Fine lines denote time series south of 40°N. Bold lines denote time series north of 40°N. Fine and bold broken lines denote time series for 39N and 43N COADS boxes, respectively.

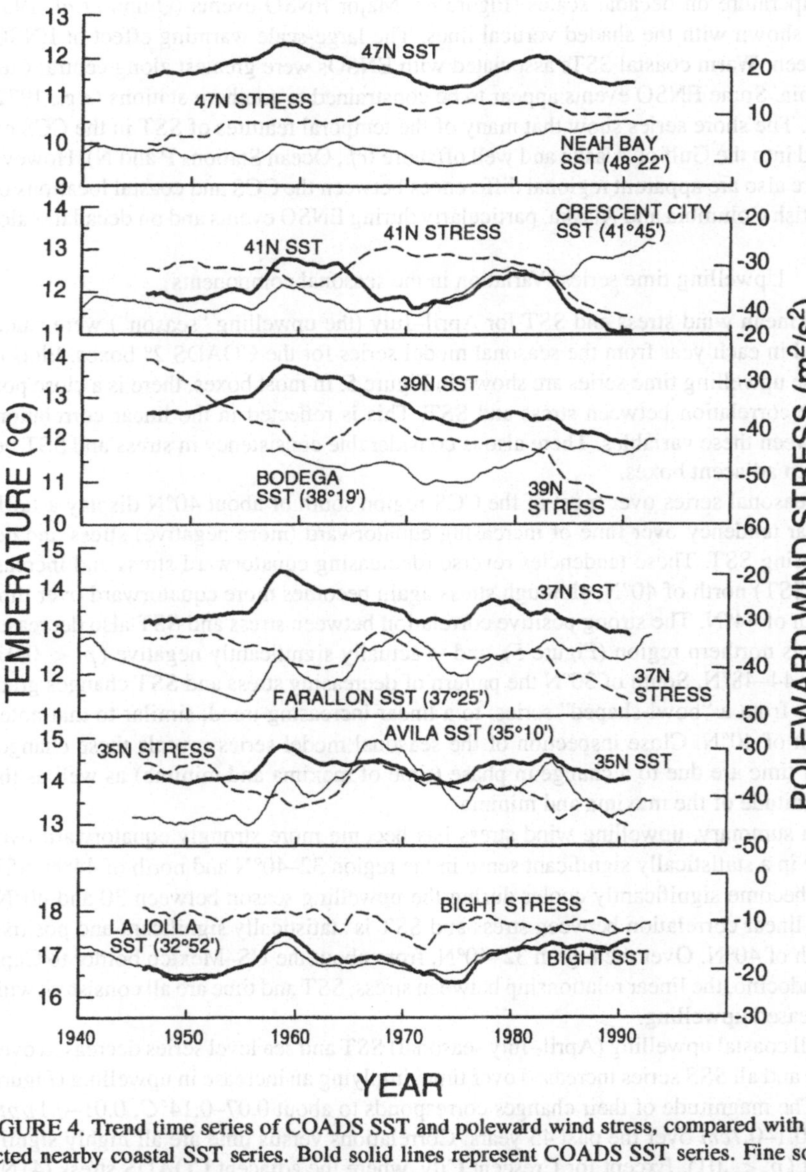

FIGURE 4. Trend time series of COADS SST and poleward wind stress, compared with selected nearby coastal SST series. Bold solid lines represent COADS SST series. Fine solid lines represent coastal SST series. Broken lines represent wind stress series. Location of time series shown in each plot. In the bottom panel, wind stress and SST series are displayed from the Southern California Bight as well as a 2° box (32–34°N, 120–122°W) immediately seaward of the Bight box.

COADS SSTs, the shore series demonstrate well-defined latitudinal regimes in ocean temperature on decadal scales (Figure 4). Major ENSO events (Quinn *et al.* 1987) are shown with the shaded vertical lines. The large-scale warming effect of ENSOs is seen. Warm coastal SSTs associated with ENSOs were greatest along central California. Some ENSO events appear to be constrained to southern stations (*e.g.*, 1972–73). The shore series show that many of the temporal features of SST in the CCS extend into the Gulf of Alaska and well offshore (*cf.*, Ocean Stations P and N). However there also are apparent regional differences between the CCS and coastal locations off British Columbia and Alaska, particularly during ENSO events and on decadal scales.

3.2 Upwelling time series: Variation in the seasonal components

The mean wind stress and SST for April–July (the upwelling "season") were calculated in each year from the seasonal model series for the COADS 2° boxes. Plots of these upwelling time series are shown in Figure 5. In most boxes, there is a close positive correlation between stress and SST. This is reflected in the linear correlations between these variables. There also is considerable consistency in stress and SST between adjacent boxes.

Seasonal series over most of the CCS region south of about 40^oN display a fairly linear tendency over time of increasing equatorward (more negative) stress and decreasing SST. These tendencies reverse (decreasing equatorward stress and increasing SST) north of 40^oN, although stress again becomes more equatorward over time north of 44^oN. The strong positive correlation between stress and SST also decreases in this northern region (Figure 5), and is actually significantly negative ($p < 0.01$) over 44–48^oN. South of 30^oN the pattern of decreasing stress and SST changes gradually from a "bowl-shaped" series, to a linear increasing trend, similar to that noted north of 40^oN. Close inspection of the seasonal model series reveals these changes over time are due to a change in phase (time of maxima and minima) as well as the magnitude of the maxima and minima.

In summary, upwelling wind stress has become more strongly equatorward over time in a statistically significant sense in the region 32–40^oN and north of 44^oN. SST has become significantly cooler during the upwelling season between 30 and 40^oN. The linear correlation between stress and SST is statistically significant and positive south of 40^oN. Over the region 32–40^oN, from about the US–Mexico border to Cape Mendocino, the linear relationship between stress, SST and time are all consistent with increased upwelling.

All coastal upwelling (April–July seasonal) SST and sea level series decreased over time and all SSS series increased over time, implying an increase in upwelling (Figure 6). The magnitude of their changes corresponds to about $0.07–0.14^oC$, $0.01–0.1ppt$, and $0.1–0.7cm$ over the past 45 years. Correlations versus time are all highly significant ($p < .01$). Except for Crescent City, where the adjacent COADS stress (41N) shows an increasing seasonal trend, SST and sea level are highly positively (negatively) correlated with local wind stress. The patterns occurring during the period covered by the COADS data (1946–90) are consistent with those seen in the full-length shore series (57–78 years). Regressions between coastal SST, SSS and sea level (the series shown in Figure 6) are highly significant and of the sign consistent with that

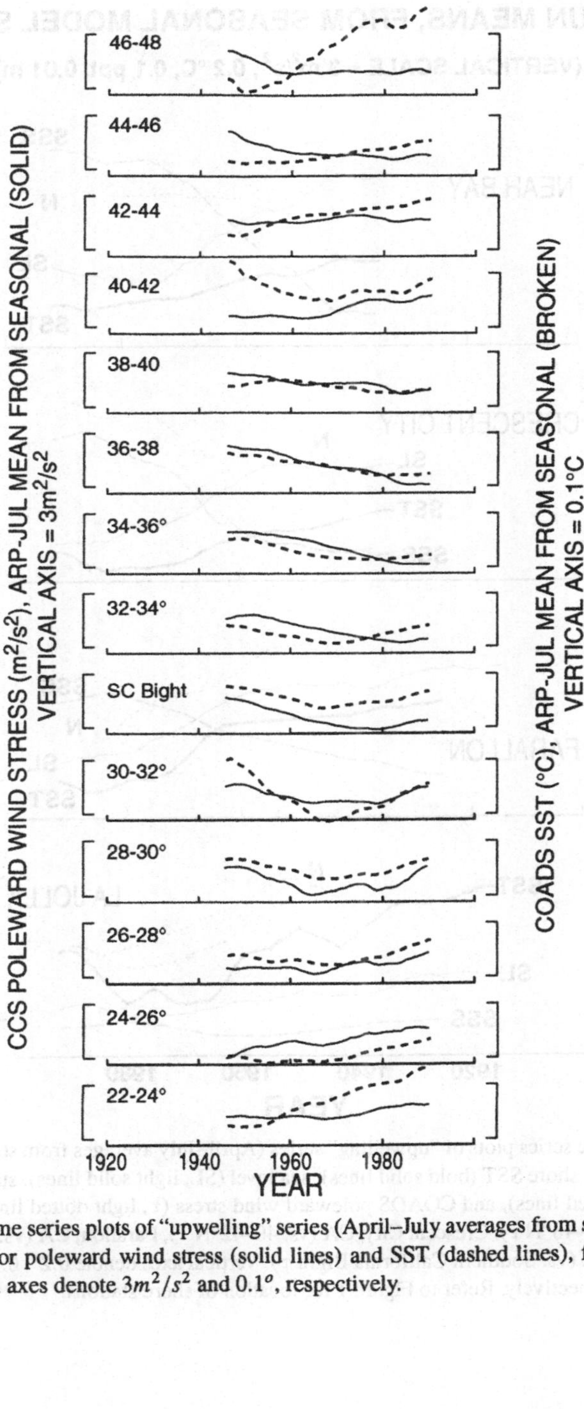

FIGURE 5. Time series plots of "upwelling" series (April–July averages from seasonal model components) for poleward wind stress (solid lines) and SST (dashed lines), for COADS 2° boxes. Vertical axes denote $3m^2/s^2$ and 0.1^o, respectively.

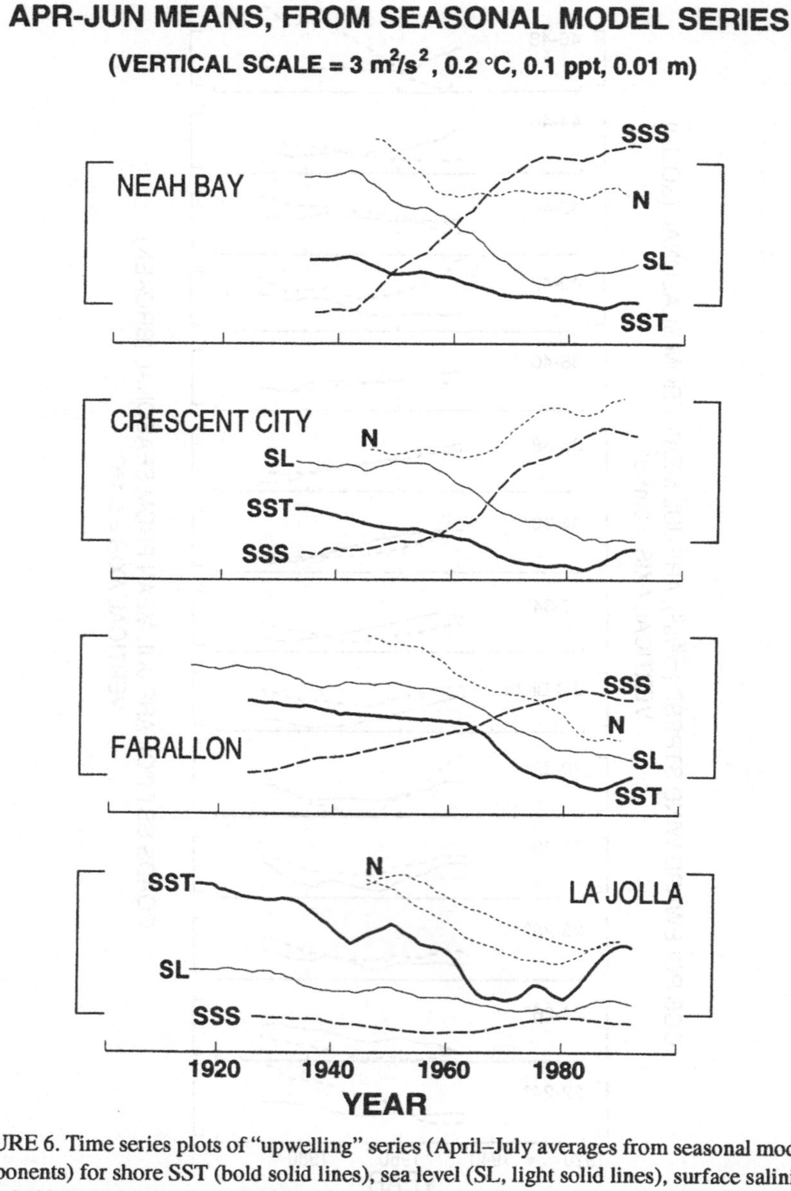

FIGURE 6. Time series plots of "upwelling" series (April–July averages from seasonal model components) for shore SST (bold solid lines), sea level (SL, light solid lines), surface salinity (SSS, bold dashed lines), and COADS poleward wind stress (τ, light dotted lines), for Neah Bay, WA (vs. 46–48°N τ), Crescent City, CA (vs. 40–42°N τ), Farallon, CA (vs. 36–38°N τ), and La Jolla, CA (vs. Southern California Bight τ). Vertical axes denote 0.2^o, 0.1ppt, 0.01m, and $3m^2/s^2$, respectively. Refer to Figure 1 for location of shore stations.

expected if upwelling is the controlling process. This is consistent with the fact that upwelling is a dominant process off much of the west coast during April–July.

3.3 Common Trends

The previous sections suggest that the dynamics of the CCS are divided into three regions, with a transition zone that both shifts its center and changes in extent. A next question is to more formally determine what aspects of the trends are global in nature, and what aspects represent more regional variation. The state-space model can be extended to this case by now making the observed series a vector of time series, one for each area, by succesively modeling for one underlying trend component, then two underlying (independent) trend components etc., and by assuming the observation matrix A, rather than being known, has an unknown coefficient for each series for each underlying component. The estimate for A can be derived from the equations in the Appendix to be:

$$\hat{A} = MS_t(0)^{-1} \tag{11}$$

where $S_t(0)$ is defined in the Appendix and M is defined as

$$M = \sum_{t=1}^{T} y(t)x(t)'. \tag{12}$$

Note that the observation equation could be rescaled so that each component has a coefficient of one for a selected observed series.

The time series for SST can be adequately represented by three underlying trend components (Figure 7). The first component shows a sharp increase due to the 1956-57 El Nino, then a decrease up to about 1975-76, and increase to the mid 1980's, then a decrease, a common pattern above. The third component exhibits what is close to a twenty year cycle. There are several papers that suggests physical reasons for long-term behavior at this frequency, and will be explored more closely in future research.

The effect of adding each component can be seen by comparing the estimated trend for two selected areas based on one, two and three underlying components with the trend estimated for that time series individually (Figure 8). The first component clearly is an underlying "mean trend", global behavior common to the entire CCS. The other components appear to be more regional in nature, thus dividing the trends into global and local components. Though not shown, the wind stress series require four underlying components to adequately model the series. On examination, the components have a similar interpretation of a mean, global trend modified by several more regional trends.

4 Summary

The state-space model appears to be a powerful tool for separating the interannual-to-interdecadal fluctuations in environmental time series from seasonal patterns of variability. The model results help provide a better understanding of the linkages between long-term variations in atmospheric forcing and the coastal ocean's response to this

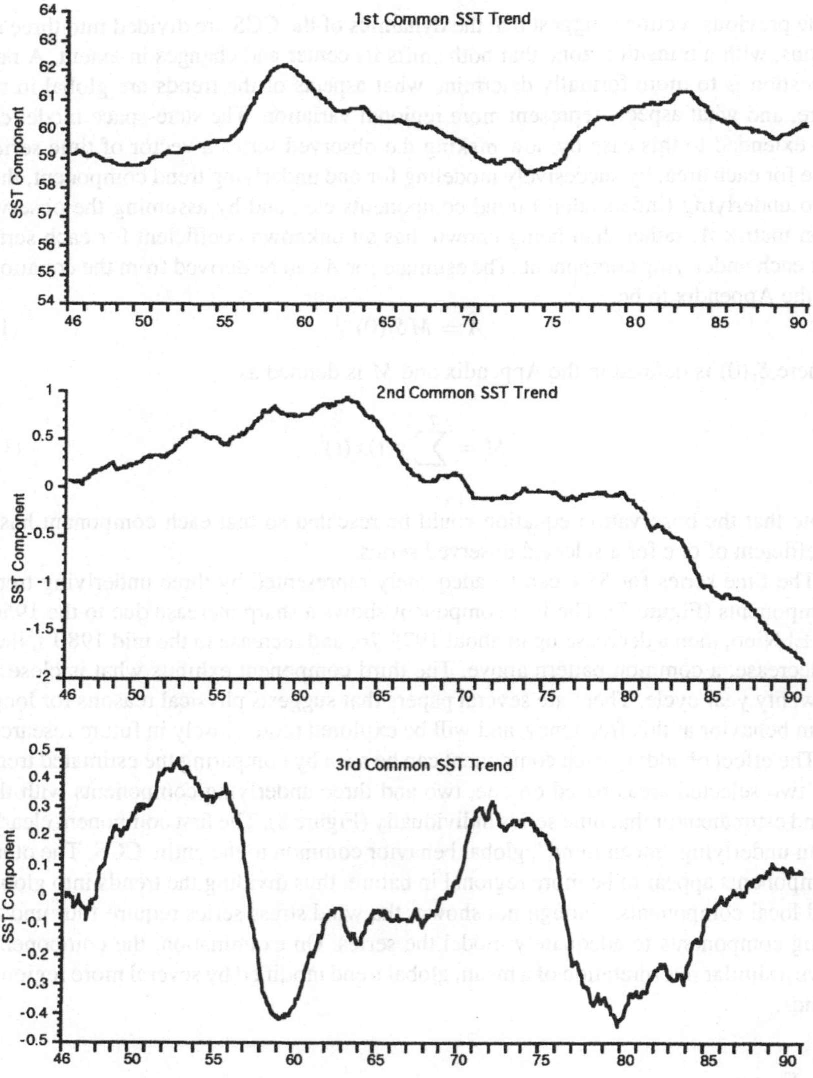

FIGURE 7. Time series of the three estimated common trends for SST.

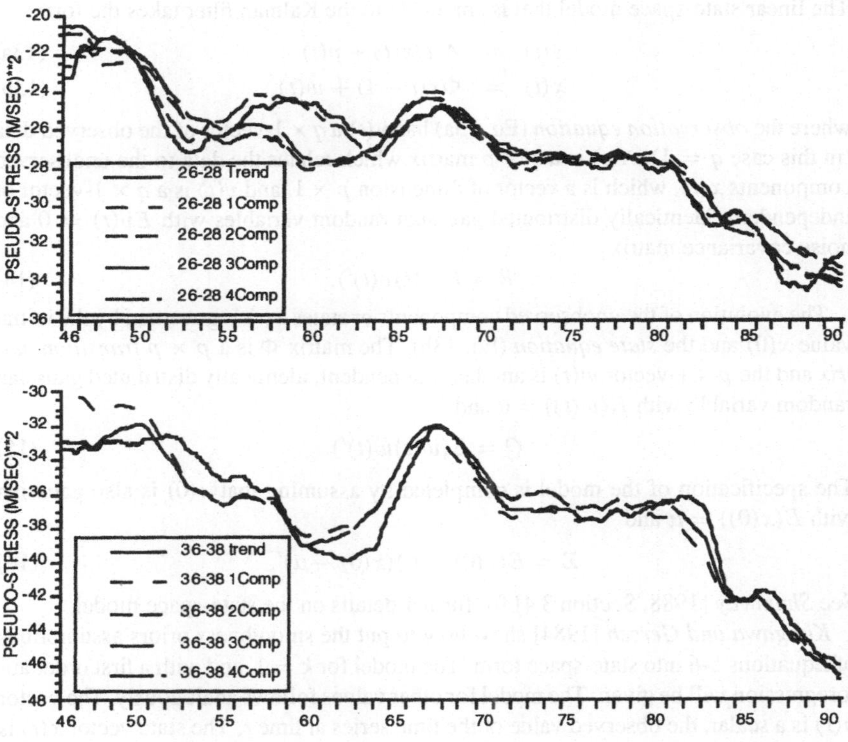

FIGURE 8. SST trend versus trend estimated from common components for two areas.

variability, as well as the potential contribution of natural and anthropogenic signals, and regional differences in these effects. The results presented here demonstrate the importance of evaluating temporal and spatial variability over the entire spectrum, rather than simply at global climate scales, when examining long-term environmental fluctuations. They also demonstrate the importance of considering independently the change in seasonal patterns versus changes in the long-term climate trend.

5 Appendix: State-Space Decomposition of a Time Series

The linear state-space model that is amenable to the Kalman filter takes the form:

$$y(t) = A(t)x(t) + v(t) \tag{13a}$$

$$x(t) = \Phi x(t-1) + w(t) \tag{13b}$$

where the *observation equation* (Eq. 13a) has $y(t)$ a $q \times 1$-vector of the observed data (in this case $q = 1$), $A(t)$ is a $q \times p$ matrix which relates the data to the unobserved components $x(t)$, which is a vector of dimension $p \times 1$, and $v(t)$ is a $q \times 1$-vector of independent, identically distributed gaussian random variables with $Ev(t) = 0$ and noise covariance matrix

$$R = E(v(t)v(t)'). \tag{14}$$

The evolution of the unobserved components or states $x(t)$ is governed by the initial value $x(0)$ and the *state equation* (Eq. 13b). The matrix Φ is a $p \times p$ *transition matrix* and the $p \times 1$-vector $w(t)$ is another independent, identically distributed gaussian random variable with $E(w(t)) = 0$ and

$$Q = E(w(t)w(t)'). \tag{15}$$

The specification of the model is completed by assuming that $x(0)$ is also gaussian with $E(x(0)) = \mu$ and

$$\Sigma = E(x(0) - \mu)(x(0) - \mu)'. \tag{16}$$

See *Shumway* [1988, Section 3.4] for further details on the state-space model.

Kitagawa and Gersch [1984] show how to put the smoothness priors assumptions of Equations 5-6 into state-space form. The model for $k = 1$, and with a first order autoregression will be given. The model for other values follows analogously. The vector $y(t)$ is a scalar, the observed value of the time series at time t. The state vector $x(t)$ is of dimension 13 and is of the form:

$$x(t)' = (T(t), S(t), \ldots, S(t-11), I(t)) \tag{17}$$

and the transition matrix Φ is given by

$$\begin{pmatrix} 1 & 0 & 0 & \ldots & 0 & 0 \\ 0 & -1 & -1 & \ldots & -1 & 0 \\ 0 & 1 & 0 & \ldots & 0 & 0 \\ 0 & 0 & 1 & \ldots & 0 & 0 \\ \vdots & \vdots & \vdots & \ddots & \vdots & \vdots \\ 0 & 0 & 0 & \ldots & 1 & 0 \\ 0 & 0 & 0 & \ldots & 0 & \phi \end{pmatrix} \tag{18}$$

where ϕ is the autoregressive parameter which is to be estimated. The observation matrix (from Eq. 1) is given by

$$A = (\,1\;\;1\;\;0\;\;0\;\;0\;\;0\;\;0\;\;0\;\;0\;\;0\;\;0\;\;0\;\;0\;\;1\,)\,. \tag{19}$$

The specification is complete by setting the observation error covariance matrix equal to $R = \sigma_e^2$ and by setting the state noise covariance matrix equal to

$$\begin{pmatrix} \sigma_T^2 & 0 & 0 & \dots & 0 & 0 \\ 0 & \sigma_S^2 & 0 & \dots & 0 & 0 \\ 0 & 0 & 0 & \dots & 0 & 0 \\ \vdots & \vdots & \vdots & \vdots & \vdots & \vdots \\ 0 & 0 & 0 & \dots & 0 & \sigma_I^2 \end{pmatrix}. \tag{20}$$

For given values of the vector of parameters $\Theta = (\phi, \sigma_T^2, \sigma_S^2, \sigma_I^2, \sigma_e^2)$, the minimum mean-square-error estimates of each of the components of the state vector can be estimated using the Kalman filter and smoother. A particularly efficient form of the algorithm was developed independently by *Ansley and Kohn* [1985, 1990], *Kohn and Ansley* [1987] and *De Jong* [1989, 1990, 1991]. Both algorithms include the case where $\Sigma \to \infty$, a diffuse or noninformative prior (see above references). Let $x(t|\tau)$ denote the expected value of the state vector given the data up to and including time τ, let $P(t|\tau)$ denote the covariance matrix of the state vector conditional on the data up to and including time τ. Let $r(t)$ be a vector of dimension p, and $R(t)$ a $p \times p$ matrix. Then the filter and smoothing steps proceed as follows:

<u>FILTERING:</u>

Initialize:

$$x(0) = \mu \tag{21a}$$
$$P(0) = \Sigma \tag{21b}$$

Iterate for $t = 1, T$

$$e(t) = y(t) - Ax(t|t-1) \tag{22a}$$
$$D(t) = AP(t|t-1)A + R \tag{22b}$$
$$K(t) = \Phi P(t|t-1)A'D(t)^{-1} \tag{22c}$$
$$x(t+1|t) = \Phi x(t|t-1) + K(t)e(t) \tag{22d}$$
$$L(t) = \Phi - K(t)A \tag{22e}$$
$$P(t+1|t) = \Phi P(t|t-1)L(t)' + Q \tag{22f}$$

<u>SMOOTHING:</u>

Initialize:

$$r(T) = 0 \tag{23a}$$
$$R(T) = 0 \tag{23b}$$

Iterate for $t = T-1, 0$

$$\begin{aligned} r(t-1) &= A'D(t)^{-1}e(t) + L(t)'r(t) && \text{(24a)} \\ R(t-1) &= A'D(t)^{-1}A + L(t)'R(t)L(t) && \text{(24b)} \\ x(t|T) &= x(t|t-1) + P(t|t-1)r(t-1) && \text{(24c)} \\ P(t|T) &= P(t|t-1) + P(t|t-1)R(t-1)P(t|t-1) && \text{(24d)} \end{aligned}$$

The log-likelihood $L(Y;\Theta)$ given all the data Y and the parameter vector Θ is given by (see for example *Shumway* [1988, page 178])

$$\ln L(Y;\Theta) = -\frac{1}{2}\sum_{t=1}^{T}\ln|D(t)| - \frac{1}{2}\sum_{t=1}^{T} e(t)'D(t)^{-1}e(t). \tag{25}$$

In order to use the EM algorithm [*Dempster et al.,* 1977] it is necessary to derive the complete data likelihood; here the components $x(t)$ are viewed as unobserved or "missing". After some manipulation, this can be shown to be [*Shumway,* 1988, page 179]

$$\begin{aligned} &-\frac{1}{2}\ln|\Sigma| - \frac{1}{2}tr\left\{\Sigma^{-1}[P(0|T) + (x(0|T)-\mu)(x(0|T)-\mu)']\right\} \\ &-\frac{T}{2}|\ln Q| \\ &-\frac{1}{2}tr\left\{Q^{-1}[S_t(0) - S_t(1)\Phi' - \Phi S_t(1)' + \Phi S_{t-1}\Phi']\right\} \\ &-\frac{T}{2}\ln|R| \\ &-\frac{1}{2}tr\left\{R^{-1}\sum_{t=1}^{T}[(y(t)-Ax(t|T))(y(t)-Ax(t|T))' + AP(t|T)A']\right\} \end{aligned}$$

where the terms $S_t(0)$,$S_t(1)$, and $S_{t-1}(0)$ are defined as:

$$\begin{aligned} S_t(0) &= \sum_{t=1}^{T}(P(t|T) + x(t|T)x(t|T)') && \text{(26a)} \\ S_{t-1}(0) &= \sum_{t=0}^{T-1}(P(t1|T) + x(t|T)x(t|T)') && \text{(26b)} \\ S_t(1) &= \sum_{t=1}^{T}(P(t,t-1|T) + x(t|T)x(t-1|T)') && \text{(26c)} \end{aligned}$$

and

$$P(t,t-1|T) = E\left[(x(t)-x(t|T))(x(t-1)-x(t-1|T))'|y(1),\ldots,y(T)\right].$$

A recursion for $P(t,t-1|T)$ is given in *Shumway and Stoffer* [1982] and *De Jong* [1990]. The complete data likelihood is maximized by setting:

$$\Phi = S_t(1)[S_{t-1}(0)]^{-1} \tag{27a}$$

$$Q = T^{-1}[S_t(0) - S_t(1)\Phi' - \Phi S_t(1)' + \Phi S_{t-1}(0)\Phi'] \quad (27b)$$

$$R = T^{-1}\sum_{t=1}^{T}\left[(y(t) - Ax(t|T))(y(t) - Ax(t|T))' + AP(t|T)A'\right]. \quad (27c)$$

In the model of this paper, since most of Φ is fixed, the new estimate of ϕ is the (13,13) element of Φ, and since most of Q is fixed to zero, the new estimates of Q are the (1,1), (2,2) and (13,13) elements of the above matrix, with all other elements set to zero. Here R is a scalar. The complete algorithm then consists of

1. Choose initial values for $x(0)$, Σ and Θ.
2. Calculate the Kalman filter and smoother for the given parameter values
3. Update the parameters by Equations 27a-27c
4. Iterate until convergence

The Kalman smoothers, calculated at the final parameter estimates, produce the component time series.

References

Akaike, H. Likelihood and the Bayes procedure, in *Bayesian Statistics*, edited by J.M. Bernardo, M.H. DeGroot, D.V. Lindley and A.F.M. Smith,pp 143-166, University Press, Valencia, 1980.

Ansley, C.F. and R. Kohn, Estimation, filtering and smoothing in state space models with incompletely specified initial conditions, *Ann. Statist. 13*, 1286-1316, 1985.

Ansley, C.F. and R. Kohn, On the equivalence of two stochastic approaches to spline smoothing, in *Time Series and Allied Processes*, edited by J. Gani and M.B. Priestly, pp. 391-405, *J. Appl. Prob. 23A*, 1986

Ansley, C.F. and R. Kohn, Filtering and Smoothing in State Space Models with Partially Diffuse Initial Conditions, *J. Time Series Anal. 11*, 275-293, 1990.

Aoki, M. State Space Models for Vector-Valued Time Series with Random Walk Components, *Invited Paper, 1988 Joint Statistical Meeting, New Orleans.*

Aoki, M., *State Space Modeling of TIme Series, 2nd Edition*, 323pp., Springer-Verlag, Berlin, 1990a.

Aoki, M. A State Space Approach to Modeling Cointegrated Times Series, in *Advances in Econometrics, Cointegration, Spurious Regressions and Unit Roots*, edited by G.F. Rhodes Jr. and T.B. Fomby, JAI Press Inc., Connecticut.

Bakun, A., Global climate change and intensification of coastal ocean upwelling, *Science 247*, 198-201, 1990.

Bakun, A., Global greenhouse effects, multi-decadal wind trends, and potential impacts on coastal pelagic fish populations, *ICES Mar. Sci. Symp. 195*, 316-325, 1992.

Bakun, A. and C.S. Nelson, Climatology of upwelling related processes off Baja California. *CalCOFI Reports 19*, 107-127, 1977.

Bakun, A. and C.S. Nelson, The seasonal cycle of wind-stress curl in subtropical eastern boundary current regions, *J. Phys. Oceanogr. 21*, 1815-1834, 1991.

Brink, K.H., D.C. Chapman, and G.R. Halliwell, Jr., A stochastic model for wind-driven currents over the continental shelf, *J. Geophys. Res. 92*, 1783-1797, 1987.

Brotherton, T. and W. Gersch, A data analytic approach to the smoothing problem and some of its variation,. *Proceedings of the 20th IEEE Conference on Decision and Control*, 1061-1069, 1981.

Cardone, V.J., J.G. Greenwood and M. Can, On Trends in the Historical Marine Wind Data, *J. Climate 3*, 113-127, 1990.

Cayan, D.R., D.R. McLain, W.D. Nichols, and J.S. DiLeo-Stevens, *Monthly Climatic Time Series Data from the Pacific Ocean and Western Americas*, 379 pp., USGS Open-File Report, 1988.

Cury, P. and C. Roy, Optimal environmental window and pelagic fish recruitment success in upwelling areas, *Can. J. Fish. Aquat. Sci. 46*, 670-680, 1989.

De Jong, P., Smoothing and interpolation with the state-space model, *Journal of the American Statistical Association 84*, 1085-1088, 1989.

De Jong, P., Stable Algorithms for the State Space Model, *J. Time Series Anal. 12*, 143-157, 1990.

De Jong, P., The diffuse Kalman filter, *Ann. Statist. 19*, 1073-1083, 1991.

Dempster, A.P., N.M. Laird and D.B. Rubin, Maximum likelihood from incomplete data via the EM algorithm, *J. Roy. Statist. Soc., Ser. B. 39*, 1-38, 1977.

Denbo, D.W. and J.S. Allen, Large-scale response to atmospheric forcing of shelf currents and coastal sea level off the west coast of North America: May-July 1981 and 1982, *J. Geophys. Res. 92*, 1757-1782, 1987.

Ebbesmeyer, C.C., D.R. Cayan, D.R. McLain, F.H. Nichols, D.H. Peterson and K.T. Redmond, 1991. 1976 step in the Pacific climate: Fourty environmental changes between 1968-1975 and 1977-1984, in *Proc. 7th Ann. Pacific Climate Workshop*, edited by J. Betancourt and V.L. Sharp, pp 129-141, Calif. Dept. of Water Resources, Interagency Ecol. Stud. Prog. Tech. Rep. 26. 1991.

Graham, N.E., Decadal scale variability in the 1970's and 1980's: observations and model results, *Clim. Dyn. 10*, 60-70, 1994.

Harvey, A.C., *Forecasting, Structural Time Series Models and the Kalman Filter*, 554 pp., Cambridge University Press, Cambridge, 1989.

Isemer, H.-J., Comparison of Estimated and Measured Marine Surface Wind Speed, in *Proceedings of the International COADS Workshop, Boulder, Colorado, 13-15 January 1992*, edited by Diaz, H.F., K. Wolter and S.D. Woodruff, pp. 142-158, U.S. Dept. of Commerce, 1992.

Kitagawa, G. and W. Gersch, A smooothness priors - state space modeling of time series with trend and seasonality, *J. Am. Statist. Assoc. 79*, 378-389, 1984.

Kitagawa, G. and W. Gersch, A smoothness priors time varying AR coefficient modeling of nonstationary time series, *IEEE Trans. on Auto. Control AC-30*, 48-56, 1985.

Kitagawa, G. and W. Gersch, Smoothness priors in time series, in *Bayesian Analysis of Time Series and Dynamic Models*, edited by J. Spall, Marcel Dekker, N.Y,

1988.

Kohn, R. and C.F. Ansley, A new algorithm for spline smoothing and interpolation based on smoothing a stochastic process, *SIAM J. Sci. Statist. Comput. 8*, 33-48, 1987.

Kohn, R. and C.F. Ansley, The equivalence between Bayesian smoothness prioirs and optimal smoothing for function estimation, in *Bayesian Analysis of Time Series and Dynamic Models*, edited by J. Spall, Marcel Dekker, N.Y, 1988

Lauritzen, S.L., Time series analysis in 1880: A discussion of contributions made by T.N. Thiele, *Int. Statist. Review 49*, 319-331, 1981.

Mann, M.E. and J. Park, Spatial correlations of interdecadal variation in global surface temperatures, *Geophys. Res. Lett., 20*, 1055-1058, 1993.

Mendelssohn, R and C. Roy, CODE (Comprehensive Ocean Data Extraction) Users Guide, *In Preparation*, 1996.

Miller, A.J., D.R. Cayan, T.P. Barnett, N.E. Graham and J.M. Oberhuber, The 1976-77 climate shift of the Pacific Ocean, *Oceanograph, 7*, 21-26, 1994..

Parrish, R.H., C.S. Nelson, and A. Bakun, Transport mechanisms and reproductive success of fishes in the California Current, *Biol. Oceanogr. 1*, 175-203, 1981.

Parrish, R.H., A. Bakun, D.M. Husby, and C.S. Nelson, Comparative climatology of selected environmental processes in relation to eastern boundary current fish production, *FAO Fish Rep 291*, 731-778, 1983.

Roemmich, D., Ocean warming and sea level rise along the southwest U.S. coast, *Science 257*, 373-375, 1992.

Rosenfeld, L.K., F.B. Schwing, N. Garfield, and D.E. Tracy, Bifurcated flow from an upwelling center: a cold water source for Monterey Bay, *Continental Shelf Research 14*, 931-964, 1994.

Sakuma, K.M., H.A. Parker, S. Ralston, F.B. Schwing, D.M. Husby and E.M. Armstrong, The physical oceanography off the central California coast during February-March and May-June, 1992: a summary of CTD data from pelagic young-of-the-year rockfish surveys, *U.S. Department of Commerce, NOAA Technical Memorandum NOAA-TM-NMFS-SWFSC-208*, 169pp., La Jolla, CA, 1994.

Schwing, F.B., Long-term and seasonal patterns in coastal temperature and salinity along the North American west coast, in *Proceedings of the Tenth Annual Pacific Climate (PACLIM) Workshop*, edited by K.T. Redmond and V.L. Tharp, pp. 129-143, Interagency Ecological Studies Program for the Sacramento-San Joaquin Estuary, Tech. Rep. 36, 1994.

Schwing, F.B., D.M. Husby, N. Garfield and D.E. Tracy, Mesoscale oceanic response to wind events off central California in spring 1989: CTD surveys and AVHRR imagery, *CalCOFI Reports 32*, 47-62, 1991.

Schwing, F.B. and R. Mendelssohn, Increased Coastal Upwelling in the California Current System, *Submitted to Journal of Geophysical Research*, 1996.

Schwing, F.B., R. Mendelssohn and R.H. Parrish, Interdecadal variability in the spatial structure of wind and SST in the California Current System, in *Proceedings of 1995 Annual Pacific Climate (PACLIM) Workshop*, in press, 1996.

Shiller, R., A distributed lag estimator derived from smoothness priors, *Econometrica 41*, 775-778, 1973.

Shumway, R.H., *Applied Statistical Time Series Analysis*, 379pp, Prentice-Hall, Englewood Cliffs, 1988.

Shumway, R.H. and D.S. Stoffer, An approach to time series smoothing and forecasting using the EM algorithm, *J. Time Series Anal. 3*, 253-264, 1982.

Skillman, J.B., *ENSO forced variations of the sea surface temperature and adjusted sea level along the west coast of the United States*, 92 pp., M.S. thesis, Naval Postgraduate School, Monterey, CA, 1993.

Slutz, R.J., S.J. Lubker, J.D. Hiscox, S.D. Woodruff, R.L. Jenne, D.H. Joseph, P.M. Steurer, and J.D. Elms, *Comprehensive Ocean-Atmosphere Data Set; Release 1*, 268pp., NOAA Environmental Research Laboratories, Climate Research Program, Boulder, CO., 1985.

Thiele, T.N., Om Anvedelse af Mindste Kvadraters Methode i Nogle Tilfaelde, Hvor en Komlikation af Visse Slags Uensartede Tilfaeldige Fejlkilder Giver Fejlene en 'Systematisk' Karakter, *Vidensk. Selsk*, French Translation: Sur la Compensation de Quelques Erreurs Quasisytematiques par la Methode des Moindres Carree,. Kobenhaven: Rietze, *Vidensk. Selsk. Skr. 5. rk., Naturvid og Mat. Afd. 12*, 381-408, 1880.

Trenberth, K.E., Recent observed interdecadal climate changes in the Northern Hemisphere, *Bull. Amer. Meteor. Soc., 71*, 988-993, 1990.

Trenberth, K.E. and J.W. Hurrell, Decadal atmospheric-ocean variations in the Pacific, *Climate Dynamics, 9*, 303-319, 1994.

Wahba, G., *Spline Functions for Observational Data*, CBMS-NSF Regional Conference Series in Applied Mathematics, Vol. 59, SIAM, Philadelphia, 1990.

Walker, P.M., D.M. Newton and A.W. Mantyla, *Surface water temperatures, salinities and densities at shore stations, United States West Coast, 1992*, 46pp., University of California, San Diego, Scripps Institution of Oceanography, La Jolla, CA, SIO Ref. 93-18, 1993.

Wecker, W.E. and C.R. Ansley, The Signal Extraction Approach to Nonlinear Regression and Spline Smoothing, *J. Am. Stat. Assoc. 78*, 81-89, 1983.

Whittaker, E.T., On a New Method of Graduation, *Proc. Edinborough Math. Assoc 78*, 81-89, 1923.

Woodruff, S.D., R.J. Slutz, R.L. Jenne, and P.M. Steurer, A Comprehensive Ocean-Atmosphere Data Set, *Bull. Amer. Meteor. Soc. 68*, 1239-1250, 1987.

Wu, Z. and R.E. Newell, The Wind Problem in COADS and its Influence on the Water Balance, in *Proceedings of the International COADS Workshop, Boulder, Colorado, 13-15 January 1992*, edited by Diaz, H.F., K. Wolter and S.D. Woodruff, pp. 189-200, U.S. Dept. of Commerce, 1992.

Xie, L. and W.W. Hsieh, The global distribution of wind-induced upwelling, *Fish. Oceanogr. 4*, 52-67, 1995.

Young, P.C., C.N. Ng, K. Lane and D. Parker, Recursive Forecasting, Smoothing and Seasonal Adjustment of Non-stationary Environmental Data, *Journal of Forecasting, 10*, 57-89, 1991.

Part III

Applications of Neural Networks

On the Equivalence Between ARMA Models and Simple Recurrent Neural Networks

Henrik Saxén
Department of Chemical Engineering, Åbo Akademi
Biskopsgatan 8, FIN–20500 Åbo, Finland. E–mail: hsaxen@abo.fi

Abstract

This paper presents analytical results for a class of linear discrete time recurrent neural networks. The networks are shown to be able to act as autoregressive moving average models. Minimal network sizes for representing ARMA(p,q) models are derived, and analogies between recurrent networks and state space models are pointed out.

1 Introduction

Recently, there has been a growing interest in recurrent neural networks for modeling dynamical systems [1, 2]. Recurrent networks can *represent time*, *learn* and *produce temporal sequences*, which makes them better suited than feed-forward networks for solving problems where temporal concepts are central. The interest in partially recurrent networks has been considerable, since constraints in the architecture may lead to more efficient and robust training of the networks. Well-known architectures are those introduced by Jordan [3] and Elman [4]. For a good review of different approaches, see ref. [5]. In this work, we derive analytical results for a class of simple partially recurrent neural networks, which arises if feedback connections are introduced into feedforward networks. The outputs from some of the units in the output layer are fed back via unity time-delay connections to *fictitious input nodes*. (Thus, the feedback weights are fixed to equal the backward shift operator, B, defined by $Bx_t = x_{t-1}$.) Also, additional feedback nodes (*fictitious outputs*) in the output layer are allowed. A recurrent net with N^{in} input nodes and N^{out} output nodes will be denoted by (N^{in},N_M^{out}), where M is the (sub)set of output feedback nodes. The network parameters to be trained are the weights and the initial states of the fictitious inputs. The reader is referred to refs. [6, 7] for a more detailed description of the architecture.

2 Some analytical results

We confine our study to networks where each unit has a "linear" (or identity) activation function, and we therefore study networks without hidden layers. Networks with non-linear activation functions are studied elsewhere [6 – 8]. The networks are shown to act as autoregressive moving average models [9]; a result which can also be derived through state space representation (see *e.g.* [10, 11]). It may be noted that Back and Tsoi [12] have shown similar analytical results for networks with more complex (multi-lag) feedback connections.

2.1 Networks with one true input and one true output

Without loss of generality, we here analyze networks with only one input and one output. Networks without true inputs, which describe fully autonomous systems, are treated elsewhere [13]. In the notation of the ARMA models that follow, ε refers to Gaussian (white) noise with zero mean.

The (2,1₃) network

The simplest (meaningful) network includes one *true* input (Figure 1a). From

$$\begin{aligned} y_{3,t} &= w_{30} + w_{31}x_{1,t} + w_{32}x_{2,t} = w_{30} + (w_{31} + w_{32})x_{1,t} - w_{32}(x_{1,t} - y_{3,t-1}) \\ &= w_{30} + (w_{31} + w_{32})x_{1,t} - w_{32}\underbrace{(x_{1,t} - \hat{x}_{1,t})}_{\varepsilon_{1,t}} \end{aligned} \tag{1}$$

it is seen that we can entertain an ARMA(1,1) model, $\hat{x}_{1,t+1} = \phi_0 + \phi_1 x_{1,t} + \theta_1 \varepsilon_{1,t}$, since $y_{3,t} = \hat{x}_{1,t+1}$ gives us $w_{30} = \phi_0$, $w_{31} = \phi_1 + \theta_1$, and $w_{32} = -\theta_1$. Note that the weight on the self-loop of the single output (Figure 1b) is $-\theta_1$, since the moving average part is equivalent to an infinite autoregressive series [9]. Naturally, pure AR(1) and MA(1) models can also be realized.

The (2,2₄) network

This network (Figure 2a) includes a feedback connection from a fictitious output unit. Since the first (true) output node gives

$$\begin{aligned} y_{3,t} &= w_{30} + w_{31}x_{1,t} + w_{32}(w_{40} + w_{41}x_{1,t-1} + w_{42}x_{2,t-1}) \\ &= (w_{30} + w_{32}w_{40}) + w_{31}x_{1,t} + w_{32}w_{41}x_{1,t-1} + w_{32}w_{42}x_{2,t-1}, \end{aligned} \tag{2}$$

it is straightforward to verify that the parameters of an AR(2) process $x_{1,t+1} = \phi_0 + \phi_1 x_{1,t} + \phi_2 x_{1,t-1} + \varepsilon_{t+1}$ can be estimated by training the network with $y_{3,t} = \hat{x}_{1,t+1}$. A possible choice of non-zero weights is $w_{30} = \phi_0$, $w_{31} = \phi_1$, $w_{32} = \phi_2$, and $w_{41} = 1$. This corresponds to a pruned network where the second output node acts as a simple lag or delay line. Naturally, higher order (AR($n \geq 3$)) models cannot be entertained.

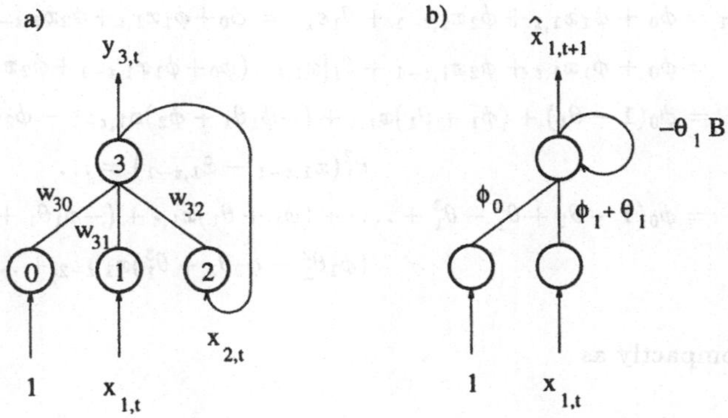

Fig. 1 a) The $(2,1_3)$ network and b) a possible realization of an ARMA(1,1) model

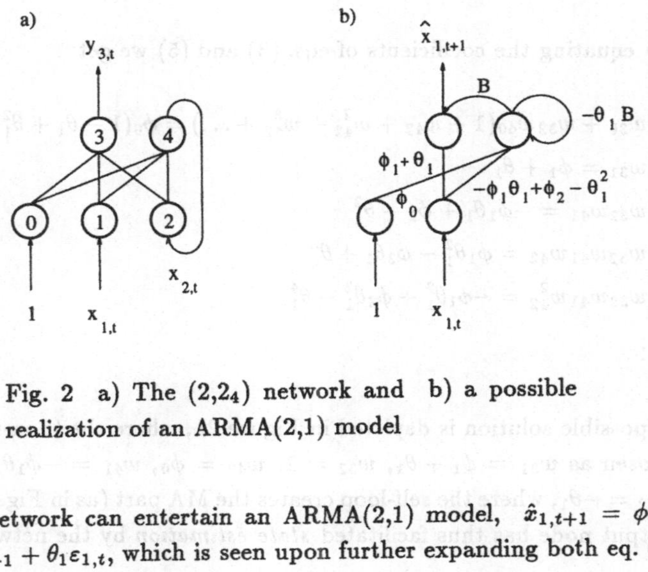

Fig. 2 a) The $(2,2_4)$ network and b) a possible realization of an ARMA(2,1) model

However, the network can entertain an ARMA(2,1) model, $\hat{x}_{1,t+1} = \phi_0 + \phi_1 x_{1,t} + \phi_2 x_{1,t-1} + \theta_1 \varepsilon_{1,t}$, which is seen upon further expanding both eq. (2)

$$
\begin{aligned}
y_{3,t} &= (w_{30} + w_{32}w_{40}) + w_{31}x_{1,t} + w_{32}w_{41}x_{1,t-1} + \\
&\qquad w_{32}w_{42}(w_{40} + w_{41}x_{1,t-2} + w_{42}x_{2,t-2}) = \cdots \\
&= (w_{30} + w_{32}w_{40} + w_{32}w_{40}w_{42} + w_{32}w_{40}w_{42}^2 + \ldots) + w_{31}x_{1,t} + \\
&\qquad w_{32}w_{41}x_{1,t-1} + w_{32}w_{41}w_{42}x_{1,t-2} + w_{32}w_{41}w_{42}^2x_{1,t-3} + \ldots \qquad (3)
\end{aligned}
$$

and the ARMA(2,1) model

$$
\begin{aligned}
\hat{x}_{1,t+1} &= \phi_0 + \phi_1 x_{1,t} + \phi_2 x_{1,t-1} + \theta_1 \varepsilon_{1,t} = \phi_0 + \phi_1 x_{1,t} + \phi_2 x_{1,t-1} + \theta_1 (x_{1,t} - \hat{x}_{1,t}) \\
&= \phi_0 + \phi_1 x_{1,t} + \phi_2 x_{1,t-1} + \theta_1 [x_{1,t} - (\phi_0 + \phi_1 x_{1,t-1} + \phi_2 x_{1,t-2} + \theta_1 \varepsilon_{1,t-1})] \\
&= \phi_0(1-\theta_1) + (\phi_1+\theta_1)x_{1,t} + (-\phi_1\theta_1 + \phi_2)x_{1,t-1} - \phi_2\theta_1 x_{1,t-2} - \\
&\qquad \theta_1^2(x_{1,t-1} - \hat{x}_{1,t-1}) = \ldots \\
&= \phi_0(1-\theta_1+\theta_1^2-\theta_1^3+\ldots) + (\phi_1+\theta_1)x_{1,t} + (-\phi_1\theta_1+\phi_2-\theta_1^2)x_{1,t-1} + \\
&\qquad (\phi_1\theta_1^2 - \phi_2\theta_1 + \theta_1^3)x_{1,t-2} + \ldots, \qquad (4)
\end{aligned}
$$

or compactly as

$$
\begin{aligned}
&\hat{x}_{1,t+1} = \varphi_0 + \varphi_1 x_{1,t} + \varphi_2 x_{1,t-1} + \varphi_3 x_{1,t-2} + \ldots \\
&\varphi_0 = \phi_0 \sum_{j=0}^{\infty} (-\theta_1)^j; \qquad \varphi_1 = \phi_1 + \theta_1; \\
&\varphi_2 = -\phi_1\theta_1 + \phi_2 - \theta_1^2; \qquad \varphi_n = -\theta_1 \varphi_{n-1}; \qquad n \geq 3.
\end{aligned} \qquad (5)
$$

By equating the coefficients of eqs. (3) and (5) we get

$$
\begin{aligned}
&w_{30} + w_{32}w_{40}(1 + w_{42} + w_{42}^2 + w_{42}^3 + \ldots) = \phi_0(1 - \theta_1 + \theta_1^2 - \theta_1^3 + \ldots) \\
&w_{31} = \phi_1 + \theta_1 \\
&w_{32}w_{41} = -\phi_1\theta_1 + \phi_2 - \theta_1^2 \\
&w_{32}w_{41}w_{42} = \phi_1\theta_1^2 - \phi_2\theta_1 + \theta_1^3 \\
&w_{32}w_{41}w_{42}^2 = -\phi_1\theta_1^3 + \phi_2\theta_1^2 - \theta_1^4 \\
&\vdots
\end{aligned} \qquad (6)
$$

A possible solution is depicted in Figure 2b, where the (non-zero) weights are chosen as $w_{31} = \phi_1 + \theta_1$, $w_{32} = 1$, $w_{40} = \phi_0$, $w_{41} = -\phi_1\theta_1 + \phi_2 - \theta_1^2$, and $w_{42} = -\theta_1$, where the self-loop creates the MA part (as in Fig. 1b). The second output node has thus facilitated *state estimation* by the network [1,14].

The (3,$2_{4,5}$) network

The outputs of the $(3, 2_{4,5})$ network (Figure 3a) are given by

$$
\begin{aligned}
y_{4,t} &= w_{40} + w_{41}x_{1,t} + w_{42}x_{2,t} + w_{43}x_{3,t} \\
y_{5,t} &= w_{50} + w_{51}x_{1,t} + w_{52}x_{2,t} + w_{53}x_{3,t}.
\end{aligned} \qquad (7a,b)
$$

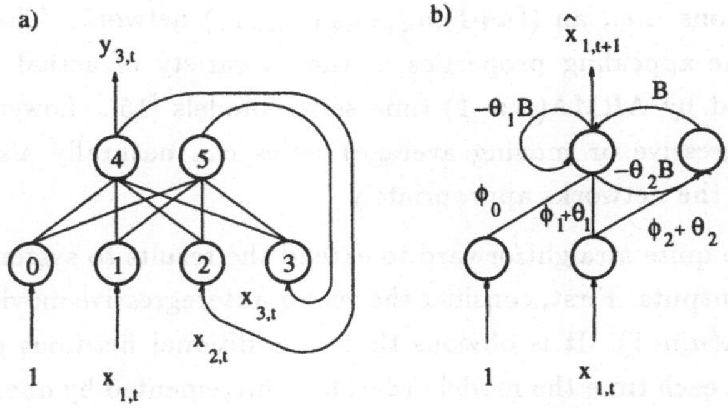

Fig. 3 a) The $(3,2_{4,5})$ network and b) a possible realization of an ARMA(2,2) model

In order to model an ARMA(2,2) process, $x_{1,t+1} = \phi_0 + \phi_1 x_{1,t} + \phi_2 x_{1,t-1} + \varepsilon_{1,t+1} + \theta_1 \varepsilon_{1,t} + \theta_2 \varepsilon_{1,t-1}$, the conditions for the weights can be derived by inserting $x_{3,t} = y_{5,t-1}$ from eq. (7b) into eq. (7a)

$$\begin{aligned} y_{4,t} &= w_{40} + w_{41}x_{1,t} + w_{42}x_{2,t} + \\ &\qquad w_{43}(w_{50} + w_{51}x_{1,t-1} + w_{52}x_{2,t-1} + w_{53}x_{3,t-1}) \\ &= (w_{40} + w_{43}w_{50}) + (w_{41} + w_{42})x_{1,t} - w_{42}(x_{1,t} - y_{4,t-1}) + \\ &\qquad w_{43}w_{51}x_{1,t-1} + w_{43}w_{52}x_{2,t-1} + w_{43}w_{53}x_{3,t-1} \\ &= (w_{40} + w_{43}w_{50}) + (w_{41} + w_{42})x_{1,t} - w_{42}(x_{1,t} - \hat{x}_{1,t}) + \\ &\qquad w_{43}(w_{51} + w_{52})x_{1,t-1} - w_{43}w_{52}(x_{1,t-1} - y_{4,t-2}) + w_{43}w_{53}x_{3,t-1} = \ldots \\ &= (w_{40} + w_{43}w_{50}) + (w_{41} + w_{42})x_{1,t} + w_{43}(w_{51} + w_{52})x_{1,t-1} - \\ &\qquad w_{42}\varepsilon_{1,t} - w_{43}w_{52}\varepsilon_{1,t-1} + w_{43}w_{53}x_{3,t-1} \end{aligned} \tag{8}$$

with $y_{4,t} = \hat{x}_{1,t+1}$. The non-zero terms in a simple weight set which satisfies the above conditions are $w_{40} = \phi_0$, $w_{41} = \phi_1 + \theta_1$, $w_{42} = -\theta_1$, $w_{43} = 1$, $w_{51} = \phi_2 + \theta_2$, and $w_{52} = -\theta_2$, which corresponds to the network of Figure 3b.

2.2 Extension of the results

Based on the results of the previous section, general formulae can be concluded for networks with only one true input and output. In order to entertain an ARMA(n,n–1) model, we require a network with n inputs and n outputs, feeding back all but the first output node. The notation for this network is

$(n, n_{n+2,n+3,\dots,2n})$. Correspondingly, an ARMA(n, n) process can be modeled by a network with one additional input node, where all outputs have feedback connections, *i.e.*, an $((n+1), n_{n+2,n+3,\dots,2n+1})$ network. The former network has some appealing properties in that a variety of actual processes can be described by ARMA$(n,n-1)$ time series models [15]. Lower orders of either (autoregressive or moving average) series can naturally also be realized by pruning the networks appropriately.

It is also quite straightforward to extend the results to systems with m inputs and m outputs. First, consider the vector autoregressive moving average model VARMA$(n,n-1)$. It is obvious that m additional fictitious output nodes are required each time the model order, n, is incremented by one. The parameters of a VARMA$(n,n-1)$ process can thus be estimated by a network with $n \times m$ inputs and outputs, where all but the first m outputs are fed back. By analogy, a network with $(n+1) \times m$ inputs and $n \times m$ feedback outputs is the minimal network for a VARMA(n, n) model.

3 A brief example

This section presents a brief example where partially recurrent neural networks are applied to a simple univariate time series studied in [15]:

$$x_{1,t+1} = -x_{1,t} - x_{1,t-1} + \varepsilon_{1,t+1} + 0.45\varepsilon_{1,t}.$$

1000 observations of the series were produced with $x_{1,0} = 0$ and $\varepsilon_1 \sim N(0,1)$. Because of the special form of the ARMA(2,1) process (with $\phi_1 = -1$ and $\phi_2 = -1$), the resulting series has a periodicity of 3 and a considerable variance of x_1. Using the first 500 observations (Fig. 4) as training instances, the results for some network configurations are given in Table 1.

Table 1 Number of parameters and prediction errors for some networks

Network	Parameters	Error
(1,1)	2	10.951
$(2,1_3)$	4	6.787
$(2,2_4)$	8	1.034
$(3,2_{4,5})$	10	1.026

A feedforward network with $x_{1,t}$ as the single input has clearly not enough capacity to describe the dynamics of the process. Even though the prediction error is considerably reduced by including a feedback connection from the output, the resulting model is still not acceptable.

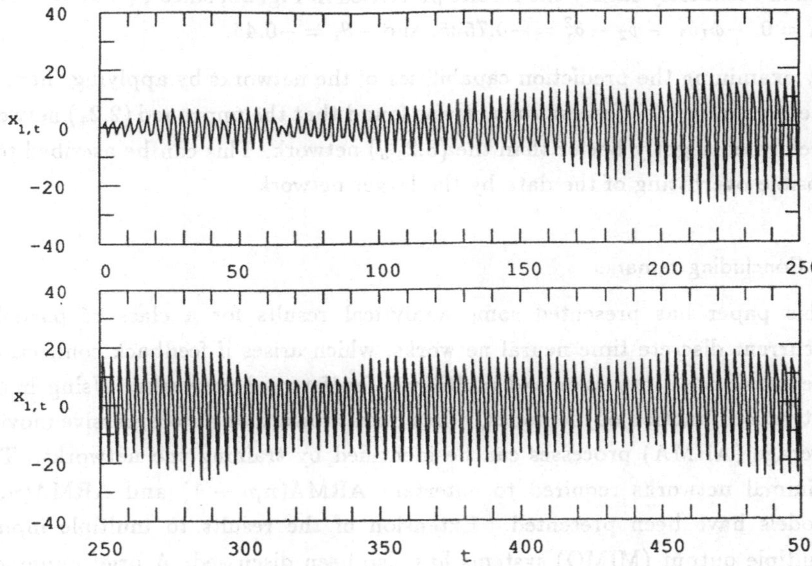

Fig. 4 500 first observations of an ARMA(2,1) process with periodicity

By using a fictitious output node instead, the error is reduced to a value which closely agrees with the actual noise standard deviation of 1.043 for the training set. A slight decrease in the prediction error is observed if also the actual output is fed back.

Training the $(2,2_4)$ network, which can be considered a parsimonious model for the process studied, yields the the initial condition $x_{2,0} = 3.0869$ and the weights

$$w_{30} = \ 0.10048 \cdot 10^{-5}; \qquad w_{31} = -0.55289; \qquad w_{32} = \ 0.80001;$$
$$w_{40} = -0.01121; \qquad w_{41} = -0.94395; \qquad w_{42} = -0.44355.$$

We may verify that the conditions of eq. (6) are approximately satisfied by these numerical values. The overparametrization can be avoided by pruning the network prior to training. By fixing $w_{30} = 0$ and $w_{32} = 1$, we get after

retraining $x_{2,0} = 2.4692$ and

$$w_{30} = 0.0; \qquad w_{31} = -0.55286; \qquad w_{32} = 1;$$
$$w_{40} = -0.00896; \qquad w_{41} = -0.75517; \qquad w_{42} = -0.44356,$$

which accurately satisfy the results presented in Fig. 2b, since $\phi_1 + \theta_1 = -0.55$, $\phi_0 = 0$, $-\phi_1\theta_1 + \phi_2 - \theta_1^2 = -0.7525$, and $-\theta_1 = -0.45$.

By examining the prediction capabilities of the networks by applying them on the remaining 500 observations, it was found that the unpruned $(2,2_4)$ network preformed slightly better than the $(3,2_{4,5})$ network. This can be ascribed to a posible overfitting of the data by the larger network.

4 Concluding remarks

This paper has presented some analytical results for a class of partially recurrent discrete time neural networks, which arises if feedback connections are introduced into standard multilayer feedforward networks. Using linear activation function of the nodes, it has been shown that autoregressive moving average (ARMA) processes can be identified by training the networks. The minimal networks required to entertain ARMA(n,$n-1$) and ARMA(n,n) models have been presented. Extension of the results to multiple input–multiple output (MIMO) systems has also been discussed. A brief numerical example has illustrated the theory.

References

1. O. Nerrand, P. Roussel-Ragot, L. Personnaz, G. Dreyfus and S. Marcos: "Neural Networks and Nonlinear Adaptive Filtering: Unifying Concepts and New Algorithms", *Neural Computation* **5** (1993) 165–199.
2. F-S. Tsung: *Modelling Dynamical Systems with Recurrent Neural Networks*, Ph.D. Dissertation, University of California, San Diego, 1994.
3. M.I. Jordan: "Attractor dynamics and parallelism in a connectionist sequential machine", Proceedings of the *Eight Annual Conference of the Cognitive Science Society*, Amherst, 1989, pp. 531–546. Hillsdale: Erlbaum.
4. J.L. Elman: "Finding Structure in Time", *Cognitive Science* **14** (1990) 179–211.
5. J. Hertz, A. Krogh and R.G. Palmer: *Introduction to the theory of neural computation*, chap. 7, Addison Wesley, 1991.

6. A.B. Bulsari and H. Saxén: "A recurrent network for modeling noisy temporal sequences", *Neurocomputing* **7** (1995) 29–40.

7. A.B. Bulsari and H. Saxén: "A Recurrent Neural Network Model", in *Artificial Neural Networks, 2* (Eds. I. Aleksander and J. Taylor) (Proceedings of the *International Conference on Artificial Neural Networks*, Brighton, UK, September 1992), Vol. 2, pp. 1091–1094.

8. A.B. Bulsari and H. Saxén: "A Partially Recurrent Connectionist Model", Proceedings of the *10th European Conference on Artificial Intelligence (ECAI 92)* August 1992, Vienna, Austria, pp. 198–202.

9. G.E.P. Box and G.M. Jenkins: *Time Series Analysis. Forecasting and Control*, Holden-Day, Oakland, CA, 1976.

10. M. Aoki: *State Space Modeling of Time Series*, Springer-Verlag, New York, 1987.

11. R.J. Bennett: *Spatial Time Series*, Pion Limited, London, 1979.

12. A.D. Back and A.C. Tsoi: "FIR and IIR Synapses, a new neural network architecture for time series modelling", *Neural Computation* **3** (1991) 375–385.

13. A.B. Bulsari and H. Saxén: "A class of partially recurrent neural networks in discrete time", Technical report 91–17, Heat Eng. Lab., Åbo Akademi, Finland, 1991.

14. O. Sørensen: "Optimal Control with Neural Networks", Proceedings of the *IASTED International Symposium on Applied Informatics*, Zurich, July 1991, pp. 90–95.

15. S.M. Pandit and S.M. Wu: *Time Series Analysis with Applications*, John Wiley and Sons, New York, 1983.

Forecasting Stock Market Indices with Recurrent Neural Networks

Maxwell J. Rhee*
BNP/Cooper Neff, Inc.
Radnor, PA 19087

Abstract

A recurrent neural network is used to forecast the out-of-sample return of a stock market index. The use of an extensive information set and a stochastic minimization algorithm distinguishes this study from prior work. The data set encompasses daily observations from 1970 through 1993, with the following forecast exercise undertaken. For a variety of model sizes, the network task is to approximate the weekly, monthly or quarterly conditional mean return. These forecasts are conditioned on a daily information set containing a number of index-specific and market-wide variables, term structure and corporate bond yields, and calendar variables. Network performance is evaluated by out-of-sample normalized mean-squared error, sample statistics describing the joint distribution of forecasted and actual returns, and a test for market-timing ability. A further performance evaluation concerns the construction of trading portfolios with transaction costs. Finally, bootstrapping techniques are applied to construct surrogate distributions of the out-of-sample statistics. Neural network models are found to perform more than adequately when compared with a benchmark linear model, and are able to generate large risk-adjusted returns over simple buy-and-hold strategies.

1 Introduction

A growing body of evidence has recently emerged suggesting the presence of forecastable components in equity returns. Much of this work has relied on constructing linear regressions between returns and past returns or a number of reasonable factors.[1] Interestingly, one strand of this literature has begun to explore the potential nonlinear structure in these returns. Tests of nonlinear independence using the procedure of Brock, Dechert and Scheinkman (1987) are one example. However, actual estimation of nonlinear relationships has only been partially investigated.[2]

---

*This chapter is based on work contained in my doctoral dissertation. A version of this paper was presented at the Second International Workshop on Neural Networks in the Capital Markets, Pasadena, CA, November 1994. I wish to thank Masanao Aoki for introducing me to this topic and for his guidance and criticism.

[1]A short list would include Fama (1981, 1990), Fama and French (1988, 1989ab, 1992), Fama and Schwert (1977), Jegadeesh (1990, 1991), and Keim and Stambaugh (1986).

[2]Forecasting equity returns and exchange rates have been the usual candidates. Some attempts using a variety of nonlinear techniques are: Diebold and Nason (1990), Hsieh (1993), Hutchinson, Lo and Poggio (1994), LeBaron (1992), Weigend, Huberman and Rumelhart (1992), and White (1988).

This intensified attention given to nonlinear techniques has been a powerful trend in recent empirical economics and finance, indeed, spurred on by dramatic increases in computing power and its cheap and readily available access. Of course, not only has this phenomenal increase allowed investigators to sift through massive amounts of data, but it has also allowed the exploration of new and computationally intensive mathematical and statistical models. Specifically, recent developments have focused on the extension of standard regression analysis to incorporate and capture important nonlinear characteristics. This extension is meant to address the unsettlingly poor performance of standard linear regression methods, especially in forecasting security returns.

In this study, we investigate the potential for accurate forecasting of a stock market index using a recurrent neural network as a flexible nonlinear model. This model is used to forecast the out-of-sample return on two stock market indices: the Dow Jones Industrial Average and Standard and Poor's 500 Composite Index. The use of an extensive, multivariate information set and simulated annealing as a global stochastic minimization algorithm distinguishes this study from prior work. The major difficulties in a study of this type are: (1) selecting a modeling strategy, (2) deciding network complexity, (3) implementing a network learning algorithm, and (4) evaluating network performance.

First, the belief is that there are "general rules" that map current inputs into future values. However, there are few theoretical insights into the nature of this mapping. Therefore, several "reasonable" variables are chosen as inputs. Furthermore, to guard against capturing spurious relationships and to allow for the estimation of a wide range of network sizes, relatively large training sets are used. Second, large networks or greater model complexity implies an increase in approximation capability; however, the likelihood of overfitting increases consequently. In view of this, network size as measured by the number of hidden units is varied across a reasonable range of values. Third, in contrast to typical methods that use online or incremental estimation, for this study, off-line or batch learning is employed, in conjunction with a simulated annealing algorithm, to minimize in-sample error. Lastly, in order to characterize and evaluate network performance, several measures are compiled: (1) out-of-sample normalized mean-squared error, (2) sample statistics describing the joint distribution of forecasted and actual, *ex post* returns, (3) a test for market-timing ability, and (4) the construction and simulation of trading portfolios with transaction costs that actively apply these forecasts. Finally, the use of bootstrapping techniques allows for examination of the sample statistic distributions and the calculation of confidence intervals.

The balance of this chapter is as follows. First, I briefly review neural network theory and introduce the basic nonlinear forecast model. Next, I describe the manner in which network performance is evaluated, with a description of the data set following. Finally, the results of a number of forecasting exercises are presented.

2 Methodology

2.1 Brief Review of Neural Network Theory

Although successfully applied in many diverse fields, artificial neural networks represent a general class of nonlinear models that has seen relatively few economic related applications.[3] Neural networks possess several interesting properties that make them extremely promising with regard to econometric applications. First, the class of multilayer neural networks can approximate well a large number of diverse functions, whereas most commonly used nonlinear time series models do not possess this property. Second, broadly speaking, polynomial, spline or trigonometric expansions require the number of parameters to grow exponentially to achieve some certain approximation capability. In contrast, neural networks require only that the parameter number grow linearly to achieve the same approximation rate.[4]

This first property is particularly interesting in that neural networks have been shown to possess a so-called *universal approximation property.* Theoretical justification provided by a number of authors show that networks with a sufficient number of hidden units and properly adjusted parameters can approximate a multitude of reasonable functions arbitrarily well.[5] Moreover, even when faced with data contaminated by noise or the inclusion of irrelevant inputs, neural networks are still capable of learning the features of a data set. Unfortunately, accurate learning in the presence of noise can be extremely difficult—hampered precisely by their very ability to approximate many functional forms. For instance, network estimation on time series of short record length likely leads to overfitting and poor out-of-sample generalization.

For example, suppose an investigator chooses some particular embedding for his data in relation to a desired mapping. If meaningful structure exists, then neural networks possess the power to perform a smooth interpolation between data points. However, they do not build an explicit description of the function in the sense of equations of motion. On the other hand, if the approximated function is a good *generalization*, then reliable out-of-sample forecasts can be made. This is the question at hand. Essentially, neural networks are a specialized case of nonlinear regression. The broad hope for these models comes from their ability to capture unforeseen regularities in a complex signal. Combinations of inputs, that perhaps at first glance seem to possess no information, may in fact prove useful.

The architecture of a neural network is defined by the placement and number of its units and connections. The number of connections or free parameters can become quite large. However, conventional wisdom concerning the number of estimable parameters and sample size can be misleading. While the number of parameters in other nonlinear models usually limits the order or degree of the input interactions, the size of a network merely limits the number of *patterns* that can be learned. Intuitively neural networks are pattern recognition devices. For example, rising volume in the number of shares traded in conjunction with falling short-term interest rates may be significant in terms of affecting the short-term return on stocks. Again, the hope is that these patterns generalize and are not simply idiosyncratic noise.

---

[3] A good text on neural network theory is Hertz, Krogh and Palmer (1991).

[4] Assuming a certain degree of smoothness for the target function, Barron (1991) provides bounds on the convergence rate of the approximation error for feedforward neural networks.

[5] This thread of the literature has been active: see for example, Chen and Chen (1993), Cybenko (1989), Hornik (1989), and Hornik, Stinchcombe and White (1990). A good review of these issues is White (1992). It has also been shown that neural networks with as little as a single hidden unit layer and mild restrictions on the hidden unit activation function are capable of arbitrarily accurate approximation of not only the unknown function but its derivatives as well (Gallant and White, 1992).

However, the number of significant patterns is not known *a priori.* Typically, one starts with a large enough network that can categorize as many of the strongest patterns. And, by a variety of *ad hoc* methods, overfitting is prevented by stopping or inhibiting training. One might also selectively estimate many networks, varying the size of the network by adding or deleting units, and then selecting some "best" architecture on the basis of a complexity measure. Although a reasonable procedure, this approach is quite computationally burdensome.

Finally, much of the enthusiasm for neural networks concerns the notion of learning, though learning in its most primitive sense: presenting new and evolving information to a "machine," with this machine acquiring knowledge and defining *rules* that work most of the time. This entails changing the connections or weights in a meaningful manner as new data is presented. In the neural network literature, this is called online learning, and much of the work in the field has been devoted to discovering adequate learning algorithms of this type. In contrast, for this study, I investigate only the empirical and statistical properties of neural networks, that is, their ability to produce a large number of rich and varied representations.

2.1.1 Network Architectures

A neural network is simply a function that receives input and returns output. Network architectures generally have the following basic structure: a *hidden unit layer* separates an *input layer* containing the input variables and an *output layer* containing the final network output units. Network output may be a vector-valued function. Two relatively straightforward architectures are the *feedforward* and *recurrent* networks, with the latter encompassing the former as a special case. For this study, the focus is on a recurrent network, but keep in mind that in general network architectures can easily become quite complicated.

The architecture of a generic, single-output, recurrent neural network consists of an input layer with n input units, a hidden unit layer with q hidden units, a second *recurrent hidden unit layer*, and finally, an output layer with a single output unit. The recurrent hidden layer is essentially a short-term memory clipboard that may contain any number of items. For example, it may contain past values of the output or input units.[6] Specifically, in this study, we will concentrate on a network that connects the current period's hidden units with the previous period's hidden unit layer, so that each current period's hidden units has access to the resultant values from all hidden units, one-step lagged. Network output is calculated as follows. Input or state variables $x_{i,t}$ simultaneously activate the q hidden units through a *hidden unit activation function* Ψ. These hidden units $h_{i,t}$ are connected both to the input units and the recurrent hidden unit layer. Finally, the single output unit is then activated through an *output unit activation function* Φ, thus producing the network output o_t.

Generally, few restrictions limit our use of neural networks as an approximation vehicle. The output unit activation function Φ needs to be continuous in its argument, and the hidden unit activation function Ψ needs to be a continuous, bounded (sigmoid) function. For example, standard usage in the literature is

[6] A network containing past values of the output or input units are called state-space reconstructions or tapped delay-line networks.

to have Φ the identity function and Ψ the logistic function or the hyperbolic tangent function.

Network architectures can become quite complicated, with the number of layers, units and connections at the mercy of the investigator's imagination. Although to date, practical considerations severely limit the size of an implemented network. In most cases, the largest networks seen in the literature do not approach more than on the order of a few hundred units or several tens of thousands of individual connections. To put this into perspective, the human brain is believed to be a massively parallel connection machine, estimated to have around 10^{11} individual processing units (neurons) and perhaps 10,000 or more connections (synapses) per neuron.

2.1.2 Architecture of a Recurrent Elman Network

In Figure 1, I depict the architecture of a single-output, multilayer recurrent neural network that will serve as the basis for the empirical work of this study. Originally studied in the speech processing literature, this particular network due to Elman (1990) has the following mathematical form:

$$h_{i,t} = \Psi(\gamma_{i0} + \sum_{j=1}^{n} \gamma_{ij} x_{j,t} + \sum_{k=1}^{q} \delta_{ik} h_{k,t-1}) \tag{1}$$

$$o_t = \Phi(\beta_0 + \sum_{i=1}^{q} \beta_i h_{i,t}). \tag{2}$$

These two equations can be equivalently written as:

$$o_t = \Phi\big(\beta_0 + \sum_{i=1}^{q} \beta_i \Psi(\gamma_{i0} + \sum_{j=1}^{n} \gamma_{ij} x_{j,t} + \sum_{k=1}^{q} \delta_{ik} h_{k,t-1})\big) = g(\boldsymbol{X}_t; \boldsymbol{\theta}) \tag{3}$$

where $\boldsymbol{X}_t = (\boldsymbol{x}_1, \ldots, \boldsymbol{x}_t)'$ is a matrix containing vectors of the input variables $\boldsymbol{x}_\tau$ for each time $\tau = 1, \ldots, t$, and where $\boldsymbol{\theta}$ is a vector containing the entire set of network parameters. An Elman network with n inputs and q hidden units has a total of $(q+1) + q(n+1) + q^2$ connections or parameters.

This recurrent network differs from a standard feedforward network in the addition of connections δ_{ik} linking a given time period's hidden units with lagged hidden units. By increasing the complexity of a feedforward architecture with a time-dependent component, network output thus becomes a function of some particular truncated *history* of the input variables. Compared to a feedforward network, the hope then is to capture more of the dynamic structure implicit in the input variables. Although the feedforward network has been the prevailing and dominant network architecture with most current applications incorporating it, results have been mixed at best. Of course, the real difficulty with the feedforward framework is its inability to adequately handle complicated dynamical behavior, e.g., situations in which the order or history of the variables is important. Indeed, for instance, this is crucial when attempting to forecast security returns. However, owing to their greater complexity and difficulties with practical implementation and estimation, currently, there has been virtually no emphasis on recurrent networks.

Thus, from a theoretical point of view, there are two key points for advocating the use a recurrent neural network as a forecasting model: approximation

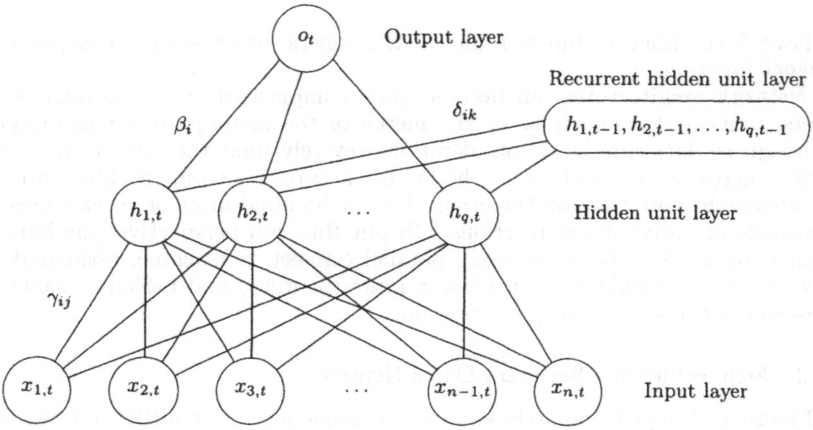

Figure 1: **Elman network architecture.** This recurrent Elman network has n inputs and q hidden units. Parameters are represented by connecting lines. Inputs and hidden units are represented by circles.

capability or flexibility and history dependence. First, due to their universal approximation property, neural networks allow for greater flexibility in accurately approximating an unknown functional relationship unlike other available models. Second, by incorporating a recurrent structure into a neural network model, we benefit from the greater degree of history dependence permitted by this formulation.

2.2 Network Implementation and Estimation

2.2.1 Statistical Interpretation

For a fixed *forecast horizon* k, suppose we observe, realizations from a stochastic process, pairs of *targets* and *input* vectors $(y_{t+k}, \boldsymbol{x}_t)$, where the target y_{t+k} is a scalar and $\boldsymbol{x}_t$ is a $(1 \times n)$ vector of input variables. We are interested in the temporal behavior of the target variable. It seems natural to find the conditional expectation of the target given the inputs: $\mathrm{E}[\tilde{y}_{t+k}|\boldsymbol{x}_t]$. From a statistical perspective, this is the best estimate or expectation of a target realization given the current information contained in the inputs. However, for some time series applications, the information contained in the inputs $\boldsymbol{x}_t$ may not be sufficient enough to accurately characterize the temporal behavior of the (stochastic) target $\tilde{y}_{t+k}$. It is possible then that the stochastic behavior of the target may be jointly determined by a history of the inputs $\boldsymbol{X}_t \equiv [\boldsymbol{x}_t, \boldsymbol{x}_{t-1}, \ldots, \boldsymbol{x}_{t-L}]$ for some large value of L. Therefore, the natural candidate for analysis is the conditional expectation: $\mathrm{E}[\tilde{y}_{t+k}|\boldsymbol{X}_t]$. We wish to approximate this function, of which, its precise formulation is not known.

This line of argument leads us to the following nonlinear regression model:

$$y_{t+k} = g(\boldsymbol{X}_t; \boldsymbol{\theta}) + \varepsilon_t \tag{4}$$

where $\boldsymbol{\theta}$ is a $(1 \times P)$ parameter vector and the function $g(\boldsymbol{X}_t; \boldsymbol{\theta})$ may be any appropriate approximation model. Since we will be approximating an unknown functional relationship, this regression model is necessarily misspecified. However, the theory of least squares for misspecified nonlinear regression models immediately applies.[7]

Under fairly general conditions, given a set of T observations of the target and input variables $\{(y_{t+k}, \boldsymbol{x}_t)\}_{t=1}^{T}$, we know that a nonlinear least-squares estimator $\tilde{\boldsymbol{\theta}}_T$ exists which solves the minimization problem:

$$\min_{\boldsymbol{\theta}} \frac{1}{T}\sum_{t=1}^{T} \left(y_{t+k} - g(\boldsymbol{X}_t;\boldsymbol{\theta})\right)^2 . \tag{5}$$

Moreover, as the sample size T becomes large, this estimator converges almost surely to $\boldsymbol{\theta}^*$, the solution to the problem:

$$\min_{\boldsymbol{\theta}} \left\{ \mathrm{E}\left[(\tilde{y}_{t+k} - g(\boldsymbol{X}_t;\boldsymbol{\theta}))^2\right] = \mathrm{E}\left[(\epsilon(\boldsymbol{X}_t) - g(\boldsymbol{X}_t;\boldsymbol{\theta}))^2\right] + \sigma^2 \right\} \tag{6}$$

where $\epsilon(\boldsymbol{X}_t) \equiv \mathrm{E}[\tilde{y}_{t+k}|\boldsymbol{X}_t]$ and $\sigma^2 \equiv \mathrm{E}\left[(\tilde{y}_{t+k} - \epsilon(\boldsymbol{X}_t))^2\right]$. Thus the estimator reduces the approximation error between the true conditional expectation and the approximation model; all that remains then is an intrinsic error σ^2 which is independent of the parameter vector. In other words, we estimate the conditional mean of the target by the approximating function $g(\boldsymbol{X}_t;\boldsymbol{\theta})$—at an optimal parameter estimate, this function returns the best L_2-forecast of the target $\tilde{y}_{t+k}$ given the inputs $\boldsymbol{X}_t$.

2.2.2 Implementation

At this point, it should clear how we will implement the recurrent neural network model: we will use the network to approximate an unknown conditional expectation function. Thus, network estimation involves searching for an optimal parameter vector that minimizes the mean-squared error:

$$\mathrm{MSE}(\boldsymbol{\theta}; \boldsymbol{y}, \boldsymbol{X}_T) = \frac{1}{T}\sum_{t=1}^{T} \left(y_{t+k} - g(\boldsymbol{X}_t;\boldsymbol{\theta})\right)^2 . \tag{7}$$

Hence, at an optimal parameter vector, the solution to the network estimation problem is our best approximation of the conditional expectation: $g(\boldsymbol{X}_t;\boldsymbol{\theta}^*) \approx \mathrm{E}[\tilde{y}_{t+k}|\boldsymbol{X}_t]$.

However, for this estimation problem, the chief difficulty is that even for networks of modest size, there is virtually no guarantee of finding a global minimum, or even second-best, "good" local minima, compounded by the limitation of finite computational time. As opposed to batch estimation in which all the data is processed simultaneously, in the neural network literature, one typically attempts to minimize the error with incremental or online learning rules.[8]

---

[7]This theory from an econometric and time series perspective is presented in Gallant and White (1988).

[8]This type of estimation algorithm is conceptually identical to the "method of stochastic approximation." The seminal article is Robbins and Munro (1951) with a recent comprehensive treatment available in Ljung and Söderström (1986). The essential concept is that the first-order conditions of an optimization problem can be approximated by incremental passes through a data set. For large data sets, the key benefit is that simultaneous storage

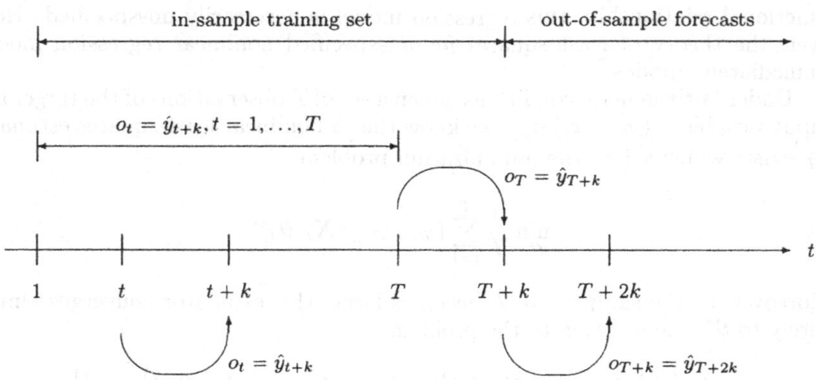

Figure 2: **Training set schematic.** A typical training set is diagrammed in which network output is a k-step ahead forecast of the target. The total number of in-sample observations is $T + k$. The first out-of-sample target k periods after the last in-sample observation is $o_{T+k} = \hat{y}_{T+2k}$.

There have been several suggestions made in order to preserve the incremental approach, also called backpropogation, for example, by Kuan (1993) and Kuan, Hornik and White (1993). However, for an Elman network, there are complications due to the presence of recurrent feedbacks, so that more than local computation and memory is required. For this study, online learning methods are completely abandoned, and off-line or batch estimation is undertaken. The approach here is to simply treat eq. (7) as a large-scale minimization problem and to use standard methods in attacking it.

2.2.3 Network Task

The recurrent neural network model of eq. (3) learns the features of this mapping over an in-sample *training set* $\mathcal{T}$. For example, a typical training set may contain input observations for dates $1, \ldots, T$, with network output produced at each of these dates, $\{o_t\}_{t=1}^{T}$. However, note that if network output is a k-step ahead forecast, then the training set $\mathcal{T}$ encompasses data for the dates $1, \ldots, T+k$. Further notice that the network output at time $T+k$ is the conditional forecast $o_{T+k} = \mathrm{E}\left[\tilde{y}_{T+2k} \mid \boldsymbol{X}_{T+k}\right]$, which is the first out-of-sample forecast k periods after the last in-sample observation in $\mathcal{T}$. Figure 2 presents the schematic of a typical training set.

We can create a sequence of training sets each with equal length T. Adjacent

---

of all the data becomes unnecessary. However, if sufficient computing resources are available, batch estimation is preferred and more efficient. Typically, the proof for viability of an online learning rule revolves around showing the equivalence between the online rule and batch estimation. Often for online rules, there are additional constraints on the underlying stochastic processes, since incremental methods usually attempt to approximate the global gradient with local information—the global gradient being defined on the entire data set. For recurrent networks, the local gradient is usually a poor (biased) estimate of the global gradient.

elements of this sequence $\{\mathcal{T}_1, \mathcal{T}_2, \ldots\}$ have the following form:

$$\begin{aligned} \mathcal{T}_i &= \{1 + (i-1)k, \ldots, T + (i-1)k\} \\ \mathcal{T}_{i+1} &= \{1 + ik, \ldots, T + ik\}, \end{aligned}$$

which produces a sequence of out-of-sample point forecasts $\{\hat{y}_{T+2k}, \hat{y}_{T+3k}, \ldots\}$, that will be the subject of our statistical tests.

Thus, given daily observations on a time series, is it possible to accurately forecast its short-term behavior? The focus here is on a stock market index. From a sequence of index values $\{S_t\}_{t=1}^{T+k}$, we can construct a sequence of k-period simple returns:

$$r_{t+k} = \frac{S_{t+k} - S_t}{S_t} \tag{8}$$

where r_{t+k} is the return or capital gain on the index at time $t + k$ over the previous k periods. We will use this sequence of returns $\{r_{t+k}\}_{t=1}^{T}$ as our target series.

At time t, the network task then is to produce a conditional forecast of the return from t to $t + k$. Equivalently, the network is asked to approximate the following function:

$$o_t \approx \hat{r}_{t+k} \equiv \mathrm{E}\left[\tilde{r}_{t+k} \mid \boldsymbol{X}_t\right],$$

which is the conditional expectation of $\tilde{r}_{t+k}$ given the information contained in the input variables $\boldsymbol{X}_t$.

2.2.4 Choice of Input Variables

Having chosen our target series, we now turn to the question of which variables to employ as inputs. Currently, there are thought to be three sources of predictable variation in stock returns; and so, the following classes of variables were considered: (1) term structure information or discount rates, (2) business conditions or shocks to expected returns, and (3) volatility measures.

First, according to classical valuation models, asset returns should vary with the discount rates that price expected cash flows. Term structure variables are likely candidates where expected short-term and long-term inflationary behavior should be captured. These variables would be for example the yields on government bonds of differing maturity.

Second, there are shocks to expected cash flows. These are business cycle-like variables that co-move with economic conditions. Typically, one might think of GDP or other measures of industrial production. However, this type of data, available at best on a monthly basis, is unlikely to proxy well for any type of daily variation. Instead, I use yields on corporate bond portfolios whose market participants are likely to be interested in these daily movements.

Lastly, recent work has focused on the possibility of predictable time-varying movements in the higher-order moments of asset returns. Examples of these models are the ARCH formulation and its variants.[9] Recall that these models posit a specific form of nonlinear relationship between the conditional expected return and the conditional variance. Although, we do not have to embrace this particular formulation, we can take it as fact that some sort of relationship exists.

[9] Autoregressive, conditionally heteroskedastic (ARCH) models are reviewed by Bollerslev, Chou and Kroner (1992).

And so, we introduce variables that might contain this kind of information. For example, we have observations of the highest and lowest achieved daily returns as well as daily closing returns. In addition, on a market level, we have daily information on trading volume and the percentage of advancing stocks. Large variability in these variables should point to increased volatility at the index and market levels.

2.2.5 Determining Network Complexity

In constructing a neural network forecasting model, one further consideration involves determining network complexity, that is, choosing the appropriate number of hidden units. An overly simple network, in terms of the number of hidden units, may not approximate the unknown conditional mean function well enough. On the other hand, an excessively complex network may overfit the data. Concentrating solely on in-sample performance, one possible discriminating measure is the Schwarz (1977) Information Criterion (SIC). This statistic is defined as:

$$\mathrm{SIC} = \log\left(\mathrm{MSE}(\widehat{\boldsymbol{\theta}}; \boldsymbol{y}, \boldsymbol{X})\right) + N\,\frac{\log(T)}{T} \tag{9}$$

where N is the number of estimated parameters and T is the sample size. Fixing N, as MSE $\rightarrow 0$ and $T \rightarrow \infty$, we have SIC $\rightarrow -\infty$. A model is chosen such that it minimizes or reduces the value of the SIC measure. Thus, small positive or large negative values indicate better model generalization for a given level of in-sample error, parameter number, and sample size.

However, this statistic is sensitive to correct model specification. One possible robust solution is the method of cross-validation. This method involves estimating many network sizes and comparing their out-of-sample performance, with the appropriate network selected that produces the best performance. Since my main interest is in out-of-sample performance, I essentially employ this method, as well as estimating over a reasonable range of network sizes. Unfortunately, one drawback is that it is computationally intensive.[10]

2.2.6 Forecast Exercises

The forecast exercise is to produce a sequence of k-step ahead forecasts of a stock market index return. The main goal of these exercises is to mechanically reproduce, as closely as possible, a real-time situation in which a portfolio manager is directed to forecast the return on an index, and shuffle his holdings accordingly—with attention paid to transaction costs—either on a weekly, monthly or quarterly basis, such that forecasts are calculated only with data available at the time portfolio holding decisions are made.

[10]For most neural network applications that employ online estimation, a typical cross-validation approach usually requires three disjoint sets of sample data. Beginning with one network that is sufficiently complex enough for the task, the first sample is a training set on which the network parameters are estimated. The second sample is a cross-validation set, where the minimum of the error over this set determines effective network complexity, usually meaning that training is stopped at a suboptimal estimate. Finally, an "untouched" test sample is then used to gauge out-of-sample generalization. In contrast, I simply estimate a range of network sizes with the goal of finding optimal estimates for a given network complexity. For examples and discussion, see Wahba (1990), Weigend, Huberman and Rumelhart (1990) or Weigend and LeBaron (1994).

First, a fixed training set length of 3000 trading days or a little less than 12 calendar years is used. Second, the forecast horizon is varied with $k = 5$, 20 and 60 trading days, or approximately one week, one month and one quarter of calendar time. For convenience, these three horizons are referred to as weekly, monthly and quarterly. Third, for each forecast horizon, four network models are estimated as indexed by the number of hidden units: $q = 6$, 9, 12 and 15. These networks have parameters numbering 133, 226, 337 and 466 respectively. Fourth, the logistic function $\Psi(x) = 1/(1 + e^{-x})$ is chosen for the hidden unit activation function. Fifth, the vector of hidden unit values for the date immediately before the start of a training set is set to the constant vector 1/2. Finally, for each training set, the target and each input series are normalized to zero mean and unit variance.

2.2.7 Estimation

It was previously argued that network estimation should be treated as a large-scale minimization problem. This involves batch estimation instead of online learning or estimation rules. However, approaching network estimation as a straightforward minimization problem still leaves us with the practical difficulty of seeking a global solution to a highly nonlinear optimization program.

With this in mind, one robust optimization method is the global stochastic minimization technique of simulated annealing.[11] This technique minimizes the objective function gradually: "downhill" steps are taken almost always, but probabilistically, some "uphill" steps are taken. The probability of steps that increase the objective is a function of a controlling "temperature," which is lowered slowly, so that the probability of these wrong steps becomes smaller. The intuition is to allow a function to escape poor local minima by forcing some moves away from these minima and explore more of the parameter space.[12] To demonstrate this technique's utility, a comparison is made with the online learning rule of Kuan (1993).

Employing batch estimation and a simulated annealing algorithm, for a given training set $\mathcal{T}_i$, an iterative procedure is used to find a local minimum of the network error. The first step is to randomly select 5000 points from a small ball around the origin, calculating the error at each of these points, and then using the point with the smallest network error as an initial parameter vector. For the second step, as the controlling temperature is slowly lowered, downhill moves follow the local gradient with uphill moves repeating the first step. Five complete passes are made through a relevant temperature range. Finally, for comparison, online estimation is also undertaken.[13]

For each training set, we attempt to fit the best functional form, using only in-sample NMSE as a ranking criterion. By looking at the overall out-of-sample

[11]See Aarts and Korst (1989) for an exposition on this subject.

[12]Of course, given an infinite amount of computational time, one is always guaranteed of finding a global minimum. Essentially with simulated annealing, the parameter space is explored in a *systematic* manner so that asymptotically the probability of finding the global minimum in a finite amount of time goes to one.

[13]Both online and batch estimation essentially employ the same stopping criterion: when the fractional decrease in the in-sample error is greater than some tolerance, then stop the parameter search. In addition, for online estimation, either this last test stops training or when 5000 training epochs are performed, whichever comes first, where an epoch is defined as one complete pass through the training set.

performance of a given network architecture across training sets, we can infer some idea on this architecture's average out-of-sample performance. Notice also that by repeated training of fixed architectures on different training sets and by looking at the time profile of the in-sample NMSE, we can ferret out "poor" minima. Since adjacent training sets do not vary much, good local minima, in terms of in-sample NMSE across training sets, should also not vary much.

Finally, notice that different sized architectures are nested, that is, a small network with a smaller number of hidden units is nested within a large network. Obviously, this implies that any network solution achieved by a small network *can be* implemented by a large network. One can think of a small network as a large one with just some of its connections set to zero (i.e. zero restrictions). This means that if we estimate different network sizes, any instance over the same training set, in which we observe a lower NMSE for a small network, indicates that the large network's current solution can surely be dominated.

2.3 Performance Evaluation

2.3.1 Normalized Mean-Squared Error

We need to evaluate the performance of a network solution independent of the record length and dynamic range of the data. Given pairs of targets y_{t+k} and forecasts $o_t = \hat{y}_{t+k}$ over a data set $\mathcal{M}$, we can define the *normalized mean-squared error* (NMSE) of a network solution as:

$$\text{NMSE}(\mathcal{M}) = \frac{\sum\limits_{t \in \mathcal{M}} (y_{t+k} - o_t)^2}{\sum\limits_{t \in \mathcal{M}} (y_{t+k} - \bar{y})^2} = \left(\frac{M}{M-1}\right) \frac{\frac{1}{M} \sum\limits_{t \in \mathcal{M}} (y_{t+k} - \hat{y}_{t+k})^2}{\hat{\sigma}_y^2} \tag{10}$$

where M is the number of data points in $\mathcal{M}$, and $\bar{y}$ and $\hat{\sigma}_y^2$ are the sample mean and variance of the targets. Except for a bias factor, this is just a ratio of the mean-squared error and sample variance. With this statistic, we can easily compare the relative performance of different network solutions. Two boundary conditions are obvious: (1) NMSE $= 0$ implies that network output is a perfect forecast of the target; and (2) NMSE $= 1$ implies that the network performs just as well as the unconditional mean of the targets. This statistic can also be interpreted as the fraction of the sample variance unexplained by the network solution. Finally note that if the network solution has the same first and second moments as the targets, this statistic is related to the sample linear correlation ($\sqrt{\rho}$) between pairs of targets and forecasts: NMSE $= 1 - \rho$.

2.3.2 Henriksson-Merton Statistic

A test for market-timing ability introduced by Henriksson and Merton (1981) is used to evaluate the network's sign predictions. This test is independent of how equilibrium returns are generated and relies solely on the forecasted sign or direction compared to an alternative investment possibility. Typically, this is a short-term bond whose return over the relevant time period is known with certainty.

Assuming that a down market means the index return is less than the certain return on a bond, and likewise, that an up market means the index return is greater than the bond return, let p_1 equal the conditional probability of a correct forecast given a realized down market, and let p_2 equal the conditional probability of a correct forecast given a realized up market. If the forecasts have no value, so that actual and forecasted returns are distributed independently, then the null hypothesis for the statistic H-M $= p_1 + p_2$ is that it equal one. Forecasts are valuable if H-M is sufficiently greater than one.

2.4 Description of the Data Set

The primary data set investigated consists of daily observations from the New York Stock Exchange (NYSE) for the period December 23, 1969, through December 2, 1993. For this period, there are 6054 observations or trading days. Two stock market indices are chosen as target series: the Dow Jones Industrial Average (DJIA) and Standard and Poor's 500 Composite Index (S&P 500). Both indices are forecasted conditional on essentially the same information set. we constructed 14 time series for use as inputs or conditioning variables. The following describes the data set construction of the DJIA information set, however, the same applies to the S&P 500 information set unless noted to the contrary.

Since many of these series are readily available at worst on a weekly basis, I established a simple procedure to deal with this problem. My main concern is in preserving the information integrity or "honesty" of the forecasting exercises; that is, only information available up to a given date can be used in making a forecast. Therefore, for example, if the value of a series is available only once during a week, the daily observation of the series between updates is simply its last value until the next update.

Tables 1 and 2 summarize the description and summary statistics of the data set used for both the DJIA and S&P 500 forecast exercises. Very little attempt is made to preprocess the data before entering the network; except that for each training set, the target series and each input series are normalized to zero mean and unit variance by subtracting and dividing by their respective in-sample mean and standard deviation.

2.4.1 Input Variables

The first three series are observations of the daily high, low and closing values of the DJIA. These index values and consequently the realized returns are *not* dividend adjusted, so that the target series consists of only the capital gains component of the index changes. These three series are transformed to daily returns using the previous day's close as a basis. The fourth series is the daily volume or number of shares traded in the various stocks comprising the DJIA.[14] This volume series is clearly trending and so was transformed to log-differences. These first four series are complete with daily observations for the entire sample period. For the fifth series, information on the average weekly dividend yield of the DJIA was gathered. This series is only available on a weekly basis, and so the daily series was constructed using the procedure outlined above. Each weekly observation is simply the DJIA dividend yield averaged over the past week.[15]

[14] For the S&P 500 information set, I substitute the daily trading volume on the NYSE.

[15] For the S&P 500 information set, I use a trailing 12-month dividend series that was available weekly. To construct the daily series, I simply calculate the dividend yield using the daily close as a basis.

Variable	Description	Transformation	Number
DJCLOSE	Daily DJIA close	Daily return	6054
DJHIGH	Daily DJIA high	Daily return from last close	6054
DJLOW	Daily DJIA low	Daily return from last close	6054
DJVOL	Daily DJIA volume	Daily log-difference	6054
DJYIELD	DJIA dividend yield	None	1249
S&PCLOSE	Daily S&P 500 close	Daily return	6054
S&PHIGH	Daily S&P 500 high	Daily return from last close	6054
S&PLOW	Daily S&P 500 low	Daily return from last close	6054
NYSEVOL	Daily NYSE volume	Daily log-difference	6054
S&PYIELD	S&P 500 dividend yield	None	1245
%NYADV	Daily % NYSE advancing stocks	None	6054
3MON	Daily 3-month T-bill yield	None	2261
6MON	Daily 6-month T-bill yield	None	2261
10YR	Daily 10-year T-note yield	None	2262
30YR	Daily 30-year T-bond yield	None	5937
Aaa	Daily Aaa ind. bond yield	None	6026
Baa	Daily Baa ind. bond yield	None	6026
MONTH	Month of the year	None	6054
DAY	Day of the week	None	6054

Table 1: **Description of the Data Set.** Transformation refers to any preprocessing performed on the raw data series. Number refers to the actual number of observations in the raw series.

Variable	Mean	Skew.	Kurt.	Max.	Min.
DJIA weekly target	0.1520 (2.24)	-0.1422	4.1889	13.0997	-18.2298
DJIA monthly target	0.6078 (4.45)	-0.3914	3.0177	18.3607	-28.5831
DJIA quarterly target	1.8617 (7.85)	-0.0996	1.5722	37.5657	-30.8573
S&P 500 weekly target	0.1579 (2.19)	-0.1992	5.0263	13.8542	-21.9089
S&P 500 monthly target	0.6324 (4.37)	-0.3572	2.8519	19.5652	-26.7443
S&P 500 quarterly target	1.9496 (7.66)	-0.1917	1.6457	40.2529	-29.0526
DJCLOSE	0.0307 (1.00)	-1.7305	47.4753	10.1488	-22.6105
DJHIGH	0.9065 (0.78)	3.7533	61.3324	18.9062	-3.6756
DJLOW	-0.8789 (0.78)	-6.4519	167.6649	6.0157	-25.3340
DJVOL	0.0477 (29.76)	0.0673	10.0722	274.1008	-254.9307
DJYIELD	4.3625 (1.08)	0.5123	-0.7375	7.1300	2.5200
S&PCLOSE	0.0317 (0.96)	-1.5392	39.2681	9.0994	-20.4669
S&PHIGH	0.7955 (0.72)	1.9961	11.8480	9.4709	-4.2411
S&PLOW	-0.7778 (0.73)	-4.4828	94.6291	6.1712	-20.4705
NYSEVOL	0.0480 (21.29)	-0.0592	14.9445	248.1464	-222.3804
S&PYIELD	4.0273 (0.94)	0.5026	-0.7521	6.7174	2.6249
%NYADV	39.1373 (12.08)	0.0940	0.0804	88.7249	2.4989
3MON	7.1378 (2.79)	0.9985	1.0758	16.7500	2.6000
6MON	7.2694 (2.69)	0.8910	0.7525	15.8400	2.7500
10YR	8.6355 (2.24)	0.9635	0.1356	15.5100	5.1700
30YR	8.6098 (2.26)	0.7603	-0.2823	15.1900	5.3200
Aaa	9.3417 (1.98)	1.0128	0.2414	15.2600	6.3400
Baa	10.6376 (2.24)	0.9803	0.2289	16.7400	7.1400
MONTH	6.5117 (3.43)	-0.0098	-1.1995	12.0000	1.0000
DAY	3.0116 (1.40)	-0.0040	-1.2801	5.0000	1.0000

Table 2: **Data Set Summary Statistics.** All sample statistics are scaled by a factor 10^{-2}, except skewness and kurtosis and all MONTH and DAY statistics. Summary statistics calculated for the transformed data set (6054 trading days). Values in parentheses are sample standard deviations.

The sixth series is the daily percentage of stocks on the NYSE advancing in price. This series was constructed from three daily series containing the number of stocks that advanced, declined and were unchanged in price. These first six series were collected from *Standard and Poor's Stock Price Record* and *The Wall Street Journal.*

The next four series are annualized yield observations of a number of US Treasury securities. Daily yields on 3-month and 6-month Treasury bills, 10-month Treasury notes, and 30-year Treasury bonds summarize the term structure. Unfortunately, for the source employed here, yields for the bills and notes were posted only on a weekly basis from the beginning of the data set through October 14, 1988. Daily observations were again filled in with the procedure outlined above. On the other hand, observations of the 30-year bond are nearly complete for the entire sample period. However, not all the observations actually denote the yield on a true 30-year bond. For some early observations, this yield is an average over a portfolio of US Treasury securities that were neither due nor callable for a period of not less than 10 years. Thus, this series should be interpreted as simply a "long government bond." These four interest rate series were collected from *Moody's Bond Survey.* Two more interest rate series were also gathered from the same source. They are the daily yields on two portfolios of corporate bonds containing debt securities for a selected sample of industrial companies rated Aaa and Baa by Moody's Investors Service. These two series are nearly complete with daily observations for the entire sample period. Finally, two more series were created to denote the month of the year and day of the week for a given trading day, so that a value of "1" denotes a January, and a value of "1" denotes a Monday.

2.4.2 Target Variables

Six target series are modeled. For both stock indices, k-period overlapping returns were constructed as in eq. (8), with forecast horizons $k = 5$ (weekly), $k = 20$ (monthly) and $k = 60$ (quarterly) trading days. These returns use the daily index close as their basis. Note that because of the way the overlapping returns were calculated, the starting and ending dates of these series are not exactly the same as the original daily data set. I report in Table 2 summary statistics for the DJIA and S&P 500 targets, and in Figure 3, I present the time behavior of these six series over the entire sample period. One interesting pattern we can see from these plots is the striking—though entirely understandable—gradation in smoothness as the forecast horizon shifts from short to long.

3 Benchmark Linear Model

A neural network is a particular nonlinear regression model. In standard regression analysis, the estimated functions are usually linear in their arguments. There are many reasons for this assumption: one primary justification being the relative ease in estimating and interpreting regressions of this type. One obvious extension is to allow the estimated functions to depend *nonlinearly* on their arguments as in the case of a neural network. For comparison, a relatively straightforward linear model is also estimated. We can compare both in-sample and out-of-sample network performance relative to the performance of this benchmark linear model.

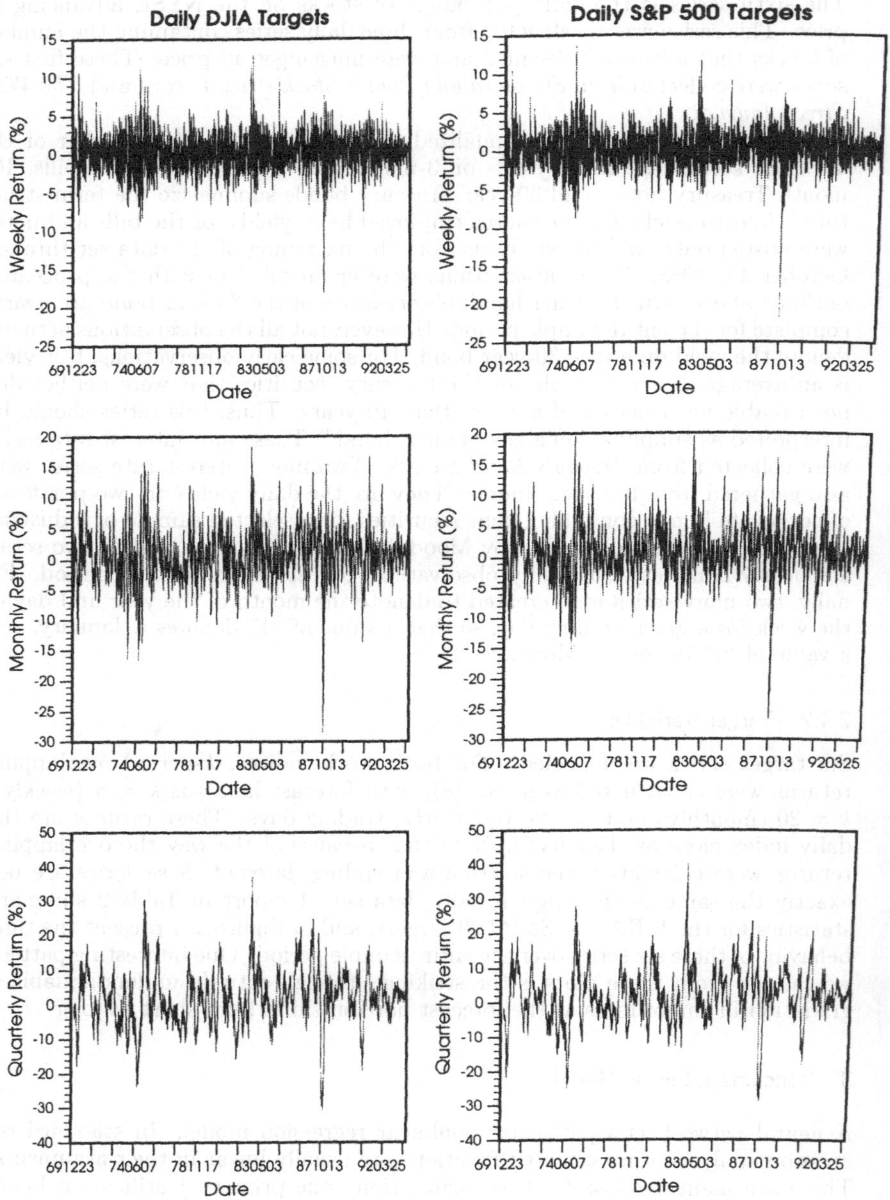

Figure 3: **DJIA and S&P 500 targets.** Six target series are modeled. For both indices, the data depicted in these graphs represent daily k-period overlapping returns, for weekly ($k = 5$ trading days), monthly ($k = 20$), and quarterly ($k = 60$) forecast horizons. The sample period is December 23, 1969, to December 2, 1993.

Analogous to the nonlinear model of eq. (3), I apply the following benchmark linear model:

$$a_t = \alpha_0 + \sum_{i=1}^{n}\sum_{\tau=0}^{L} \alpha_{i\tau} x_{i,t-\tau} = \ell(\boldsymbol{X}_{t(L)}; \mathbf{A}) \tag{11}$$

where $\boldsymbol{X}_{t(L)} = (\boldsymbol{x}_{t-L}, \ldots, \boldsymbol{x}_t)'$ is a matrix containing vectors of the input variables $\boldsymbol{x}_\tau$ for each time $\tau = t - L, \ldots, t$, and where $\mathbf{A}$ is a parameter vector. Eq. (11) is simply a linear multivariate, distributed-lag model with fixed lag depth L and parameter vector $\mathbf{A}$. This is a simple framework in that each input enters the model with a fixed and equal lag depth. Interestingly, notice the difference between the recurrent neural network and this linear model: theoretically, the network model has no restriction on its memory, while the linear model has a fixed memory controlled by the lag depth. Adjusting this lag depth establishes the memory as well as model size in terms of the number of parameters. Finally, estimation of this model involves minimizing with respect to the parameter vector, the mean-squared error between targets and the output of the linear model. For completeness, I use the same procedures outlined for the network batch estimation to minimize linear in-sample error.

3.1 Linear Estimation Results

For each of the three forecast horizons, I estimate the benchmark linear model with four values of the lag depth: $L = 9$, 16, 23 and 33, resulting in models with parameters numbering 141, 239, 337, and 477 respectively. These lag depths are chosen so that the parameter numbers are about the same as the four network sizes. In Table 3, I report the in-sample performance of the four linear models for both indices and all forecast horizons. Correspondingly, I report in Table 4 out-of-sample summary statistics, and in Tables 5 and 6, out-of-sample simulated trading portfolio performance.[16]

Several observations of note can be made concerning these results.

(1) For a given target series, the average in-sample error monotonically decreases with the lag depth, so that the in-sample fit becomes better as the memory of the linear model increases.

(2) However, the in-sample error does not decrease rapidly enough to counterbalance the increase in the number of estimated parameters as illustrated by the SIC values, which only progressively worsen with the lag depth.

(3) Turning to out-of-sample performance, the results are uniformly poor. Substantive ability to forecast both magnitude and sign is absent with no other clear patterns present.[17]

[16] See Section 4.3 for an explanation of the simulated trading portfolios. For comparison, later in Table 15, one can find benchmark performance of the two indices and a short-term bond.

[17] For quarterly forecasting, it appears that the linear models are doing a good job in sign prediction since so many of the fractions seem large, e.g. around 60 percent or more. However, remember that significant market-timing ability exists only relative to the actual distribution of returns. For example, if a model always forecasts up markets and the fraction of up markets is 75 percent, then sign prediction will be likewise. Substantial market-timing ability means accurate forecasting for both up *and* down markets.

DJIA Average In-Sample Performance

Forecast Horizon	Lag Depth	NMSE	SIC
Weekly	9	0.9077 (0.01)	0.2794 (0.01)
	16	0.8750 (0.01)	0.5043 (0.01)
	23	0.8418 (0.01)	0.7272 (0.01)
	33	0.7887 (0.01)	1.0356 (0.01)
Monthly	9	0.8171 (0.02)	0.1741 (0.02)
	16	0.7887 (0.02)	0.4003 (0.02)
	23	0.7596 (0.02)	0.6242 (0.02)
	33	0.7088 (0.02)	0.9284 (0.03)
Quarterly	9	0.6987 (0.05)	0.0156 (0.07)
	16	0.6690 (0.05)	0.2334 (0.07)
	23	0.6392 (0.05)	0.4486 (0.08)
	33	0.5922 (0.06)	0.7447 (0.09)

S&P 500 Average In-Sample Performance

Forecast Horizon	Lag Depth	NMSE	SIC
Weekly	9	0.9030 (0.01)	0.2742 (0.01)
	16	0.8698 (0.01)	0.4983 (0.01)
	23	0.8351 (0.01)	0.7190 (0.02)
	33	0.7857 (0.01)	1.0316 (0.02)
Monthly	9	0.8147 (0.02)	0.1710 (0.03)
	16	0.7876 (0.02)	0.3986 (0.03)
	23	0.7617 (0.03)	0.6265 (0.04)
	33	0.7268 (0.03)	0.9531 (0.04)
Quarterly	9	0.7052 (0.05)	0.0242 (0.08)
	16	0.6816 (0.05)	0.2520 (0.07)
	23	0.6564 (0.05)	0.4758 (0.07)
	33	0.6126 (0.05)	0.7802 (0.07)

Table 3: **Benchmark Linear In-Sample Performance.** The benchmark linear model is a multivariate, distributed-lag model with equal lags for each input variable as in eq. (11). In-sample NMSE and SIC values are averaged over all training sets. SIC refers to the Schwarz Information Criterion: small positive or large negative values indicate better model generalization for a given level of in-sample error, parameter number, and training set size which is fixed at 3000 daily observations. Values in parentheses are sample standard deviations. There are 608 weekly, 150 monthly and 48 quarterly training sets.

DJIA Out-of-Sample Performance

Forecast Horizon	Lag Depth	NMSE	Forecast Correlation	Sign Prediction	H-M
Weekly	9	1.0982	0.0146	0.5313	1.0354
	16	1.1721	-0.0393	0.5000	0.9799
	23	1.2145	-0.0319	0.5099	1.0070
	33	1.2951	0.0022	0.5329	1.0348
Monthly	9	1.1476	0.0460	0.5400	1.0379
	16	1.2491	-0.0192	0.5467	1.0411
	23	1.1857	0.0740	0.5667	1.0552
	33	1.4226	0.0888	0.5400	1.0563
Quarterly	9	1.5868	0.0127	0.6667	1.2071
	16	1.7958	-0.0609	0.6667	*1.3429*
	23	1.7174	-0.0086	0.6458	1.2071
	33	1.9475	-0.0633	0.6250	1.1429

S&P 500 Out-of-Sample Performance

Forecast Horizon	Lag Depth	NMSE	Forecast Correlation	Sign Prediction	H-M
Weekly	9	1.0841	0.0134	0.5066	0.9786
	16	1.1486	-0.0253	0.4967	0.9450
	23	1.1644	-0.0011	0.5082	1.0098
	33	1.1954	0.0253	0.5329	1.0385
Monthly	9	1.0631	0.1311	0.5733	1.1050
	16	1.1303	0.0674	0.5667	1.0425
	23	1.1493	0.0816	0.5533	1.0581
	33	1.2115	0.1005	0.5267	1.1126
Quarterly	9	1.4539	0.0146	0.5833	1.1005
	16	1.5695	-0.0367	0.5833	1.0423
	23	1.4955	0.0470	0.5833	1.0794
	33	1.7171	0.0391	0.6250	1.0794

Table 4: **Benchmark Out-of-Sample Linear Performance.** Using the benchmark linear model, this table tabulates out-of-sample performance by forecast horizon and lag depth. Correlation is between forecasted and actual returns. Sign prediction is the fraction of forecasted returns that have the same sign as the actual returns. H-M refers to the Henriksson-Merton statistic for market-timing ability: *italic* type indicates statistically significant ability at the 95 percent level. There are 608 weekly, 150 monthly and 48 quarterly forecasts. The out-of-sample forecast period is approximately January 1982 to December 1993.

Average Per Period Percentage Returns

Forecast Horizon	Lag Depth	P1	P2	P3	P4
Weekly	9	0.1705 **(2.16)**	0.2004 **(2.07)**	0.1683 (4.41)	0.2011 (4.25)
	16	0.1591 **(2.22)**	0.1354 **(2.14)**	0.1101 (4.40)	0.0892 (4.32)
	23	0.1432 **(2.24)**	0.1536 **(2.16)**	0.0445 (4.39)	0.0384 (4.32)
	33	0.1418 **(2.18)**	0.1965 **(1.73)**	0.2560 (3.91)	0.2210 (3.78)
Monthly	9	0.9212 **(4.00)**	0.8668 **(3.77)**	**1.2492** (7.96)	0.9696 (6.28)
	16	0.8803 **(4.08)**	0.8794 **(3.76)**	**1.2463** (8.04)	0.8242 (6.49)
	23	1.0614 **(4.03)**	0.8662 **(3.73)**	**1.2921** (8.05)	0.9822 (6.66)
	33	0.9604 **(3.87)**	0.8327 **(3.78)**	**1.2415** (7.99)	**1.1061** (5.80)
Quarterly	9	3.1180 **(6.18)**	2.4419 **(4.81)**	3.0880 (11.16)	2.5815 (8.50)
	16	**3.3687** **(6.20)**	2.4532 **(4.81)**	3.1214 (11.15)	2.4284 (8.37)
	23	**3.3842** **(6.37)**	2.9585 **(5.80)**	**3.8117** (12.26)	**3.3177** (9.06)
	33	**3.3446** **(6.40)**	2.9919 **(5.94)**	**3.4914** (12.63)	3.0646 (9.50)

Terminal Portfolio Values

Forecast Horizon	Lag Depth	P1	P2	P3	P4
Weekly	9	2.4207	2.9366	1.3601	1.7299
	16	2.2406	1.9616	0.9557	0.8588
	23	2.0298	2.1832	0.6437	0.6301
	33	2.0284	3.0153	3.0075	2.4933
Monthly	9	3.5040	3.2758	3.8827	3.0604
	16	3.2814	3.3398	3.8311	2.3928
	23	**4.3110**	3.2804	4.0928	2.9851
	33	3.7442	3.1117	3.8233	4.0434
Quarterly	9	4.0107	3.0170	3.1934	2.8302
	16	**4.5047**	3.0333	3.2454	2.6443
	23	**4.5152**	3.7609	4.2963	3.9065
	33	**4.4286**	3.8059	3.6345	3.4168

Table 5: **DJIA Benchmark Out-of-Sample Linear Portfolio Performance.** Using the benchmark linear model for the DJIA, this table tabulates out-of-sample portfolio performance by forecast horizon and network size. Portfolios are constructed as described in Table 12. Portfolios P1 and P2 allow no short selling, while portfolios P3 and P4 allow limited short selling. Beginning portfolio values are normalized to one. Values in parentheses are sample standard deviations. **Bold** type indicates better than benchmark index performance: either a larger average return, a smaller sample variance or a larger terminal value.

Average Per Period Percentage Returns

Forecast Horizon	Lag Depth	P1	P2	P3	P4
Weekly	9	0.2069 (**2.13**)	0.2201 (**1.62**)	0.2020 (3.69)	0.2495 (3.30)
	16	0.0841 (**2.06**)	0.0829 (**1.95**)	-0.0731 (4.26)	-0.0390 (4.19)
	23	0.1565 (**2.14**)	0.1968 (**1.72**)	0.1602 (4.37)	0.1500 (4.06)
	33	0.0990 (**2.07**)	0.0791 (**2.01**)	0.2434 (3.79)	**0.2729** (3.68)
Monthly	9	0.9041 (**3.91**)	0.9778 (**3.68**)	**1.6843** (7.70)	0.9823 (**4.50**)
	16	0.8224 (**4.03**)	0.8329 (**3.77**)	**1.2124** (7.89)	**1.0720** (4.70)
	23	0.8299 (**3.97**)	0.8883 (**3.70**)	**1.0222** (7.75)	1.0043 (4.58)
	33	0.8280 (**3.84**)	0.6935 (**3.65**)	0.9661 (7.79)	0.7135 (5.32)
Quarterly	9	2.4151 (**5.12**)	2.1958 (**4.14**)	1.9583 (10.07)	1.6977 (7.08)
	16	2.5005 (**5.75**)	2.4664 (**4.44**)	2.8715 (10.41)	2.2779 (7.91)
	23	**3.3842** (**6.37**)	2.9585 (**5.80**)	**3.8117** (12.26)	**3.3177** (9.06)
	33	2.9097 (**5.62**)	2.6816 (**5.21**)	**3.2117** (11.60)	2.5004 (8.54)

Terminal Portfolio Values

Forecast Horizon	Lag Depth	P1	P2	P3	P4
Weekly	9	3.0375	3.5179	2.2799	3.2847
	16	1.4545	1.4634	0.3362	0.4212
	23	2.2332	3.0243	1.3582	1.4515
	33	1.5904	1.4199	2.8559	3.4909
Monthly	9	3.4397	**3.8890**	**7.7321**	3.7473
	16	3.0241	3.1179	3.7592	**4.2401**
	23	3.0688	3.4004	2.8780	**3.8573**
	33	3.0828	2.5498	2.6296	2.3698
Quarterly	9	3.3689	3.0004	2.0881	2.2027
	16	2.9606	2.7252	1.9853	1.9815
	23	3.0413	3.0770	3.0067	2.5211
	33	3.6961	3.3539	3.3820	2.7368

Table 6: **S&P 500 Benchmark Out-of-Sample Linear Portfolio Performance.** Using the benchmark linear model for the S&P 500, this table tabulates out-of-sample portfolio performance by forecast horizon and network size. Portfolios are constructed as described in Table 12. Portfolios P1 and P2 allow no short selling, while portfolios P3 and P4 allow limited short selling. Beginning portfolio values are normalized to one. Values in parentheses are sample standard deviations. **Bold** type indicates better than benchmark index performance: either a larger average return, a smaller sample variance or a larger terminal value.

(4) Likewise for out-of-sample portfolio performance, results are poor. Only in a handful of cases is there any instance that using the linear model forecasts improves on benchmark index performance.

4 Network Estimation Results

4.1 In-Sample Performance

The first phase of the network forecast exercises involves estimating, for fixed model size and forecast horizon, a parameter vector for each training set. Two estimation methods are used. The first method attempts to minimize in-sample error using the online learning rule of Kuan (1993). And the second method applies batch minimization in conjunction with a simulated annealing algorithm.

4.1.1 Summary Statistics

In Table 7, I present in-sample results for both estimation methods. For the two stock indices, this table tabulates the average in-sample NMSE and SIC estimates across the training sets by forecast horizon and network size.[18]

From Table 7, we can draw several conclusions. (1) In all cases, batch estimation provides the *smallest* in-sample error. (2) For each forecast horizon, there is a smooth or monotonic decrease in the in-sample error as we increase the number of hidden units, matching well with our intuition. (3) For both indices, using the SIC estimates as a complexity measure, the 9 and 12 hidden unit networks generally seem to provide the best tradeoff between network size and in-sample error—except for weekly forecasting where no such apparent conclusion can be made.

Later when out-of-sample performance results are presented, we will see that this last inference is generally correct: for monthly and quarterly forecasting and networks with 9 and 12 hidden units, we observe both good out-of-sample fit and generalization. Underfitting seems to occur with 6 hidden units and overfitting with 15 hidden units. On the other hand, accurate weekly forecasting appears to be a nearly impossible task regardless of network size.

4.1.2 In-Sample NMSE Profiles

We can verify the competence of batch minimization by comparing the in-sample NMSE profiles from both online and batch estimation. In Figure 4, I present these profiles for monthly and quarterly forecasts and 9 and 12 hidden unit networks.

Again we observe that batch estimation dominates online estimation: amazingly, the improvement in in-sample error for some training sets approaches nearly 50 percent. Also, we observe that the time profile of in-sample error across the training sets is not a smooth plot. As we look across adjacent training sets, there is much variation in the average error. There are two possible sources for this variability. (1) We are observing sampling variation across the training sets because the estimation for each training set is independent. (2)

[18] Each training set $\mathcal{T}_i$ has an associated NMSE and SIC estimate. I simply calculate an average over all the training sets.

DJIA Average In-Sample Performance

Forecast Horizon	Hidden Units	Online NMSE	Batch NMSE	Online SIC	Batch SIC
Weekly	6	0.7054 (0.03)	0.7033 (0.03)	0.0050 (0.04)	0.0020 (0.04)
	9	0.6341 (0.04)	0.6297 (0.04)	0.1450 (0.07)	0.1384 (0.07)
	12	0.5757 (0.03)	0.5693 (0.02)	0.3462 (0.05)	0.3352 (0.04)
	15	0.5073 (0.02)	0.5031 (0.02)	0.5639 (0.05)	0.5556 (0.04)
Monthly	6	0.4735 (0.07)	0.4694 (0.06)	-0.4019 (0.13)	-0.4101 (0.13)
	9	0.3442 (0.05)	0.3112 (0.02)	-0.4723 (0.13)	-0.5671 (0.07)
	12	0.2628 (0.03)	0.2516 (0.02)	-0.4444 (0.12)	-0.4849 (0.09)
	15	0.2496 (0.04)	0.2413 (0.04)	-0.1551 (0.15)	-0.1891 (0.15)
Quarterly	6	0.2108 (0.03)	0.1977 (0.03)	-1.2119 (0.14)	-1.2742 (0.13)
	9	0.1173 (0.02)	0.1026 (0.01)	-1.5531 (0.13)	-1.6816 (0.13)
	12	0.0927 (0.02)	0.0764 (0.01)	-1.5015 (0.21)	-1.6873 (0.17)
	15	0.0740 (0.02)	0.0680 (0.01)	-1.3792 (0.20)	-1.4609 (0.18)

S&P 500 Average In-Sample Performance

Forecast Horizon	Hidden Units	Online NMSE	Batch NMSE	Online SIC	Batch SIC
Weekly	6	0.6881 (0.03)	0.6832 (0.03)	-0.0196 (0.04)	-0.0267 (0.04)
	9	0.6132 (0.03)	0.6097 (0.02)	0.1133 (0.04)	0.1075 (0.04)
	12	0.5464 (0.03)	0.5365 (0.03)	0.2937 (0.05)	0.2755 (0.05)
	15	0.4879 (0.03)	0.4849 (0.03)	0.5236 (0.07)	0.5177 (0.07)
Monthly	6	0.4281 (0.04)	0.4199 (0.04)	-0.4972 (0.09)	-0.5164 (0.09)
	9	0.3224 (0.04)	0.2866 (0.03)	-0.5367 (0.12)	-0.6505 (0.09)
	12	0.2441 (0.04)	0.1973 (0.03)	-0.5259 (0.18)	-0.7385 (0.18)
	15	0.1979 (0.04)	0.1865 (0.03)	-0.3919 (0.18)	-0.4496 (0.17)
Quarterly	6	0.2143 (0.04)	0.2046 (0.04)	-1.2013 (0.18)	-1.2476 (0.18)
	9	0.1189 (0.02)	0.1040 (0.01)	-1.5379 (0.15)	-1.6687 (0.13)
	12	0.0803 (0.02)	0.0682 (0.01)	-1.6542 (0.24)	-1.8068 (0.20)
	15	0.0587 (0.01)	0.0518 (0.01)	-1.6123 (0.20)	-1.7328 (0.18)

Table 7: **Online and Batch Estimation In-Sample Performance.** In-sample NMSE and SIC values are averaged over all training sets. SIC refers to the Schwarz Information Criterion: small positive or large negative values indicate better model generalization for a given level of in-sample error, number of parameters, and training set size which is fixed at 3000 daily observations. Values in parentheses are sample standard deviations. There are 608 weekly, 150 monthly and 48 quarterly training sets.

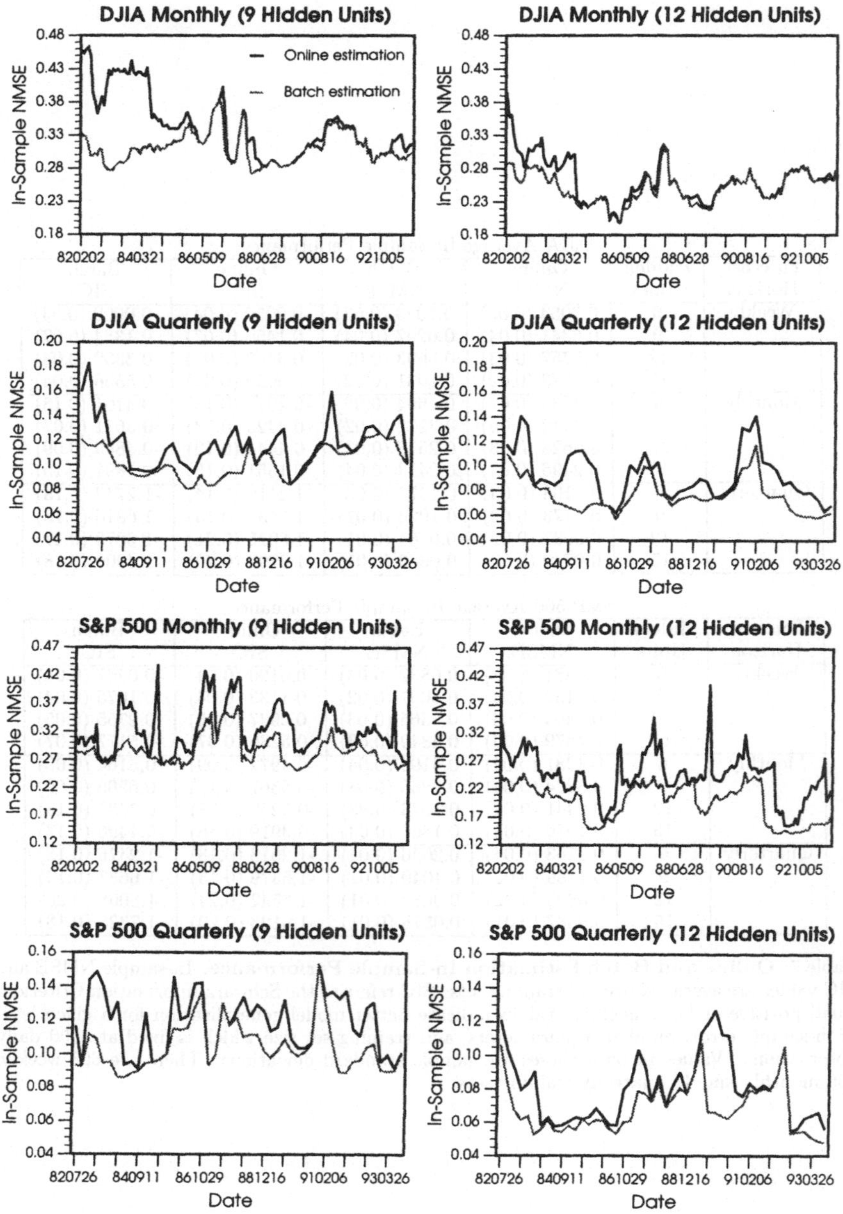

Figure 4: **DJIA and S&P 500 in-sample NMSE profiles.** For monthly and quarterly forecast horizons, graphs depict the in-sample NMSE over all training sets using 9 and 12 hidden unit networks. Training sets are dated by their last in-sample observation. On-line estimation is denoted by dark lines and batch estimation by gray lines.

Or more likely, there are significant time stationarity problems in the data. Recall how the training sets were constructed: two consecutive sets only differ in the subtraction and addition of k observations at the beginning and end of the training sets. Time stationarity will be a problem if the underlying processes generating the data are significantly different between these two intervals of data. Since training sets are about 12 years of calendar time in length, it is likely that this is the major source of the in-sample time variation in the error. Moreover, we can also point to the fact that, by comparing the error profiles between network sizes, there appears to be a good deal of co-movement in the in-sample error. That is, when the error increases at a particular training set for one network size, the same increase also occurs for a different network size on the same training set.

In Figure 5, I compare the batch estimates between the 9 and 12 hidden unit networks for both monthly and quarterly forecasts. As stated earlier, networks of different sizes are nested. Thus a large network can implement any solution by a small network. We expect that in general a large network should produce a smaller in-sample error. Indeed, the error is usually smaller for a large network with violations rare. This leads us to infer two points: (1) the estimated parameter vector of the large network for those points that the error is larger is *definitely* not optimal, since we can simply implement the solution of the small network and achieve a lower error; and (2), however, these violations are rare, implying that we are on average doing a reasonable job of searching for good minima.

4.2 Out-of-Sample Performance

4.2.1 Forecast Summary Statistics

In Table 8, I present out-of-sample performance for all forecast horizons and network sizes. As opposed to in-sample performance, out-of-sample error does not monotonically decrease with network size. Instead, the error usually first decreases and then increases as we move through the range of different network sizes. This confirms previous observations by other researchers that out-of-sample generalization is a concave function of network complexity: at first decreasing and then increasing. However, since we only make discrete changes in the network size, we cannot say by and larger whether generalization changes smoothly or continuously.[19]

Besides out-of-sample error, in Table 8, I also calculate several other sample statistics. Again, we can confirm that the best out-of-sample performance occurs with the 9 and 12 hidden unit networks for both monthly and quarterly forecasts; on the other hand, results for weekly forecasting are very poor for all network sizes.

Remarkably, not only is this substantial forecasting ability observed in both indices, but we also observe it even though the same information set is used to forecast different horizons. Recall that although different target series are estimated, varying only in forecast horizon, the *same* daily conditioning set is applied. Thus, the information contained in the daily inputs proves useful not

[19]Strictly speaking, network size should be defined by the number of connections and not by the number of hidden units. For example, in a feedforward network, the real difference between a q hidden unit and a $q+1$ hidden unit network is the discrete addition of n connections between the input and hidden unit layers.

DJIA Out-of-Sample Performance

Forecast Horizon	Hidden Units	NMSE	Forecast Correlation	Sign Prediction	H-M
Weekly	6	1.1504	0.0209	0.5625	1.0655
	9	1.6246	-0.0778	0.5280	1.0538
	12	1.2244	0.1777	0.5395	1.0523
	15	1.7802	0.0317	0.5148	1.0131
Monthly	6	1.4608	0.0153	0.5400	0.9740
	9	1.0712	**0.4018**	0.5333	1.0476
	12	1.1980	**0.3851**	0.6067	**1.2749**
	15	1.4512	0.2802	0.5933	1.2002
Quarterly	6	2.4100	-0.0464	0.6042	1.1286
	9	1.4615	0.2486	0.6875	**1.3500**
	12	1.6427	**0.5274**	0.7292	**1.5143**
	15	1.2753	0.4273	0.6042	1.3786

S&P 500 Out-of-Sample Performance

Forecast Horizon	Hidden Units	NMSE	Forecast Correlation	Sign Prediction	H-M
Weekly	6	1.5041	-0.0131	0.5066	0.9657
	9	1.4091	0.1110	0.5230	1.0210
	12	1.3372	0.0749	0.5115	1.0268
	15	1.6679	0.0417	0.5197	1.0243
Monthly	6	2.1362	-0.0306	0.5133	0.9797
	9	1.3596	**0.3533**	0.6000	1.0821
	12	1.2960	**0.3295**	0.5867	1.1515
	15	2.0924	0.2965	0.6133	**1.2373**
Quarterly	6	1.9458	0.0087	0.4792	1.1587
	9	1.3733	0.1974	0.7708	**1.4074**
	12	0.8425	**0.5958**	0.7292	**1.3968**
	15	1.5993	0.1454	0.5625	1.1270

Table 8: **DJIA and S&P 500 Out-of-Sample Network Performance.** Using batch estimates, this table tabulates out-of-sample performance by forecast horizon and network size. Correlation is between forecasted and actual returns: **bold** type indicates a bootstrap probability of less than 1 percent for a zero or negative correlation. Sign prediction is the fraction of forecasted returns that have the same sign as the actual returns. H-M refers to the Henriksson-Merton statistic for market-timing ability: **bold** type indicates statistically significant ability at the 99 percent level. There are 608 weekly, 150 monthly and 48 quarterly forecasts. The out-of-sample forecast period is approximately January 1982 to December 1993.

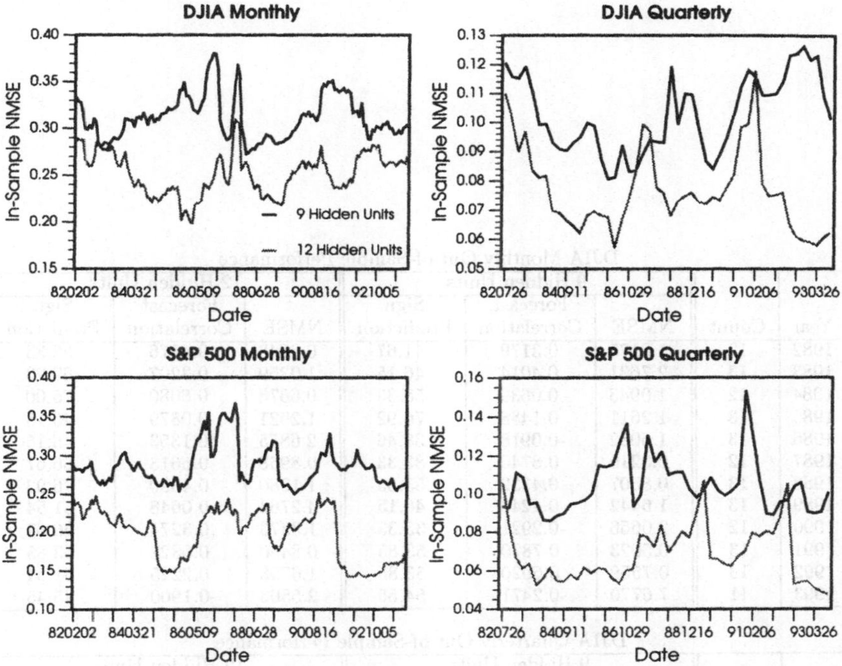

Figure 5: **DJIA and S&P 500 in-sample NMSE profile comparisons.** For monthly and quarterly forecasts, batch in-sample NMSE profiles are compared between 9 (dark line) and 12 (gray line) hidden unit networks. Training sets are dated by their last in-sample observation.

only in forecasting more than one horizon, but also in forecasting relatively long horizon returns.

Lastly, in Tables 9 and 10, I break down out-of-sample performance by calendar year for monthly and quarterly forecasting using the 9 and 12 hidden unit networks. For the most part, there is a good deal of variation in out-of-sample performance, although there does not appear to be any obvious pattern. Interestingly, there seems to be a tendency for the larger network to outperform the smaller one; however, this does not certainly hold on a year-by-year basis.

4.2.2 Bootstrap Sample Statistic Distributions

To get an idea of their distributions, we can bootstrap some of the out-of-sample forecast statistics. For a chosen statistic, a bootstrap sample with the same size as the original, empirical distribution is created by sampling with replacement from the original distribution. We then calculate the statistic on this bootstrap sample. Repeating these steps 5000 times, we in turn build up a distribution of the statistic. Asymptotically, this bootstrap distribution should approach the true distribution of the underlying statistic.[20]

[20]These techniques were introduced by Efron (1982). An introductory treatment is provided by Efron and Tibshirani (1993).

DJIA Monthly Out-of-Sample Performance

Year	Count	9 Hidden Units			12 Hidden Units		
		NMSE	Forecast Correlation	Sign Prediction	NMSE	Forecast Correlation	Sign Prediction
1982	12	1.0175	0.3179	41.67	0.8848	0.4176	58.33
1983	13	2.7631	0.4014	46.15	1.0359	0.3207	69.23
1984	12	1.0943	0.0639	58.33	0.6678	0.6080	75.00
1985	13	1.2611	0.1488	76.92	1.2921	-0.0879	46.15
1986	13	1.6082	-0.0918	38.46	2.6875	0.1353	46.15
1987	12	0.3210	0.8745	83.33	0.8968	0.6613	66.67
1988	13	0.8507	0.4332	53.85	1.1280	0.4090	76.92
1989	13	1.6442	0.4248	46.15	4.2763	0.0648	61.54
1990	12	2.0656	-0.2921	33.33	1.3373	0.3274	66.67
1991	13	0.5023	0.7820	53.85	0.8460	0.6322	53.85
1992	13	0.7559	0.6020	53.85	1.6798	0.2225	61.54
1993	11	7.6770	-0.2471	54.55	2.5503	-0.1900	45.45

DJIA Quarterly Out-of-Sample Performance

Year	Count	9 Hidden Units			12 Hidden Units		
		NMSE	Forecast Correlation	Sign Prediction	NMSE	Forecast Correlation	Sign Prediction
1982	2	0.2453	1.0000	50.00	0.0900	1.0000	100.00
1983	5	18.7714	0.1756	40.00	14.0994	0.5417	40.00
1984	4	0.2666	0.9391	100.00	0.1720	0.9783	100.00
1985	4	4.1765	-0.7408	75.00	0.2580	0.8883	100.00
1986	4	0.4763	0.8856	100.00	4.1327	0.5064	50.00
1987	4	1.3580	-0.1105	75.00	11.7822	0.9826	75.00
1988	5	2.8124	-0.3194	60.00	1.0653	0.8136	60.00
1989	4	2.6747	0.5729	75.00	1.2301	0.6796	100.00
1990	4	1.2291	0.4146	50.00	3.0244	0.9033	50.00
1991	4	0.3584	0.8060	100.00	0.5605	0.7152	75.00
1992	5	7.8175	-0.3944	60.00	0.1351	0.9440	80.00
1993	3	8.8618	0.2914	33.33	52.7156	-0.7270	66.67

Table 9: **DJIA Out-of-Sample Network Performance: Breakdown by Year.** For monthly and quarterly forecasting and 9 and 12 hidden unit networks, this table tabulates out-of-sample network performance by calendar year. Count refers to the number of forecasts made within a given year.

S&P 500 Monthly Out-of-Sample Performance

		9 Hidden Units			12 Hidden Units		
Year	Count	NMSE	Forecast Correlation	Sign Prediction	NMSE	Forecast Correlation	Sign Prediction
1982	12	0.5977	0.7387	41.67	0.8629	0.4502	50.00
1983	13	3.5392	0.3569	61.54	1.8988	0.5840	69.23
1984	12	1.6816	0.2477	75.00	0.6821	0.6145	83.33
1985	13	2.2948	-0.0070	69.23	1.5695	0.4238	69.23
1986	13	2.0854	0.4856	61.54	3.1357	-0.0142	53.85
1987	12	0.8633	0.6330	58.33	0.4793	0.7445	75.00
1988	13	1.1637	0.3284	46.15	3.2855	0.1494	30.77
1989	13	1.4664	-0.0566	53.85	2.1090	0.0382	46.15
1990	12	1.9445	0.0476	50.00	1.6186	-0.1467	41.67
1991	13	0.9106	0.4701	69.23	0.3197	0.8655	69.23
1992	13	0.8378	0.6937	61.54	0.9983	0.5129	53.85
1993	11	2.1737	-0.0987	72.73	5.0068	0.0268	63.64

S&P 500 Quarterly Out-of-Sample Performance

		9 Hidden Units			12 Hidden Units		
Year	Count	NMSE	Forecast Correlation	Sign Prediction	NMSE	Forecast Correlation	Sign Prediction
1982	2	1.0736	1.0000	100.00	0.0186	1.0000	100.00
1983	5	0.8482	0.8521	80.00	2.2975	0.3780	80.00
1984	4	5.5557	-0.9184	75.00	0.7350	0.6166	75.00
1985	4	1.3358	0.2189	75.00	1.5215	-0.7169	75.00
1986	4	1.0679	0.4818	100.00	0.2513	0.9882	75.00
1987	4	0.6890	0.9657	100.00	4.2177	0.3018	75.00
1988	5	2.3168	-0.1716	60.00	1.6913	0.8856	60.00
1989	4	0.4381	0.7585	100.00	0.2781	0.8607	100.00
1990	4	1.2273	0.3432	50.00	1.6081	0.4427	25.00
1991	4	3.6163	-0.4424	50.00	0.9540	0.5004	75.00
1992	5	0.5670	0.7665	80.00	0.4459	0.8844	60.00
1993	3	1.2582	0.0590	66.67	1.4807	0.4256	100.00

Table 10: **S&P 500 Out-of-Sample Network Performance: Breakdown by Year.** For monthly and quarterly forecasting and 9 and 12 hidden unit networks, this table tabulates out-of-sample network performance by calendar year. Count refers to the number of forecasts made within a given year.

DJIA Bootstrapped Out-of-Sample NMSE

Forecast Horizon	Hidden Units	Mean	Max.	Min.	Pr[$x \leq 1$]
Monthly	9	1.1177 (0.25)	2.1326	0.4989	0.3428
	12	1.2485 (0.29)	2.6267	0.5057	0.2076
Quarterly	9	1.5224 (0.45)	3.9100	0.4330	0.1118
	12	1.8008 (0.83)	7.6978	0.3115	0.1482

DJIA Bootstrapped Forecast Correlation

Forecast Horizon	Hidden Units	Mean	Max.	Min.	Pr[$x \leq 0$]
Monthly	9	0.3826 (0.13)	0.7292	-0.0401	0.0010
	12	0.3694 (0.12)	0.7326	-0.0230	0.0016
Quarterly	9	0.2604 (0.18)	0.7546	-0.2848	0.0820
	12	0.5252 (0.11)	0.8543	0.0374	0.0000

S&P 500 Bootstrapped Out-of-Sample NMSE

Forecast Horizon	Hidden Units	Mean	Max.	Min.	Pr[$x \leq 1$]
Monthly	9	1.4138 (0.32)	3.1410	0.6037	0.0868
	12	1.3390 (0.28)	2.8334	0.6938	0.0954
Quarterly	9	1.4115 (0.44)	4.0232	0.3832	0.1748
	12	0.9241 (0.34)	3.1822	0.1896	0.6524

S&P 500 Bootstrapped Forecast Correlation

Forecast Horizon	Hidden Units	Mean	Max.	Min.	Pr[$x \leq 0$]
Monthly	9	0.3366 (0.13)	0.7441	-0.1087	0.0058
	12	0.3289 (0.09)	0.6570	-0.0318	0.0002
Quarterly	9	0.2230 (0.17)	0.8120	-0.2645	0.0812
	12	0.5793 (0.14)	0.8917	0.0447	0.0000

Table 11: **DJIA and S&P 500 Bootstrap Sample Statistics.** Forecasts are based on batch estimates for monthly and quarterly forecasts and 9 and 12 hidden unit networks. Two panels for each index: (1) summary statistics for 5000 bootstrapped out-of-sample NMSEs, and (2) summary statistics for 5000 bootstrapped forecast correlations. Correlation is between forecasted and actual returns. Values in parentheses are bootstrap sample standard deviations.

By using bootstrapping techniques, I address the possible difficulty that out-of-sample performance is too dependent on the particular sample period. For example, during this period, there were several rare occurrences of large declines in the indices. Whether out-of-sample point estimates are driven by accurate or inaccurate forecasting of these declines is the question.[21]

For monthly and quarterly forecasts and 9 and 12 hidden unit networks, I bootstrapped two statistics: out-of-sample NMSE and the correlation between forecasted and actual returns. Since this technique gives us a view of the underlying distribution, we should be able to generate confidence intervals. I report in Table 11 the results of this procedure, and in Figures 6 and 7, I offer histograms of these bootstrapped distributions.

For the out-of-sample NMSE, bootstrapping reveals that its distribution is very wide, suggesting that this statistic is very sensitive to the particular sample. Accordingly, we should give it less weight as a good measure of out-of-sample

[21]Weigend and LeBaron (1994) provide bootstrapping and simulation evidence showing that the splitting of data sets between in-sample and out-of-sample is the primary variation in network performance, and not particular network initial conditions (e.g. architecture and initial weights).

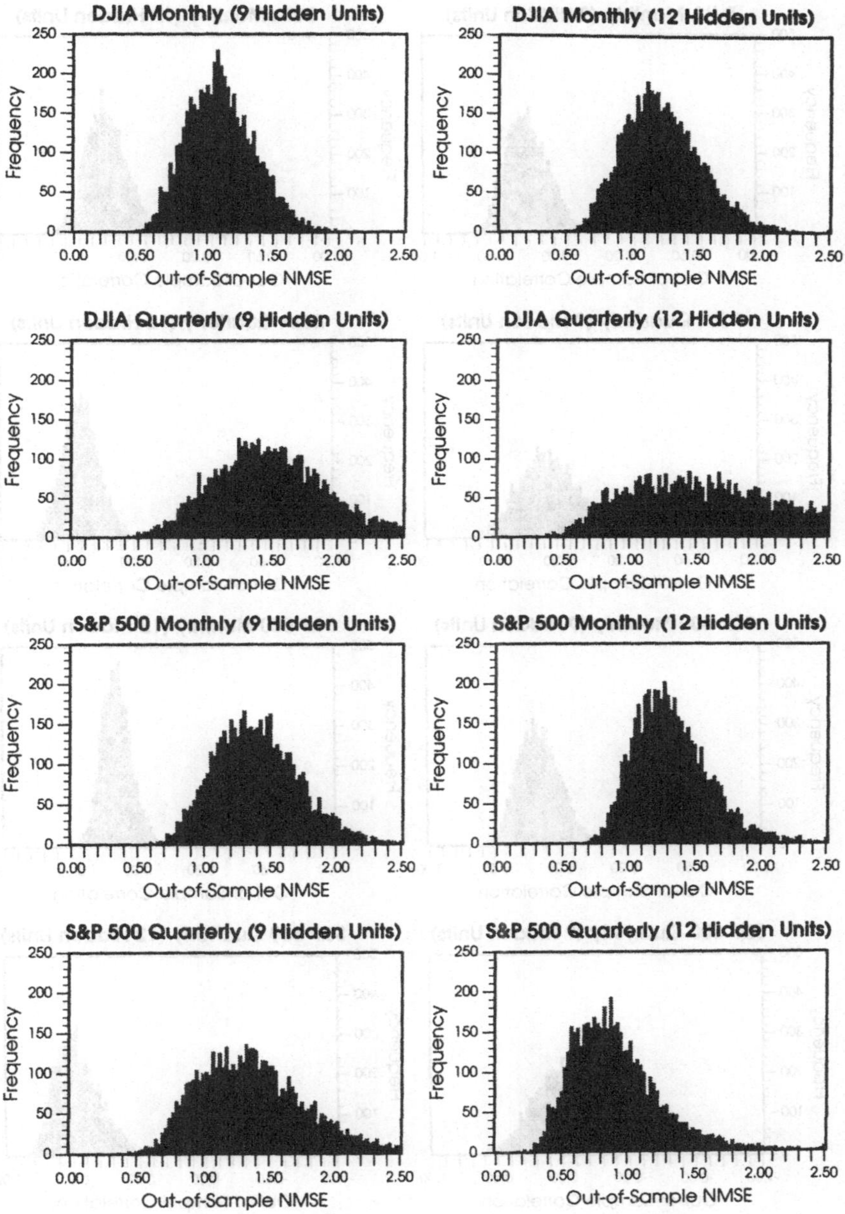

Figure 6: **Bootstrap out-of-sample NMSE distributions.** For monthly and quarterly forecasts, graphs depict histograms of 5000 bootstrapped out-of-sample NMSEs using 9 and 12 hidden unit networks.

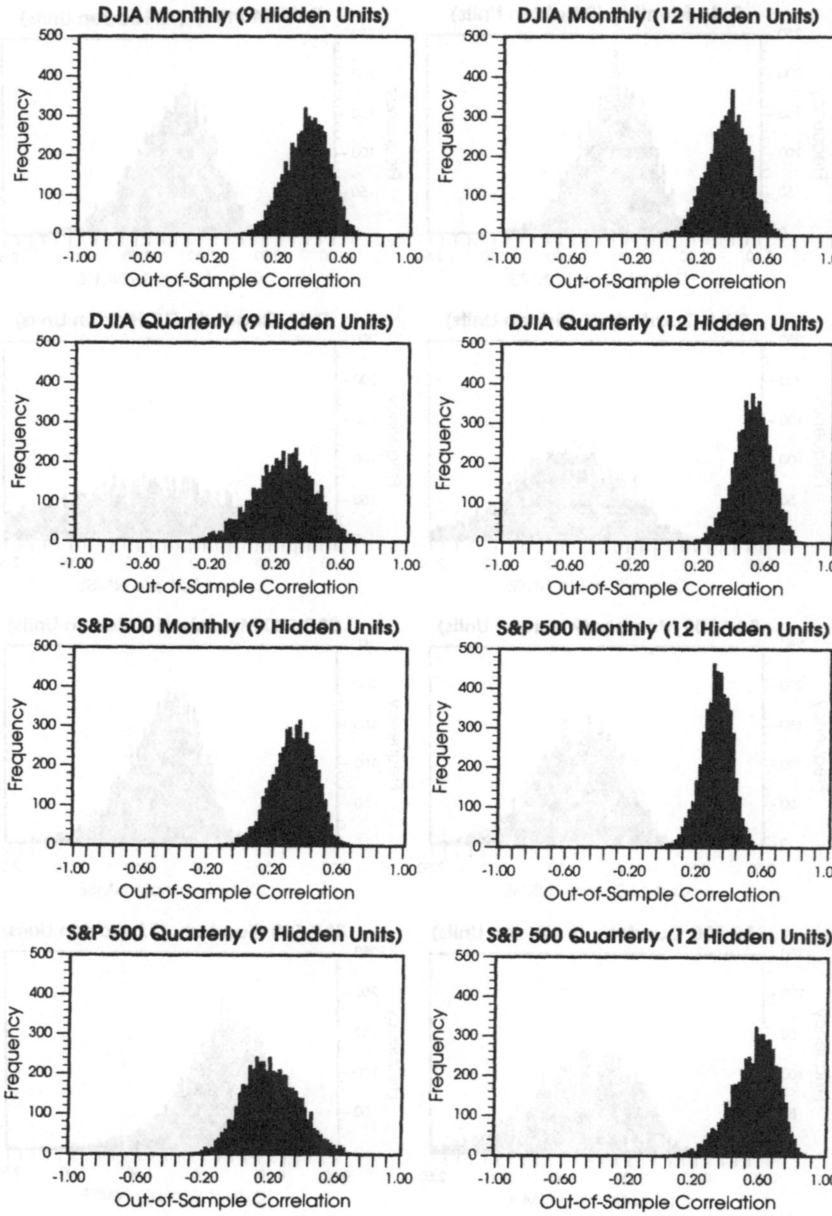

Figure 7: **Bootstrap forecast correlation distributions.** For monthly and quarterly forecasts, graphs depict histograms of 5000 bootstrapped correlations using 9 and 12 hidden unit networks. Correlation is between forecasted and actual returns.

Portfolio	Rule Description
P1	Assuming in the previous period the portfolio is invested in the bond, if the forecasted index return net of transaction costs is greater than zero, then invest 100% in the index. Otherwise, invest in the bond. Reverse the rule if in the previous period the portfolio is invested in the index.
P2	Same as P1, except use the bond return as the threshold.
P3	Assuming in the previous period the portfolio is invested in the bond, if the forecasted index return net of transaction costs is greater than the bond return, then invest 200% of the portfolio in the index and short the bond 100%. Otherwise, invest 200% in the bond and short the index. Reverse the rule if in the previous period the portfolio is invested in the index.
P4	Same as P3, except that the portfolio weights are a linear function of the forecasted return if the forecasted return on the index lies within a specified interval. For weekly returns this interval is $(-1\%, 1\%)$, for monthly returns $(-5\%, 5\%)$, and for quarterly returns $(-10\%, 10\%)$.

Table 12: **Portfolio Rules.** For all portfolios, a fixed percentage transaction cost of 0.5 percent on the value of the portfolio is deducted whenever there is portfolio turnover.

fit. Also this statistic seems to be very sensitive to the size of the sample, since the distributions for the quarterly forecasts are all uniformly wider than for the monthly forecasts, with three times as many monthly forecasts (150) as there are quarterly forecasts (48). Lastly, these distributions have long and thin righthand tails, suggesting that probably several relatively poor forecasts are responsible for inflating the NMSE point estimates.

For the forecast correlation, bootstrapping reveals extremely small probabilities of zero or negative correlations. Moreover, the distributions are fairly sharp-peaked and resting to the positive side. For example, in the case of monthly DJIA forecasts, the probabilities of a correlation between forecasted and actual returns of less than zero are 0.10 and 0.16 percent for the 9 and 12 hidden unit networks respectively. Thus, with high confidence, the true forecast correlation is large and positive. This suggests that the networks are doing a credible job in forecasting the magnitude of future returns.

4.3 Out-of-Sample Portfolio Performance

4.3.1 Portfolio Statistics

As a final evaluation of out-of-sample network performance, I constructed four simulated trading portfolios based on simple rules that use the forecasts as triggers which are summarized in Table 12. The portfolio strategies were very simple: using the forecasted return on the index, decide whether to invest in the index or a safe cash-equivalent security, in this case, a 3-month Treasury bill. The major differences among the strategies are in the calculation of the portfolio holding weights and whether short sales are allowed. Basically, for portfolios P1 and P2, short selling is not allowed, while in portfolios P3 and P4, some limited short selling is allowed. Lastly, a fixed transaction cost of 0.5 percent is deducted from realized returns whenever there is portfolio turnover.

In Tables 13 and 14, I tabulate out-of-sample portfolio performance for all forecast horizons and network sizes. For comparison, in Table 15, I provide corresponding actual benchmark performance for the two indices and the short-term bond. Finally, in Figure 8, I offer graphs of the portfolio performance for

Average Per Period Percentage Returns

Forecast Horizon	Hidden Units	P1	P2	P3	P4
Weekly	6	0.1898 (**2.28**)	0.1934 (**2.16**)	**0.3851** (4.63)	0.2367 (4.34)
	9	0.1292 (**2.10**)	0.1257 (**2.07**)	0.2418 (4.37)	0.1690 (4.28)
	12	0.2133 (**1.77**)	0.2102 (**1.72**)	**0.4620** (3.88)	**0.4415** (3.81)
	15	0.1352 (**2.12**)	0.1313 (**2.09**)	0.0142 (4.42)	0.0906 (4.29)
Monthly	6	0.9671 (**4.10**)	0.8992 (**3.89**)	**1.3020** (8.12)	0.9762 (6.62)
	9	**1.1123 (3.58)**	**1.1156 (3.38)**	**1.9103** (7.43)	**2.0278** (6.27)
	12	**1.2905 (3.09)**	**1.2673 (2.79)**	**2.4347** (6.55)	**1.8617** (5.66)
	15	**1.1849 (3.35)**	**1.2380 (3.15)**	**2.4489** (6.99)	**2.0594** (6.18)
Quarterly	6	2.9688 (**6.49**)	2.5463 (**5.53**)	2.9883 (11.99)	2.8679 (9.71)
	9	**3.4771 (6.19)**	**3.6254 (6.07)**	**6.2991** (12.30)	**5.3005** (10.96)
	12	**4.1054 (5.41)**	**3.7487 (5.01)**	**6.6849** (10.73)	**6.1415** (9.29)
	15	3.2789 (**5.47**)	**3.4767 (5.15)**	**5.8652** (11.09)	**5.6146** (10.22)

Terminal Portfolio Values

Forecast Horizon	Hidden Units	P1	P2	P3	P4
Weekly	6	2.6774	2.7809	**4.7693**	2.1008
	9	1.8962	1.8624	2.1299	1.3997
	12	3.3249	3.2818	**10.6423**	**9.5465**
	15	1.9630	1.9230	0.5280	0.8725
Monthly	6	3.7310	3.4150	4.1119	3.0849
	9	**4.7926**	**4.8640**	**11.5708**	**15.5071**
	12	**6.3917**	**6.2548**	**27.3213**	**12.7286**
	15	**5.3996**	**5.8974**	**26.8301**	**16.3785**
Quarterly	6	3.7061	3.1151	2.9142	3.0909
	9	**4.7379**	**5.0927**	**13.4260**	**9.1228**
	12	**6.4821**	**5.5481**	**17.7307**	**14.7674**
	15	**4.4150**	**4.8758**	**12.0256**	**11.1974**

Table 13: **DJIA Out-of-Sample Network Portfolio Performance.** Using batch estimates for the DJIA, this table tabulates out-of-sample portfolio performance by forecast horizon and network size. Portfolios are constructed as described in Table 12. Portfolios P1 and P2 allow no short selling, while portfolios P3 and P4 allow limited short selling. Beginning portfolio values are normalized to one. Values in parentheses are sample standard deviations. **Bold** type indicates better than benchmark index performance: either a larger average return, a smaller sample variance or a larger terminal value.

two selected horizons and networks. Again, the best out-of-sample performance is provided by the 9 and 12 hidden unit networks, and weekly performance for all network sizes is poor.

For the 9 and 12 hidden unit networks, all simulated portfolios are able to outperform their respective benchmark index for both monthly and quarterly forecasts, even when taking into account transaction costs, by a wide margin. Moreover, as we can see, the inherent risk in the portfolio strategies is more than moderate as measured by the *ex post* variance of the portfolio returns. In almost all cases, the magnitude of the ratio of the portfolio returns over the benchmark returns is substantially larger than the ratio of the portfolio variance over the benchmark variance. In some cases, not only are the average portfolio returns larger than the benchmark return, but the portfolio variances are smaller than the benchmark variance. This clearly indicates the possibility of above-market returns with less or moderate risk. I will address this trade-off between risk and return in the next section.

Next, I characterize portfolio turnover in Table 16. This table tabulates the

Average Per Period Percentage Returns

Forecast Horizon	Hidden Units	P1	P2	P3	P4
Weekly	6	0.1671 (**2.17**)	0.1677 (**2.12**)	0.0233 (4.48)	0.0729 (4.43)
	9	0.1899 (**1.51**)	0.2210 (**1.44**)	0.2180 (3.47)	**0.2985** (3.50)
	12	0.1528 (**1.94**)	0.1506 (**1.89**)	0.1943 (4.10)	0.2493 (4.08)
	15	0.2352 (**1.66**)	0.2237 (**1.63**)	**0.3078** (3.65)	**0.3762** (3.61)
Monthly	6	0.7604 (**3.99**)	0.6677 (**3.85**)	0.3733 (8.09)	0.6953 (6.01)
	9	**1.0989 (3.37)**	**1.0425 (3.07)**	**1.7261** (7.03)	**1.5096** (5.90)
	12	**1.0897 (3.05)**	**1.0899 (2.98)**	**2.0185** (6.79)	**1.7998** (5.73)
	15	**1.2108 (3.29)**	**1.3097 (3.06)**	**2.5437** (6.83)	**2.3690** (6.15)
Quarterly	6	2.1454 (**4.93**)	2.4590 (**4.09**)	2.5037 (9.97)	2.1612 (7.99)
	9	**3.6798 (5.70)**	3.0273 (**4.67**)	**5.8395** (11.48)	**3.5753** (7.12)
	12	**3.8527 (5.35)**	**3.7094 (4.91)**	**7.3886** (10.35)	**5.7691** (9.01)
	15	2.5684 (**6.14**)	2.6043 (**5.51**)	2.9588 (12.07)	3.2757 (9.57)

Terminal Portfolio Values

Forecast Horizon	Hidden Units	P1	P2	P3	P4
Weekly	6	2.3733	2.3977	0.5722	0.7833
	9	2.9595	3.5977	2.6308	**4.2789**
	12	2.2386	2.2234	1.7826	2.5016
	15	3.0157	2.5380	**5.5586**	**4.4503**
Monthly	6	2.7644	2.4252	1.0436	2.1758
	9	**4.7510**	**4.4304**	**9.2327**	**7.4353**
	12	**4.7545**	**4.7720**	**14.5194**	**11.5644**
	15	**5.6298**	**6.5898**	**31.4602**	**25.9603**
Quarterly	6	2.6189	3.0871	2.5832	2.3862
	9	**5.2747**	**3.9797**	**11.4013**	**4.8103**
	12	**5.7764**	**5.4607**	**24.7631**	**12.5750**
	15	3.1045	3.2114	2.9244	3.8981

Table 14: **S&P 500 Out-of-Sample Network Portfolio Performance.** Using batch estimates for the S&P 500, this table tabulates out-of-sample portfolio performance by forecast horizon and network size. Portfolios are constructed as described in Table 12. Portfolios P1 and P2 allow no short selling, while portfolios P3 and P4 allow limited short selling. Beginning portfolio values are normalized to one. Values in parentheses are sample standard deviations. **Bold** type indicates better than benchmark index performance: either a larger average return, a smaller sample variance or a larger terminal value.

Average Per Period Percentage Returns

Forecast Horizon	DJIA	S&P 500	3-Month T-Bill
Weekly	0.2772 (2.57)	0.2537 (2.50)	0.1284 (0.00)
Monthly	1.0732 (4.59)	1.0049 (4.50)	0.5575 (0.17)
Quarterly	3.3171 (6.97)	3.1294 (6.84)	1.6755 (0.52)

Terminal Index Values

Forecast Horizon	DJIA	S&P 500	3-Month T-Bill
Weekly	4.3610	3.8361	2.1812
Monthly	4.2326	3.8539	2.3019
Quarterly	4.2972	3.9562	2.2188

Table 15: **Benchmark Index Performance.** Beginning index values are normalized to one. No transaction cost was applied for the T-bill rollover. Values in parentheses are sample standard deviations. There are 608 weekly, 150 monthly and 48 quarterly observations. The sample period is approximately January 1982 to December 1993.

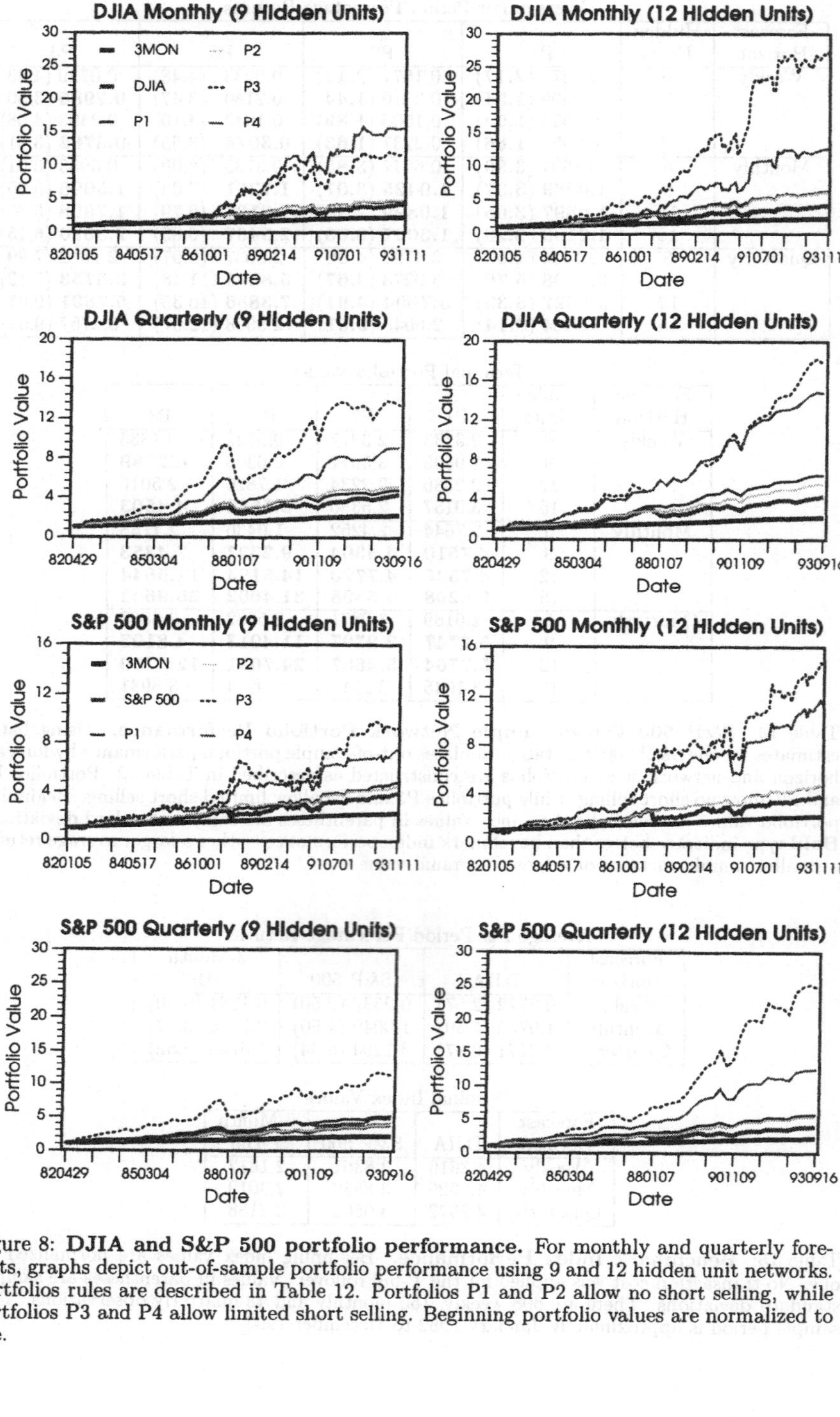

Figure 8: **DJIA and S&P 500 portfolio performance.** For monthly and quarterly forecasts, graphs depict out-of-sample portfolio performance using 9 and 12 hidden unit networks. Portfolios rules are described in Table 12. Portfolios P1 and P2 allow no short selling, while portfolios P3 and P4 allow limited short selling. Beginning portfolio values are normalized to one.

DJIA Portfolio Turnover

Forecast Horizon	Hidden Units	P1		P2		P3	
		Index	TO	Index	TO	Index	TO
Weekly (54.44)	6	75.16	9.38	60.03	9.38	65.62	19.24
	9	64.47	11.68	62.50	12.34	61.35	17.60
	12	58.55	17.11	52.80	17.43	54.77	27.30
	15	58.39	18.26	54.93	17.60	53.62	23.85
Monthly (56.00)	6	72.00	14.67	63.33	14.67	60.67	17.33
	9	62.00	22.67	52.00	24.00	53.33	28.00
	12	60.00	30.00	44.00	28.67	42.67	36.67
	15	67.33	28.67	56.00	32.67	56.67	36.67
Quarterly (58.33)	6	72.92	29.17	62.50	33.33	62.50	33.33
	9	70.83	33.33	58.33	37.50	62.50	37.50
	12	64.58	29.17	50.00	41.67	50.00	41.67
	15	66.67	43.75	52.08	41.67	52.08	41.67

S&P 500 Portfolio Turnover

Forecast Horizon	Hidden Units	P1		P2		P3	
		Index	TO	Index	TO	Index	TO
Weekly (55.26)	6	73.36	6.09	64.31	6.09	65.95	9.70
	9	44.08	8.06	38.32	7.73	40.46	12.01
	12	54.61	17.76	48.52	18.09	46.88	22.20
	15	52.96	16.94	49.84	16.28	50.66	21.88
Monthly (57.33)	6	66.00	15.33	61.33	15.33	61.33	16.67
	9	62.00	26.00	52.00	30.00	46.67	37.33
	12	60.67	30.00	53.33	27.33	54.00	31.33
	15	60.67	27.33	52.00	28.67	53.33	32.67
Quarterly (56.25)	6	52.08	22.92	39.58	22.92	37.50	22.92
	9	72.92	27.08	58.33	39.58	56.25	41.67
	12	68.75	25.00	58.33	41.67	60.42	41.67
	15	66.67	39.58	47.92	47.92	50.00	52.08

Table 16: **Out-of-Sample Portfolio Turnover.** Using batch estimates, this table tabulates out-of-sample portfolio turnover by forecast horizon and network size. Index refers to the percentage of times the portfolio is long the index. TO refers to the percentage of times the portfolio was turned over. There are no comparable values for portfolio P4. Values in parentheses are the percentage of *ex post* net positive index returns for the given forecast horizon (index return greater than bond return).

fraction of instances that a portfolio is long in the index and the fraction out of all possible trades that the an actual trade is initiated or the portfolio turned over. This turnover is by and large moderate with the fraction of initiated trades averaging about 25–35 percent. However, the most striking aspect is that for the most part the portfolios are only in the index an average of about 55–60 percent of the time, and in the short-term bond the remaining time, i.e. those instances that the network forecasts *negative* expected excess returns. In other words, although the unconditional mean return is always greater than the bond, the conditional mean return is at times negative. This seems to be the key to good out-of-sample portfolio performance: accurate forecasting of down moves in order to avoid them entirely or to take advantage of them.

Interestingly, only a relatively small amount of accurate forecasting ability was required to generate large portfolio returns—regardless of the fact that the employed portfolio strategies are by no means exhaustive or optimal. In fact, a great many simplifications were made in performing these portfolio simulations: (1) the difficulty in buying and selling baskets of securities such as indices, (2) the ignorance of true bid and ask prices, (3) the complete ignorance of dividend payments for both long and short positions, and (4) the assumption of a naive transaction costs structure. These considerations will undoubtedly hamper real-world performance.

4.3.2 Risk and Return

In Table 17, I provide a simple measure of the trade-off between risk and return by calculating the ratio of the (arithmetic) mean return of a portfolio over its standard deviation. For comparison, similar ratios are calculated for the benchmark indices and the short-term bond.[22]

Again, the best performances are turned in by the 9 and 12 hidden unit networks for monthly and quarterly forecasting. The most interesting pattern is the uniform difference between the sets of ratios for the small and large networks. All of the 12 hidden unit network values are larger in magnitude than the corresponding 9 hidden unit network values. This seems to imply that the sequence of forecasts generated by the large network incurs a smaller cost in terms of risk than the small network. However, I caution that this might be too broad a statement since this specific return-risk measure is rather simplistic.

4.3.3 Portfolio Bootstrapping

To provide some notion of confidence or significance, I bootstrapped the portfolios themselves. This is accomplished by: (1) resampling from the joint distribution of actual, forecasted and bond returns, and (2) applying the portfolio rules, thereby creating a bootstrap distribution of portfolios. This is the same procedure outlined in Section 4.2.2 for the forecast summary statistics. These efforts generate three items: bootstrap sample paths for the actual return, the forecasted return, and the bond return. In turn, this provides bootstrap distributions for the index, the portfolios, and the bond.

With these distributions, we can also calculate a variety of other distributions. Two of particular interest are: (1) the ratio of a portfolio terminal value over the corresponding benchmark index terminal value, and (2) the difference between a portfolio return-risk ratio and the corresponding benchmark return-risk ratio. For a bootstrap sample path j, these two quantities are written: P_j/I_j and $(\bar{r}^P_j/\hat{\sigma}^P_j - \bar{r}^I_j/\hat{\sigma}^I_j)$, where P_j is a portfolio terminal value, I_j is an index terminal value, and $\bar{r}_j$ and $\hat{\sigma}_j$ are the sample mean and standard variation of either the portfolio or index returns.

In Table 18, I present summary statistics for these bootstrapped quantities. These numbers can be translated as follows: (1) if the ratio of the terminal values is greater than one, then the portfolio outperforms the benchmark; (2) if the difference in return-risk ratios is positive, then the portfolio outperforms the benchmark. As we can see in terms of the mean, except for two cases, all of the estimates point towards the existence of profitable, risk-adjusted network forecasting ability.

To further bolster this inference, I quote two probabilities from the distributions: the probability of observing the original estimated value or larger, and the probability of observing a value below a natural threshold. These probabilities

[22] This ratio is basically the Sharpe (1966) index of mean-variance efficiency. I use average unadjusted returns instead of average excess returns.

DJIA Return-Risk Ratios

Forecast Horizon	Hidden Units	P1	P2	P3	P4
Weekly	6	0.0832	0.0895	0.0831	0.0545
	9	0.0617	0.0608	0.0554	0.0395
	12	**0.1202**	**0.1224**	**0.1191**	**0.1159**
	15	0.0639	0.0627	0.0032	0.0211
Monthly	6	**0.2359**	0.2314	0.1604	0.1474
	9	**0.3110**	**0.3300**	**0.2570**	**0.3232**
	12	**0.4176**	**0.4541**	**0.3715**	**0.3288**
	15	**0.3534**	**0.3925**	**0.3505**	**0.3331**
Quarterly	6	0.4578	0.4602	0.2492	0.2952
	9	**0.5619**	**0.5976**	**0.5120**	**0.4838**
	12	**0.7586**	**0.7475**	**0.6229**	**0.6609**
	15	**0.5999**	**0.6755**	**0.5289**	**0.5492**

S&P 500 Return-Risk Ratios

Forecast Horizon	Hidden Units	P1	P2	P3	P4
Weekly	6	0.0768	0.0791	0.0052	0.0164
	9	**0.1258**	**0.1536**	0.0627	0.0853
	12	0.0788	0.0798	0.0474	0.0611
	15	**0.1416**	**0.1372**	0.0844	**0.1042**
Monthly	6	0.1908	0.1733	0.0461	0.1156
	9	**0.3265**	**0.3395**	**0.2454**	**0.2560**
	12	**0.3571**	**0.3662**	**0.2973**	**0.3143**
	15	**0.3683**	**0.4276**	**0.3722**	**0.3852**
Quarterly	6	0.4356	**0.6008**	0.2511	0.2705
	9	**0.6461**	**0.6484**	**0.5086**	**0.5018**
	12	**0.7197**	**0.7559**	**0.7141**	**0.6405**
	15	0.4180	**0.4728**	0.2451	0.3424

Benchmark Return-Risk Ratios

Forecast Horizon	DJIA	S&P 500	3-Month T-Bill
Weekly	0.1078	0.1016	∞
Monthly	0.2339	0.2235	3.2187
Quarterly	0.4756	0.4573	3.2245

Table 17: **Out-of-Sample Return-Risk Ratios.** Using batch estimates, this table tabulates out-of-sample return-risk ratios by forecast horizon and network size. Top and middle panels: DJIA and S&P 500 ratios; bottom panel: benchmark indices and short-term bond ratios. **Bold** type indicates better than benchmark index performance, i.e. a smaller ratio than the corresponding benchmark index ratio.

DJIA Bootstrapped Ratio of Terminal Values

Forecast Horizon	Hidden Units	P1 Mean	P2 Mean	P3 Mean	P4 Mean
Monthly	9	1.0684 (0.44)	0.9706 (0.42)	3.3003 (3.51)	3.9805 (3.42)
	12	1.5184 (0.68)	1.5111 (0.72)	8.7315 (9.72)	5.2697 (4.77)
Quarterly	9	1.0734 (0.22)	1.2125 (0.28)	3.7002 (2.24)	1.9009 (0.87)
	12	1.4564 (0.45)	1.3288 (0.47)	5.0989 (3.96)	3.8093 (2.39)

DJIA Bootstrapped Difference in Return-Risk Ratio

Forecast Horizon	Hidden Units	P1 Mean	P2 Mean	P3 Mean	P4 Mean
Monthly	9	0.0521 (0.07)	0.0420 (0.07)	0.0017 (0.01)	0.1041 (0.09)
	12	0.1645 (0.08)	0.1929 (0.09)	0.1216 (0.12)	0.1559 (0.10)
Quarterly	9	0.0721 (0.09)	0.1405 (0.11)	0.0509 (0.14)	-0.0004 (0.13)
	12	0.2590 (0.14)	0.2671 (0.16)	0.1377 (0.21)	0.1766 (0.18)

S&P 500 Bootstrapped Ratio of Terminal Values

Forecast Horizon	Hidden Units	P1 Mean	P2 Mean	P3 Mean	P4 Mean
Monthly	9	1.1770 (0.51)	1.1807 (0.54)	3.8321 (4.38)	3.5949 (3.22)
	12	1.1665 (0.51)	1.2623 (0.56)	5.1031 (5.30)	6.8300 (5.95)
Quarterly	9	1.3474 (0.32)	1.1429 (0.32)	3.5372 (2.32)	1.7400 (0.69)
	12	1.5726 (0.41)	1.5019 (0.46)	6.8353 (4.48)	3.4739 (1.86)

S&P 500 Bootstrapped Difference in Return-Risk Ratio

Forecast Horizon	Hidden Units	P1 Mean	P2 Mean	P3 Mean	P4 Mean
Monthly	9	0.0824 (0.07)	0.0998 (0.08)	0.0266 (0.11)	0.0851 (0.10)
	12	0.0949 (0.07)	0.1250 (0.08)	0.0631 (0.11)	0.1945 (0.10)
Quarterly	9	0.2053 (0.12)	0.2030 (0.14)	0.0702 (0.16)	0.1805 (0.15)
	12	0.2735 (0.13)	0.3218 (0.15)	0.2387 (0.19)	0.2474 (0.18)

Table 18: **DJIA and S&P 500 Portfolio Bootstrapping Summary Statistics.** Using batch estimates, this table tabulates out-of-sample portfolio bootstrapping for monthly and quarterly forecasts and 9 and 12 hidden unit networks. The ratio of terminal values is the ratio of the terminal portfolio value over the corresponding terminal benchmark index value; thus, a value greater than one indicates better than benchmark performance. The difference in return-risk ratios is the difference between the portfolio risk-return ratio with the corresponding benchmark index ratio; thus, a positive value indicates better than benchmark performance. Values in parentheses are bootstrap sample standard deviations.

are presented in Table 19. For both probabilities, small values indicate superior network performance. On the whole, these results are extremely pleasing. The probability of the network outperforming the benchmark is generally large to moderate with some cases indicating a near certainty. This, coupled with the high likelihood of positive differences in the return-risk ratios, supports the claim of superlative, risk-adjusted network performance.

4.4 Comparison with the Benchmark Linear Model

Referring back to Section 3.1, we can compare network in-sample and out-of-sample performance with the benchmark linear model. Tables 3–6 for the linear results correspond with Tables 7, 8, 13 and 14 for the network results. Recall that the four linear models roughly pair up with the four estimated network models in terms of parameter number.

First, the in-sample error of the network models is uniformly and dramatically smaller than the corresponding linear model error for all target series—even though the parameter number is roughly about the same. Thus, if the only criterion is this number, a network model surely provides better in-sample fits. Second, out-of-sample performance is also decidedly superior. For all measures of performance evaluation, the network grades out much higher in these forecasting exercises than the linear model, in some cases, orders of magnitude higher.

5 Conclusions

In this study, we have applied neural network models to several difficult forecasting problems. It appears that increased attention given to nonlinear methods and techniques seems to be justified. Using a flexible nonlinear approximator, I have shown that it is possible to reasonably estimate an unknown function and accurately forecast the short-term behavior of a time series with complicated dynamics. Since linear methods generally fail in this regard, there appears then to be some degree of forecastable nonlinear structure in financial time series data.

The potential existence of forecastable nonlinear structure in asset prices has several interesting implications with regard to market efficiency. Current evidence seems to point in the direction of limited or marginal forecastability—that is, in the sense of *linear* structure. Since linear tools are available, even under a loose reading of market efficiency, much of the linear features in market movements should be incorporated in asset prices. However, since applicable nonlinear techniques are fairly recent developments, there ought to be greater room for improvement by incorporating and capturing nonlinear structure. This seems to be the case.

Still, several issues remain open. Although, we might believe there is structure present, its exact nature is clouded. On this, the neural network models investigated here are unable to supply much detail. Due to their nonparametric nature, no explicit equations of motion can be studied and interpreted, with perhaps only circumstantial understanding available. Moreover, as a nonlinear technique, practical implementation continues to be difficult: considerable computational resources and data are required that depending on the problem area may be a hindrance or simply too burdensome.

DJIA Bootstrap Probabilities: Ratio of Terminal Values

Forecast Horizon	Hidden Units	P1 Obs. Val.	P1 One	P2 Obs. Val.	P2 One	P3 Obs. Val.	P3 One	P4 Obs. Val.	P4 One
Monthly	9	31.86	58.90	24.80	63.00	31.56	27.02	38.24	17.48
	12	39.10	27.84	41.08	25.60	38.34	12.40	59.12	**3.88**
Quarterly	9	38.04	45.18	47.16	28.00	46.14	**4.28**	32.44	14.52
	12	33.92	13.86	27.92	26.18	40.18	**3.34**	37.00	**3.32**

DJIA Bootstrap Probabilities: Difference in Return-Risk Ratios

Forecast Horizon	Hidden Units	P1 Obs. Val.	P1 Zero	P2 Obs. Val.	P2 Zero	P3 Obs. Val.	P3 Zero	P4 Obs. Val.	P4 Zero
Monthly	9	**0.00**	25.26	**0.00**	31.98	0.38	52.08	**0.40**	14.64
	12	**0.02**	**2.12**	**0.12**	**1.46**	**1.40**	16.96	**4.32**	8.24
Quarterly	9	**0.00**	22.86	**0.18**	7.80	**0.50**	42.84	**0.22**	53.66
	12	**0.06**	**2.44**	**0.60**	**3.84**	**1.28**	28.56	**0.70**	17.34

S&P 500 Bootstrap Probabilities: Ratio of Terminal Values

Forecast Horizon	Hidden Units	P1 Obs. Val.	P1 One	P2 Obs. Val.	P2 One	P3 Obs. Val.	P3 One	P4 Obs. Val.	P4 One
Monthly	9	34.64	48.66	40.00	46.80	48.32	24.12	66.26	11.12
	12	31.40	47.74	37.64	40.00	40.80	7.98	76.54	6.38
Quarterly	9	47.06	12.06	37.00	37.00	44.78	7.96	74.46	13.22
	12	53.26	**3.30**	50.02	13.18	36.76	**0.42**	41.26	**1.74**

S&P 500 Bootstrap Probabilities: Difference in Return-Risk Ratios

Forecast Horizon	Hidden Units	P1 Obs. Val.	P1 Zero	P2 Obs. Val.	P2 Zero	P3 Obs. Val.	P3 Zero	P4 Obs. Val.	P4 Zero
Monthly	9	**0.04**	15.26	**0.04**	10.82	**1.60**	43.26	**3.84**	21.98
	12	**0.00**	10.56	**0.08**	5.74	**1.40**	29.18	11.98	**2.60**
Quarterly	9	**0.20**	**1.94**	**0.68**	6.16	**0.84**	38.64	**2.58**	13.30
	12	**0.16**	**1.04**	**0.38**	**1.00**	**0.78**	12.42	**1.70**	8.16

Table 19: **DJIA and S&P 500 Portfolio Bootstrap Probabilities.** Using batch estimates, this table tabulates out-of-sample portfolio bootstrap probabilities for monthly and quarterly forecasts and 9 and 12 hidden unit networks. All probabilities are in percent. For the terminal value ratios, two probabilities are calculated: (1) the probability of observing the estimated value or greater; (2) the probability of values one or less. For the return-risk ratios, two probabilities are calculated: (1) the probability of observing the estimated value or greater; (2) the probability of values zero or less. For all probabilities, small values indicate higher likelihood of better portfolio than benchmark performance. **Bold** type indicates probabilities of 5 percent or smaller. Values in parentheses are bootstrap sample standard deviations.

However, the techniques outlined in this study are exciting and promising. For the data sets investigated here, there exists considerable and convincing evidence that accurate forecasting is feasible—an ability that cannot be attributed to chance. Of course, if it is possible to accurately forecast, it is not surprising that one may profitably exploit this advantage to a large degree. Nevertheless, it is important to reiterate how closely connected these techniques are to classical regression analysis. Indeed, factor models and the like all derive from the same lineage. And yet the real novelty is that, with increased computational power and new statistical tools, practical and powerful nonlinear estimation and forecasting has become a strong and distinct possibility.

References

Aarts, Emile H. L., and Jan Korst, 1989, *Simulated Annealing and Boltzmann Machines.* New York: J. Wiley & Sons, Inc.

Barron, A. R., 1991, Universal approximation bounds for superpositions of a sigmodial function, Technical Report No. 58, Department of Statistics, University of Illinois, Urbana-Champaign.

Bollerslev, Tim, Ray Y. Chou, and Kenneth F. Kroner, 1992, ARCH modeling in finance: a review of theory and empirical evidence, *Journal of Econometrics*, **52**, 5–59.

Brock, William A., W. Davis Dechert, and Jose A. Scheinkman, 1987, A test for independence based on the correlation dimension, Department of Economics, University of Wisconsin, mimeo.

———, David A. Hsieh, and Blake LeBaron, 1991, *Nonlinear Dynamics, Chaos, and Instability: Statistical Theory and Economic Evidence.* Cambridge, MA: MIT Press.

Chen, Tianping, and Hong Chen, 1993, Approximations of continuous functionals by neural networks with application to dynamic systems, *IEEE Transactions on Neural Networks*, **4**, 910–918.

Cybenko, G., 1989, Approximations by superpositions of a sigmodial function, *Mathematics of Control, Signals and Systems*, **2**, 303–314.

Diebold, Francis X., and J. M. Nason, 1990, Nonparametric exchange rate prediction?, *Journal of International Economics*, **28**, 315–332.

Efron, Bradley, 1982, *The Jackknife, the Bootstrap and Other Resampling Plans.* CBMS-NSF Regional Conference Series in Applied Mathematics, Philadelphia: SIAM.

———, and Robert J. Tibshirani, 1993, *An Introduction to the Bootstrap*, London: Chapman & Hall.

Elman, Jeffrey L., 1990, Finding structure in time, *Cognitive Science*, **14**, 179–212.

Fama, Eugene F., 1981, Stocks, real activity, inflation, and money, *American Economic Review*, **71**, 545–565.

———, 1990, Stock returns expected returns, and real activity, *Journal of Finance*, **45**, 1089–1108.

———, and Kenneth R. French, 1989, Business conditions and expected stock returns on stocks and bonds, *Journal of Financial Economics*, **25**, 23–49.

———, and Kenneth R. French, 1988a, Dividend yields and expected stock returns, *Journal of Financial Economics*, **22**, 3–25.

———, and Kenneth R. French, 1988b, Permanent and temporary components of stock prices, *Journal of Political Economy*, **96**, 247–273.

———, and G. William Schwert, 1977, Asset returns and inflation, *Journal of Financial Economics*, **5**, 115–146.

Gallant, A. Ronald, and Halbert White, 1988, *A Unified Theory of Estimation and Inference for Nonlinear Dynamic Models*. Oxford: Basil Blackwell.

———, 1992, On learning the derivatives of an unknown mapping with multilayer feedforward networks, *Neural Networks*, **5**, 128–138.

Henriksson, Roy D., and Robert C. Merton, 1981, On market timing and investment performance, II: Statistical procedures for evaluating forecasting skills, *Journal of Business*, **54**, 513–533.

Hertz, John, Anders Krogh, and Richard G. Palmer, 1991, *Introduction to the Theory of Neural Computation*. Santa Fe Institute Studies in the Sciences of Complexity, Lecture Notes Vol. I. Redwood City, CA: Addison-Wesley Publishing.

Hornik, Kurt, Maxwell B. Stinchcombe, and Halbert White, 1989, Multilayer feedforward networks are universal approximators, *Neural Networks*, **2**, 359–366.

Hsieh, David A., 1993, Implications of nonlinear dynamics for financial risk management, *Journal of Financial and Quantitative Analysis*, **28**, 41–64.

Hutchinson, James M., Andrew W. Lo, and Tomaso Poggio, 1994, A nonparametric approach to pricing and hedging derivative securities via learning networks, *Journal of Finance*, **49**, 851–890.

Jegadeesh, Narasimham, 1990, Evidence of predictable behavior of security returns, *Journal of Finance*, **45**, 881–898.

———, 1991, Seasonality in stock price mean reversion: Evidence from the US and the UK, *Journal of Finance*, **46**, 1427–1444.

Keim, Donald B., and Robert F. Stambaugh, 1986, Predicting returns in the stock and bond markets, *Journal of Financial Economics*, **17**, 357–390.

Kuan, Chung-Ming., 1993, A recurrent Newton algorithm and its convergence properties, Department of Economics, University of Illinois, Faculty Working Paper No. 93-0139.

———, Kurt Hornik, and Halbert White, 1993, A convergence result for learning in recurrent neural networks, Department of Economics, University of California, San Diego, mimeo.

LeBaron, Blake, 1992, Some relations between volatility and serial correlation in stock market returns, *Journal of Business*, **65**, 199–219.

Ljung, L., and T. Söderström, 1986, *Theory and Practice of Recursive Estimation*. Cambridge, MA: MIT Press.

Robbins, H., and S. Munro, 1951, A stochastic approximation model, *Annals of Mathematics*, **22**, 400–407.

Rumelhart, David E., J. L. McClelland, and the PDP Research Group, 1986, *Parallel Distributed Processing: Explorations in the Microstructure of Cognition, Volume 1: Foundations*. Cambridge, MA: MIT Press.

Schwarz, G., 1978, Estimating the dimension of a model, *Annals of Statistics*, **6**, 461–464.

Sharpe, William F., 1966, Mutual fund performance, *Journal of Business*, **39**, 119–138.

Wahba, Grace, 1990, *Spline Models for Observational Data*. CBMS-NSF Regional Conference Series in Applied Mathematics, Philadelphia: SIAM.

Weigend, Andreas S., Bernardo A. Huberman, and David E. Rumelhart, 1990, Predicting the future: a connectionist approach, *International Journal of Neural Systems*, **1**, 193–209.

———, Bernardo A. Huberman, and David E. Rumelhart, 1992, Predicting sunspots and exchange rates with connectionist networks, in *Nonlinear Modeling and Forecasting*, edited by M. Casdagli and S. Eubank, Santa Fe Institute Studies in the Sciences of Complexity, Proceedings Volume XII, 395–432.

———, and Blake LeBaron, 1994, Evaluating neural network predictors by bootstrapping, Department of Computer Science, University of Colorado, mimeo.

White, Halbert, 1988, Economic prediction using neural networks: the case of IBM stock prices, *Proceedings of the IEEE Second Conference on Neural Networks*, **II**, 451–458, San Diego, CA: SOS Printing.

———, 1992, *Artificial Neural Networks: Approximation and Learning Theory*. Cambridge, MA: Blackwell Publishers.

Lecture Notes in Statistics

For information about Volumes 1 to 43 please contact Springer-Verlag

Vol. 44: D.L. McLeish, Christopher G. Small, The Theory and Applications of Statistical Inference Functions. vi, 124 pages, 1987.

Vol. 45: J.K. Ghosh (Editor), Statistical Information and Likelihood. 384 pages, 1988.

Vol. 46: H.-G. Müller, Nonparametric Regression Analysis of Longitudinal Data. vi, 199 pages, 1988.

Vol. 47: A.J. Getson, F.C. Hsuan, {2}-Inverses and Their Statistical Application. viii, 110 pages, 1988.

Vol. 48: G.L. Bretthorst, Bayesian Spectrum Analysis and Parameter Estimation. xii, 209 pages, 1988.

Vol. 49: S.L. Lauritzen, Extremal Families and Systems of Sufficient Statistics. xv, 268 pages, 1988.

Vol. 50: O.E. Barndorff-Nielsen, Parametric Statistical Models and Likelihood. vii, 276 pages, 1988.

Vol. 51: J. Hüsler, R.-D. Reiss (Editors). Extreme Value Theory, Proceedings, 1987. x, 279 pages, 1989.

Vol. 52: P.K. Goel, T. Ramalingam, The Matching Methodology: Some Statistical Properties. viii, 152 pages, 1989.

Vol. 53: B.C. Arnold, N. Balakrishnan, Relations, Bounds and Approximations for Order Statistics. ix, 173 pages, 1989.

Vol. 54: K.R. Shah, B.K. Sinha, Theory of Optimal Designs. viii, 171 pages, 1989.

Vol. 55: L. McDonald, B. Manly, J. Lockwood, J. Logan (Editors), Estimation and Analysis of Insect Populations. Proceedings, 1988. xiv, 492 pages, 1989.

Vol. 56: J.K. Lindsey, The Analysis of Categorical Data Using GLIM. v, 168 pages, 1989.

Vol. 57: A. Decarli, B.J. Francis, R. Gilchrist, G.U.H. Seeber (Editors), Statistical Modelling. Proceedings, 1989. ix, 343 pages, 1989.

Vol. 58: O.E. Barndorff-Nielsen, P. Bl¾sild, P.S. Eriksen, Decomposition and Invariance of Measures, and Statistical Transformation Models. v, 147 pages, 1989.

Vol. 59: S. Gupta, R. Mukerjee, A Calculus for Factorial Arrangements. vi, 126 pages, 1989.

Vol. 60: L. Gyorfi, W. Härdle, P. Sarda, Ph. Vieu, Nonparametric Curve Estimation from Time Series. viii, 153 pages, 1989.

Vol. 61: J. Breckling, The Analysis of Directional Time Series: Applications to Wind Speed and Direction. viii, 238 pages, 1989.

Vol. 62: J.C. Akkerboom, Testing Problems with Linear or Angular Inequality Constraints. xii, 291 pages, 1990.

Vol. 63: J. Pfanzagl, Estimation in Semiparametric Models: Some Recent Developments. iii, 112 pages, 1990.

Vol. 64: S. Gabler, Minimax Solutions in Sampling from Finite Populations. v, 132 pages, 1990.

Vol. 65: A. Janssen, D.M. Mason, Non-Standard Rank Tests. vi, 252 pages, 1990.

Vol 66: T. Wright, Exact Confidence Bounds when Sampling from Small Finite Universes. xvi, 431 pages, 1991.

Vol. 67: M.A. Tanner, Tools for Statistical Inference: Observed Data and Data Augmentation Methods. vi, 110 pages, 1991.

Vol. 68: M. Taniguchi, Higher Order Asymptotic Theory for Time Series Analysis. viii, 160 pages, 1991.

Vol. 69: N.J.D. Nagelkerke, Maximum Likelihood Estimation of Functional Relationships. V, 110 pages, 1992.

Vol. 70: K. Iida, Studies on the Optimal Search Plan. viii, 130 pages, 1992.

Vol. 71: E.M.R.A. Engel, A Road to Randomness in Physical Systems. ix, 155 pages, 1992.

Vol. 72: J.K. Lindsey, The Analysis of Stochastic Processes using GLIM. vi, 294 pages, 1992.

Vol. 73: B.C. Arnold, E. Castillo, J.-M. Sarabia, Conditionally Specified Distributions. xiii, 151 pages, 1992.

Vol. 74: P. Barone, A. Frigessi, M. Piccioni, Stochastic Models, Statistical Methods, and Algorithms in Image Analysis. vi, 258 pages, 1992.

Vol. 75: P.K. Goel, N.S. Iyengar (Eds.), Bayesian Analysis in Statistics and Econometrics. xi, 410 pages, 1992.

Vol. 76: L. Bondesson, Generalized Gamma Convolutions and Related Classes of Distributions and Densities. viii, 173 pages, 1992.

Vol. 77: E. Mammen, When Does Bootstrap Work? Asymptotic Results and Simulations. vi, 196 pages, 1992.

Vol. 78: L. Fahrmeir, B. Francis, R. Gilchrist, G. Tutz (Eds.), Advances in GLIM and Statistical Modelling: Proceedings of the GLIM92 Conference and the 7th International Workshop on Statistical Modelling, Munich, 13-17 July 1992. ix, 225 pages, 1992.

Vol. 79: N. Schmitz, Optimal Sequentially Planned Decision Procedures. xii, 209 pages, 1992.

Vol. 80: M. Fligner, J. Verducci (Eds.), Probability Models and Statistical Analyses for Ranking Data. xxii, 306 pages, 1992.

Vol. 81: P. Spirtes, C. Glymour, R. Scheines, Causation, Prediction, and Search. xxiii, 526 pages, 1993.

Vol. 82: A. Korostelev and A. Tsybakov, Minimax Theory of Image Reconstruction. xii, 268 pages, 1993.

Vol. 83: C. Gatsonis, J. Hodges, R. Kass, N. Singpurwalla (Editors), Case Studies in Bayesian Statistics. xii, 437 pages, 1993.

Vol. 84: S. Yamada, Pivotal Measures in Statistical Experiments and Sufficiency. vii, 129 pages, 1994.

Vol. 85: P. Doukhan, Mixing: Properties and Examples. xi, 142 pages, 1994.

Vol. 86: W. Vach, Logistic Regression with Missing Values in the Covariates. xi, 139 pages, 1994.

Vol. 87: J. Müller, Lectures on Random Voronoi Tessellations.vii, 134 pages, 1994.

Vol. 88: J. E. Kolassa, Series Approximation Methods in Statistics.viii, 150 pages, 1994.

Vol. 89: P. Cheeseman, R.W. Oldford (Editors), Selecting Models From Data: AI and Statistics IV. xii, 487 pages, 1994.

Vol. 90: A. Csenki, Dependability for Systems with a Partitioned State Space: Markov and Semi-Markov Theory and Computational Implementation. x, 241 pages, 1994.

Vol. 91: J.D. Malley, Statistical Applications of Jordan Algebras. viii, 101 pages, 1994.

Vol. 92: M. Eerola, Probabilistic Causality in Longitudinal Studies. vii, 133 pages, 1994.

Vol. 93: Bernard Van Cutsem (Editor), Classification and Dissimilarity Analysis. xiv, 238 pages, 1994.

Vol. 94: Jane F. Gentleman and G.A. Whitmore (Editors), Case Studies in Data Analysis. viii, 262 pages, 1994.

Vol. 95: Shelemyahu Zacks, Stochastic Visibility in Random Fields. x, 175 pages, 1994.

Vol. 96: Ibrahim Rahimov, Random Sums and Branching Stochastic Processes. viii, 195 pages, 1995.

Vol. 97: R. Szekli, Stochastic Ordering and Dependence in Applied Probability. viii, 194 pages, 1995.

Vol. 98: Philippe Barbe and Patrice Bertail, The Weighted Bootstrap. viii, 230 pages, 1995.

Vol. 99: C.C. Heyde (Editor), Branching Processes: Proceedings of the First World Congress. viii, 185 pages, 1995.

Vol. 100: Wlodzimierz Bryc, The Normal Distribution: Characterizations with Applications. viii, 139 pages, 1995.

Vol. 101: H.H. Andersen, M.Højbjerre, D. Sørensen, P.S.Eriksen, Linear and Graphical Models: for the Multivariate Complex Normal Distribution. x, 184 pages, 1995.

Vol. 102: A.M. Mathai, Serge B. Provost, Takesi Hayakawa, Bilinear Forms and Zonal Polynomials. x, 378 pages, 1995.

Vol. 103: Anestis Antoniadis and Georges Oppenheim (Editors), Wavelets and Statistics. vi, 411 pages, 1995.

Vol. 104: Gilg U.H. Seeber, Brian J. Francis, Reinhold Hatzinger, Gabriele Steckel-Berger (Editors), Statistical Modelling: 10th International Workshop, Innsbruck, July 10-14th 1995. x, 327 pages, 1995.

Vol. 105: Constantine Gatsonis, James S. Hodges, Robert E. Kass, Nozer D. Singpurwalla(Editors), Case Studies in Bayesian Statistics, Volume II. x, 354 pages, 1995.

Vol. 106: Harald Niederreiter, Peter Jau-Shyong Shiue (Editors), Monte Carlo and Quasi-Monte Carlo Methods in Scientific Computing. xiv, 372 pages, 1995.

Vol. 107: Masafumi Akahira, Kei Takeuchi, Non-Regular Statistical Estimation. vii, 183 pages, 1995.

Vol. 108: Wesley L. Schaible (Editor), Indirect Estimators in U.S. Federal Programs. iix, 195 pages, 1995.

Vol. 109: Helmut Rieder (Editor), Robust Statistics, Data Analysis, and Computer Intensive Methods. xiv, 427 pages, 1996.

Vol. 110: D. Bosq, Nonparametric Statistics for Stochastic Processes. xii, 169 pages, 1996.

Vol. 111: Leon Willenborg, Ton de Waal, Statistical Disclosure Control in Practice. xiv, 152 pages, 1996.

Vol. 112: Doug Fischer, Hans-J. Lenz (Editors), Learning from Data. xii, 450 pages, 1996.

Vol. 113: Rainer Schwabe, Optimum Designs for Multi-Factor Models. viii, 124 pages, 1996.

Vol. 114: C.C. Heyde, Yu. V. Prohorov, R. Pyke, and S. T. Rachev (Editors), Athens Conference on Applied Probability and Time Series Analysis Volume I: Applied Probability In Honor of J.M. Gani. viii, 424 pages, 1996.

Vol. 115: P.M. Robinson, M. Rosenblatt (Editors), Athens Conference on Applied Probability and Time Series Analysis Volume II: Time Series Analysis In Memory of E.J. Hannan. viii, 448 pages, 1996.

Vol. 116: Genshiro Kitagawa and Will Gersch, Smoothness Priors Analysis of Time Series. x, 261 pages, 1996.

Vol. 117: Paul Glasserman, Karl Sigman, David D. Yao (Editors), Stochastic Networks. xii, 298, 1996.

Vol. 118: Radford M. Neal, Bayesian Learning for Neural Networks. xv, 183, 1996.

Vol. 119: Masanao Aoki, Arthur M. Havenner, Applications of Computer Aided Time Series Modeling, ix, 329 pages, 1997.

GPSR Compliance

The European Union's (EU) General Product Safety Regulation (GPSR) is a set of rules that requires consumer products to be safe and our obligations to ensure this.

If you have any concerns about our products, you can contact us on ProductSafety@springernature.com

In case Publisher is established outside the EU, the EU authorized representative is:

Springer Nature Customer Service Center GmbH
Europaplatz 3
69115 Heidelberg, Germany

Batch number: 10741234

Printed by Printforce, the Netherlands